LONDON MATHEMATICAL SOCIETY LECTURE NOTE SERIES

Managing Editor: Professor J.W.S. Cassels, Department of Pure Mathematics and Mathematical Statistics, University of Cambridge, 16 Mill Lane, Cambridge CB2 1SB, England

The books in the series listed below are available from booksellers, or, in case of difficulty, from Cambridge University Press.

4 Algebraic topology, J.F. ADAMS
5 Commutative algebra, J.T. KNIGHT
16 Topics in finite groups, T.M. GAGEN
17 Differential germs and catastrophes, Th. BROCKER & L. LANDER
18 A geometric approach to homology theory, S. BUONCRISTIANO, C.P. ROURKE & B.J. SANDERSON
20 Sheaf theory, B.R. TENNISON
21 Automatic continuity of linear operators, A.M. SINCLAIR
23 Parallelisms of complete designs, P.J. CAMERON
24 The topology of Stiefel manifolds, I.M. JAMES
25 Lie groups and compact groups, J.F. PRICE
27 Skew field constructions, P.M. COHN
29 Pontryagin duality and the structure of LCA groups, S.A. MORRIS
30 Interaction models, N.L. BIGGS
31 Continuous crossed products and type III von Neumann algebras, A. VAN DAELE
34 Representation theory of Lie groups, M.F. ATIYAH *et al*
36 Homological group theory, C.T.C. WALL (ed)
38 Surveys in combinatorics, B. BOLLOBAS (ed)
39 Affine sets and affine groups, D.G. NORTHCOTT
40 Introduction to H_p spaces, P.J. KOOSIS
41 Theory and applications of Hopf bifurcation, B.D. HASSARD, N.D. KAZARINOFF & Y-H. WAN
42 Topics in the theory of group presentations, D.L. JOHNSON
43 Graphs, codes and designs, P.J. CAMERON & J.H. VAN LINT
44 Z/2-homotopy theory, M.C. CRABB
45 Recursion theory: its generalisations and applications, F.R. DRAKE & S.S. WAINER (eds)
46 p-adic analysis: a short course on recent work, N. KOBLITZ
48 Low-dimensional topology, R. BROWN & T.L. THICKSTUN (eds)
49 Finite geometries and designs, P. CAMERON, J.W.P. HIRSCHFELD & D.R. HUGHES (eds)
50 Commutator calculus and groups of homotopy classes, H.J. BAUES
51 Synthetic differential geometry, A. KOCK
52 Combinatorics, H.N.V. TEMPERLEY (ed)
54 Markov processes and related problems of analysis, E.B. DYNKIN
55 Ordered permutation groups, A.M.W. GLASS
56 Journees arithmetiques, J.V. ARMITAGE (ed)
57 Techniques of geometric topology, R.A. FENN
58 Singularities of smooth functions and maps, J.A. MARTINET
59 Applicable differential geometry, M. CRAMPIN & F.A.E. PIRANI
60 Integrable systems, S.P. NOVIKOV *et al*
61 The core model, A. DODD
62 Economics for mathematicians, J.W.S. CASSELS
63 Continuous semigroups in Banach algebras, A.M. SINCLAIR
64 Basic concepts of enriched category theory, G.M. KELLY
65 Several complex variables and complex manifolds I, M.J. FIELD
66 Several complex variables and complex manifolds II, M.J. FIELD
67 Classification problems in ergodic theory, W. PARRY & S. TUNCEL
68 Complex algebraic surfaces, A. BEAUVILLE
69 Representation theory, I.M. GELFAND *et al*

70 Stochastic differential equations on manifolds, K.D. ELWORTHY
71 Groups - St Andrews 1981, C.M. CAMPBELL & E.F. ROBERTSON (eds)
72 Commutative algebra: Durham 1981, R.Y. SHARP (ed)
73 Riemann surfaces: a view towards several complex variables, A.T. HUCKLEBERRY
74 Symmetric designs: an algebraic approach, E.S. LANDER
75 New geometric splittings of classical knots, L. SIEBENMANN & F. BONAHON
76 Spectral theory of linear differential operators and comparison algebras, H.O. CORDES
77 Isolated singular points on complete intersections, E.J.N. LOOIJENGA
78 A primer on Riemann surfaces, A.F. BEARDON
79 Probability, statistics and analysis, J.F.C. KINGMAN & G.E.H. REUTER (eds)
80 Introduction to the representation theory of compact and locally compact groups, A. ROBERT
81 Skew fields, P.K. DRAXL
82 Surveys in combinatorics, E.K. LLOYD (ed)
83 Homogeneous structures on Riemannian manifolds, F. TRICERRI & L. VANHECKE
84 Finite group algebras and their modules, P. LANDROCK
85 Solitons, P.G. DRAZIN
86 Topological topics, I.M. JAMES (ed)
87 Surveys in set theory, A.R.D. MATHIAS (ed)
88 FPF ring theory, C. FAITH & S. PAGE
89 An F-space sampler, N.J. KALTON, N.T. PECK & J.W. ROBERTS
90 Polytopes and symmetry, S.A. ROBERTSON
91 Classgroups of group rings, M.J. TAYLOR
92 Representation of rings over skew fields, A.H. SCHOFIELD
93 Aspects of topology, I.M. JAMES & E.H. KRONHEIMER (eds)
94 Representations of general linear groups, G.D. JAMES
95 Low-dimensional topology 1982, R.A. FENN (ed)
96 Diophantine equations over function fields, R.C. MASON
97 Varieties of constructive mathematics, D.S. BRIDGES & F. RICHMAN
98 Localization in Noetherian rings, A.V. JATEGAONKAR
99 Methods of differential geometry in algebraic topology, M. KAROUBI & C. LERUSTE
100 Stopping time techniques for analysts and probabilists, L. EGGHE
101 Groups and geometry, ROGER C. LYNDON
102 Topology of the automorphism group of a free group, S.M. GERSTEN
103 Surveys in combinatorics 1985, I. ANDERSON (ed)
104 Elliptic structures on 3-manifolds, C.B. THOMAS
105 A local spectral theory for closed operators, I. ERDELYI & WANG SHENGWANG
106 Syzygies, E.G. EVANS & P. GRIFFITH
107 Compactification of Siegel moduli schemes, C-L. CHAI
108 Some topics in graph theory, H.P. YAP
109 Diophantine Analysis, J. LOXTON & A. VAN DER POORTEN (eds)
110 An introduction to surreal numbers, H. GONSHOR
111 Analytical and geometric aspects of hyperbolic space, D.B.A.EPSTEIN (ed)
112 Low-dimensional topology and Kleinian groups, D.B.A. EPSTEIN (ed)
113 Lectures on the asymptotic theory of ideals, D. REES
114 Lectures on Bochner-Riesz means, K.M. DAVIS & Y-C. CHANG
115 An introduction to independence for analysts, H.G. DALES & W.H. WOODIN
116 Representations of algebras, P.J. WEBB (ed)
117 Homotopy theory, E. REES & J.D.S. JONES (eds)
118 Skew linear groups, M. SHIRVANI & B. WEHRFRITZ
119 Triangulated categories in the representation theory of finite-dimensional algebras, D. HAPPEL
120 Lectures on Fermat varieties, T. SHIODA
121 Proceedings of *Groups - St Andrews 1985*, E. ROBERTSON & C. CAMPBELL (eds)
122 Non-classical continuum mechanics, R.J. KNOPS & A.A. LACEY (eds)
123 Surveys in combinatorics 1987, C. WHITEHEAD (ed)
124 Lie groupoids and Lie algebroids in differential geometry, K. MACKENZIE
125 Commutator theory for congruence modular varieties, R. FREESE & R. MCKENZIE

London Mathematical Society Lecture Note Series. 122

Non-Classical Continuum Mechanics

Proceedings of the London Mathematical Society
Symposium, Durham, July 1986

Edited by R.J. KNOPS and A.A. LACEY
Department of Mathematics, Heriot-Watt University

The right of the University of Cambridge to print and sell all manner of books was granted by Henry VIII in 1534. The University has printed and published continuously since 1584.

CAMBRIDGE UNIVERSITY PRESS
Cambridge
New York New Rochelle Melbourne Sydney

CAMBRIDGE UNIVERSITY PRESS
Cambridge, New York, Melbourne, Madrid, Cape Town, Singapore, São Paulo

Cambridge University Press
The Edinburgh Building, Cambridge CB2 8RU, UK

Published in the United States of America by Cambridge University Press, New York

www.cambridge.org
Information on this title: www.cambridge.org/9780521349352

First published 1987
Re-issued in this digitally printed version 2007

A catalogue record for this publication is available from the British Library

ISBN 978-0-521-34935-2 paperback

FOREWORD

This volume contains most of the invited lectures delivered as part of the Symposium on "Non-classical Continuum Mechanics, Abstract Techniques and Applications" held in the University of Durham from 2 - 12th July, 1986. The meeting was under the auspices of the London Mathematical Society with financial support generously provided by the Science and Engineering Research Council of Great Britain. To these organisations and also to the University of Durham (including staff of the Department of Mathematics and of Grey College) grateful thanks are extended on behalf of all participants and members of the Organising Committee. Sincere thanks are also due to Mrs P. Hampton (Heriot-Watt University) and Mrs S. Cuttle (University of Durham) who admirably coped with the secretarial and administrative burdens.

One objective of the Symposium was to provide the opportunity for interaction between two broad trends now discernible in continuum mechanics which respectively emphasise applications and rigorous mathematical analysis. The mutual enrichment of these two developments requires frequent exchange of information and thanks to the right note struck in the formal addresses this certainly occurred at the Durham meeting. Indeed, all contributors are to be commended for ensuring that their presentations were readily accessible to both groups thus helping to further the aims of the organisers. This in turn has resulted in the present volume containing articles which besides being accounts of recent progress should also be of wide interest to the specialist and non-specialist alike.

Edinburgh
March 1987

R.J. Knops
A.A. Lacey

LIST OF CONTENTS

LMS SYMPOSIUM ON "NON-CLASSICAL CONTINUUM MECHANICS: ABSTRACT TECHNIQUES AND APPLICATIONS"
2 - 12 JULY, 1986, UNIVERSITY OF DURHAM

LIST OF PARTICIPANTS

ORGANISING COMMITTEE

Professor R J KNOPS, (Chairman), Department of Mathematics, Heriot-Watt University, Riccarton, Edinburgh EH14 4AS

Dr A A LACEY, (Secretary), Department of Mathematics, Heriot-Watt University, Riccarton, Edinburgh EH14 4AS

Professor F M LESLIE, Department of Mathematics, University of Strathclyde, Livingstone Tower, 26 Richmond Street, Glasgow G1 1XH

Professor A J M SPENCER, Department of Theoretical Mechanics, University of Nottingham, University Park, Nottingham NG7 2RD

Professor J R WILLIS, School of Mathematics, University of Bath, Claverton Down, Bath BA2 7AY

KEY SPEAKERS

Professor I MÜLLER, Hermann Föttinger-Institut Für Thermo und Fluiddynamik, Technische Universität Berlin, Strasse des 17 June 135, 1000 Berlin 12, West Germany

Professor O A OLEINIK, Department of Mathematics, University of Moscow, kozp «k" App 133, Moscow B-234, USSR

Professor P J OLVER, University of Minnesota, 127 Vincent Hall, 206 Church Street SE, Minneapolis, Minnesota 55455, USA

Professor G C PAPANICOLAOU, Courant Institute of Mathematical Sciences, New York University, 251 Mercer Street, New York NY 10012, USA

Professor L E PAYNE, Department of Mathematics, White Hall, Cornell University, Ithaca NY 14853, USA

Professor L TARTAR, Centre d'Etudes de Limeil-Valenton, Service MA, BP 27, 94190 Villeneuve-St-Georges, France

PARTICIPANTS

Professor L ANAND, Department of Mechanical Engineering, Room 1.130, Massachusetts Institute of Technology, 77 Massachusetts Avenue, Cambridge MA 02139, USA

Dr C ATKINSON, Department of Mathematics, Imperial College, Queen's Gate, London SW7 2BZ

Professor M M AVELLANEDA, Courant Institute, New York University, 251 Mercer Street, New York 10012, USA

Professor J M BALL, Department of Mathematics, Heriot-Watt University, Riccarton, Edinburgh EH14 4AS

Professor D J BERGMAN, School of Physics and Astronomy, Tel-Aviv University, Tel-Aviv 69978, Israel.

Dr J CARR, Department of Mathematics, Heriot-Watt University, Riccarton, Edinburgh EH14 4AS

Dr M G CLARK, GEC Research, East Lane, Wembley, Middlesex HA9 7PP

Professor S C COWIN, Department of Biomedical Engineering, Tulane University, New Orleans LA 70018, USA

Professor E N DANCER, Department of Mathematics, Statistics & Computer Science, University of New England, Armidale, NSW 2351, Australia

Professor J N FLAVIN, Department of Mathematical Physics, University College, Galway, Ireland

Professor I FONSECA, Centro de Matematica e Applicacoes Fundamentais, Ave Prof Gama Pinto 2, 1699 (Codex) Portugal

Professor G A FRANCFORT, Laboratoire Central des Ponts et Chaussees, 58 Bld Lefebvre, 75732 Paris, Cedex 15, France

Professor A E GREEN, 20 Lakeside, Oxford OX2 8JG

Dr J M HILL, Department of Mathematics, University of Wollongong, Wollongong, New South Wales, Australia

Dr S D HOWISON, Mathematical Institute, University of Oxford, 24-29 St Giles, Oxford OX1 3LB

Professor R JAMES, Department of Aerospace Engineering & Mechanics, 107 Akerman Hall, University of Minnesota, Minneapolis MN 55455, USA

Professor J T JENKINS, Department of Theoretical and Applied Mechanics, Thurston Hall, Cornell University, Ithaca, New York 14853, USA

Professor R V KOHN, Courant Institute of Mathematical Sciences, New York University, 251 Mercer Street, New York 10012, USA

Professor E H LEE, Department of Mechanical Engineering, Aero Engineering and Mechanics, Rensselaer Polytechnic Institute, Troy, New York 12180, USA

Professor J H MADDOCKS, Department of Mathematics, University of Maryland, College Park, Maryland 20742, USA

Professor G A MAUGIN, Universite Pierre et Marie Curie, Laboratoire de Mecanique Theorique, Tour 66, 4 Place Jussieu, 75230 Paris Cedex 05, France

Dr G MILTON, Department of Physics, 405-47, California Institute of Technology, Pasadena, CA 91125, USA

Mr S MULLER, Department of Mathematics, Heriot-Watt University, Riccarton, Edinburgh EH14 4AS

Professor F MURAT, Laboratoire d'Analyse Numerique, Tour 55-65, Universite Paris VI, 75230 Paris Cedex 05, France

Professor A NOVICK-COHEN, Department of Mathematical Sciences, Rensselaer Polytechnic Institute, Troy, New York 12180, USA

Dr J R OCKENDON, Mathematical Institute, University of Oxford, 24-29 St Giles, Oxford OX1 3LB

Professor R W OGDEN, Department of Mathematics, University of Glasgow, Glasgow G1 8QW

Professor E T ONAT, Department of Mechanical Engineering, Yale University, Becton Center, PO Box 2157, Yale Station, New Haven, Connecticut 06520, USA

Dr D F PARKER, Department of Theoretical Mechanics, Nottingham University, Nottingham NG7 2RD

Dr G P PARRY, School of Mathematics, University of Bath, Claverton Down, Bath BA2 7AY

Dr R L PEGO, Department of Mathematics, Heriot-Watt University, Riccarton, Edinburgh EH14 4AS

Professor M RASCLE, Analyse Numerique, Universite de St Etienne, 42023 St Etienne Cedex, France

Dr J F RODRIGUES, Centro de Matematica e Fundamentais Aplicacoes, Av Prof Gama Pinto 2, 1699 Lisboa Codex, Portugal

Dr J RUBINSTEIN, Institute for Mathematics and its Applications, University of Minnesota, 514 Vincent Hall, Minneapolis MN 55455, USA

Professor E SANCHEZ-PALENCIA, Department of Mechanics, Universite Pierre et Marie Curie, 4 Place Jussieu, 75230 Paris, France

Dr C M SAYERS, School of Mathematics, University of Bath, Claverton Down, Bath BA2 7AY

Dr M Shillor, Department of Mathematics, Imperial College, Queen's College, London SW7 2BZ

Dr J SIVALOGANATHAN, School of Mathematics, University of Bath, Claverton Down, Bath, BA2 7AY

Professor M SLEMROD, Department of Mathematical Sciences, Rensselaer Polytechnic Institute, Troy, New York 12180-3590, USA

Professor I N SNEDDON, Department of Mathematics, University of Glasgow, University Gardens, Glasgow G2 8QW

Professor Dr J SPREKELS, Institute of Mathematics, Universität Augsburg, Memminger Strasse 6, D-8900 Augsburg, West Germany

Professor G STRANG, Department of Mathematics, Massachusetts Institute of Technologoy, CAmbridge, Massachusetts 02119, USA

Dr B STRAUGHAN, Department of Mathematics, University of Glasgow, University Gardens, Glasgow G12 8QW

Dr D R S TALBOT, Department of Mathematics, Coventry (Lanchester), Polytechnic, Priory Street, Coventry CV1 5FB

Dr TANG QI, Department of Mathematics, Heriot-Watt University, Riccarton, Edinburgh EH14 4AS

Dr A B TAYLER, Mathematical Institute, Oxford University, 24-29 St Giles, Oxford, OX1 3LB

Professor J F TOLAND, School of Mathematics, University of Bath, Claverton Down, Bath BA2 7AY

Dr K WALTON, School of Mathematics, University of Bath, Claverton Down, Bath BA2 tAY

Dr J-R L WEBB, Department of Mathematics, University of Glasgow, Glasgow G12 8QW

Dr H T WILLIAMS, Department of Mathematics, Heriot-Watt University, Edinburgh EH14 4AS

PART I

PRINCIPAL LECTURES

Pulse Reflection by a Random Medium

R. Burridge

Schlumberger-Doll Research, Ridgefield, CT 06877

G. Papanicolaou

Courant Institute, New York University, 251 Mercer Street, New York, NY 10012

P. Sheng and B. White

Exxon Research & Engineering Company, Route 22 East, Clinton Township, Annandale, NJ 08801

1. Introduction

The study of pulse propagation in one dimensional random media arises in many applied contexts. While reflection and transmission of monochromatic waves was studied extensively some time ago [1-6 and references therein], new and perhaps surprising results emerge in the study of pulses that cannot be understood simply from the single frequency analysis by Fourier synthesis. The numerical study of Richards and Menke [7] drew our attention to these questions and led to [8] and [9]. Here we extend and simplify the analysis of [8] and give several new results. The computations are at a formal level comparable to the one in [8].

In [8] we analyzed the reflection of a pulse that is broad compared to the size of the inhomogeneities of the random medium. The random functions characterizing the medium properties were statistically homogeneous. We gave a rather complete description of the reflected signal process in a well defined asymptotic limit in which it has a canonical structure. We introduced the notion of a windowed process and showed that the canonical reflection process is windowed and Gaussian. We found a scaling law for the power spectral density but not its explicit form. All this was subjected to extensive numerical simulations in [9] where an intrinsic scaling, localization length scaling, was introduced that makes comparison to the theory much more reliable. This intrinsic scaling idea is not fully understood theoretically but seems to be very promising.

In this paper we extend the analysis to random media that are not statistically homogeneous. The incident pulse is now broad compared to the size of the inhomogeneities but short compared to the scale of variation of the mean properties. The pulse can thus resolve the mean structure while the fluctuations affect the reflected signal in a canonical way. The problem is formulated in section 2. The calculations are done in the frequency domain as in [8] and at the level of second moments (power spectra) they differ little from similar calculations in [1] for example. In section 3 we state the results which include a new equation (3.4) for the canonical power spectral density in a statistically inhomogeneous random medium. In the special statistically homogeneous case of [8] they can actually be solved explicitly (formula (3.7)). We were not able to do this in [8]. In section 4 we show how the results are obtained, including formula (3.7). Appendix A contains a brief outline of the main result in the asymptotic analysis of stochastic equations that we need here (cf also [8]).

Since all calculations here are at the level of the single (or finite) frequency results of [1] why is the analysis of pulse statistics so different? A careful look at what follows shows that we have frequently interchanged limits in the course of taking Fourier transforms and doing the small parameter asymptotics. To justify these interchanges one must do the small parameter asymptotics in an infinite dimensional setting (simultaneously for all frequencies) which is much more involved. If this seems pedantic, given that our results are correct, consider showing that the limit pulse statistics are Gaussian (this is not attempted here). In [8] we gave a finite dimensional argument for this that was incomplete and not very transparent. In the more general setting [10] the Gaussian property comes out much more naturally. It is worth noting that even though the limit law is Gaussian, the usual central limit methods do not apply because the necessary asymptotic independence (in the frequency domain) is very weak and controlled largely by the geometrical optics limit, not the mixing properties of the random medium.

2. Formulation and Scaling

We consider a one-dimensional acoustic wave propagating in a random slab of material occupying the half space $x < 0$. We will analyze in detail the backscatter at $x = 0$.

Let $p(t,x)$ be the pressure and $u(t,x)$ velocity. The linear conservation laws of momentum and mass governing acoustic wave propagation are

$$\rho(x)\frac{\partial}{\partial t}u(t,x) + \frac{\partial}{\partial x}p(t,x) = 0 \tag{2.1}$$

$$\frac{1}{K(x)}\frac{\partial}{\partial t}p(t,x) + \frac{\partial}{\partial x}u(t,x) = 0$$

where ρ is density and K the bulk modulus. We define means of ρ and $\frac{1}{K}$ as

$$\rho_o = E\,[\rho] \tag{2.2}$$

$$\frac{1}{K_o} = E\,[\frac{1}{K}].$$

In the special case that ρ and K are stationary random functions of position x, ρ_o, K_o are the constant parameters of effective medium theory. That is, a pulse of long wavelength will propagate over distances that are not too large as if in a homogeneous medium with "effective" constant parameters ρ_o, K_o, and hence with propagation speed

$$c_o = \sqrt{K_o / \rho_o}\ . \tag{2.3}$$

We consider here the case where ρ_o, K_o, c_o are not constant, but vary slowly compared to the spatial scale, l_o, of a typical inhomogeneity. We may take the "microscale" l_o to be the correlation length of ρ and $\frac{1}{K}$. We introduce a "macroscale", l_o/ε^2, where $\varepsilon > 0$ is a small parameter. It is on this macroscale that ρ_o, K_o, and other statistics of ρ and K are allowed to vary. We thus write the density and bulk modulus on the macroscale in the following scaled form.

$$\rho(x) = \rho_o\,(\frac{x}{l_o})\left[1 + \eta(\frac{x}{l_o}, \frac{x}{\varepsilon^2 l_o})\right] \tag{2.4}$$

$$\frac{1}{K(x)} = \frac{1}{K_o(x/l_o)}\left[1 + \nu(\frac{x}{l_o}, \frac{x}{\varepsilon^2 l_o})\right]$$

where the random fluctuations η and ν have mean zero and slowly varying statistics. The mean density ρ_o and the mean bulk modulus K_o are assumed to be differentiable functions of x.

Equations (2.1) are to be supplemented with boundary conditions at $x = 0$ corresponding to different ways in which the pulse is generated at the interface. In the cases analyzed below the pulse width is assumed to be on a scale intermediate between the microscale and the macroscale. That is, the pulse is broad compared to the size of the random inhomogeneities, but short compared to the non-random variations. Thus the small scale structure will introduce only random effects which the pulse is too broad to probe in detail. In contrast, the pulse is chosen to probe the non-random macroscale, from which it reflects and refracts in the manner of ray theory (geometrial optics). We will recover macroscopic variations of the medium by examination of reflections at $x = 0$.

Let typical values of ρ_o, K_o be $\bar{\rho}$, $\bar{K}$ with $\bar{c} = \sqrt{\bar{K}/\bar{\rho}}$. Then for $f(t)$ a smooth function of compact support in $[0,\infty)$ we define the incident pulse by

$$f^{\varepsilon}(t) = \frac{1}{\varepsilon^{1/2}} f\left(\frac{\bar{c}\,t}{\varepsilon\, l_o}\right). \tag{2.5}$$

This pulse, f^{ε}, will be convolved with the appropriate Green's function depending on how the wave is excited at the interface. The pre-factor $\varepsilon^{-1/2}$ is introduced to make the energy of the pulse independent of the small parameter ε.

We consider here the "matched medium" boundary condition. It is assumed that the wave is incident on the random medium occupying $x < 0$ from a homogeneous medium occupying $x > 0$ and characterized by the constant parameters $\rho_o(0)$, $K_o(0)$. One may similarly consider an unmatched medium where ρ_o, K_o are discontinuous at $x = 0$, but we do not carry this out here. To obtain the Green's function for this problem we introduce the initial-boundary condition for a left-travelling wave which strikes $x = 0$ a time $t = 0$

$$u = l_o\, \delta\left(t + \frac{x}{c_o(0)}\right) \tag{2.6}$$

$$p = -\, l_o\, \rho_o(0)\, c_o(0)\, \delta\left(t + \frac{x}{c_o(0)}\right)$$

The Green's function G will then be a right-going wave in $x > 0$ and as $x \downarrow 0$

$$G = \frac{1}{2}\left[u(t,0) - \frac{p(t,0)}{(\rho_o(0)c_o(0))}\right] \tag{2.7}$$

We non-dimensionalize by setting

$$x' = x/l_o \quad p' = p/\bar{\rho}\,\bar{c}^2 \tag{2.8}$$

$$t' = \bar{c}t/l_o \quad u' = u/\bar{c}$$

By inserting (2.8) into the above equations, and dropping primes, it can be shown that without loss of generality $\bar{K}$, $\bar{\rho}$, $\bar{c}$, l_o may be taken equal to unity, after K, ρ, c are replaced by their normalized forms.

We will determine the statistics of the Green's function convolved with the pulse f^{ε}. Let

$$G^{\varepsilon}_{t,f}(\sigma) = (G * f^{\varepsilon})(t + \varepsilon\sigma) \tag{2.9}$$

$$= \int_0^{t+\varepsilon\sigma} G(t + \varepsilon\sigma - s) f^{\varepsilon}(s)\, ds\,.$$

We consider the above expression as a stochastic process in σ, with t held fixed. That is, for each t we consider a "time window" centered at t, and of duration on the order of a pulse width, with the parameter

σ measuring time within this window.

For the analysis of this problem, we Fourier transform in time, choosing a frequency scale appropriate to the pulse $f^{\varepsilon}(t)$. Thus, letting

$$\hat{f}(\omega) = \int_{-\infty}^{\infty} e^{i\omega t} f(t)\, dt \tag{2.10}$$

we transform (2.1) by

$$\hat{u}\,(\omega, x) = \int e^{i\omega t/\varepsilon}\, u\,(t,x)\, dt \tag{2.11}$$

$$\hat{p}\,(\omega, x) = \int e^{i\omega t/\varepsilon}\, p\,(t,x)\, dt$$

so that

$$G^{\varepsilon}_{t,f}\,(\sigma) = \frac{1}{2\,\pi\,\varepsilon^{1/2}} \int_{-\infty}^{\infty} e^{-i\omega[t+\varepsilon\sigma]/\varepsilon}\, \hat{f}\,(\omega)\, \hat{G}\,(\omega) d\omega\,. \tag{2.12}$$

In (2.12) $\hat{G}$ is the appropriate combination of $\hat{u}$, $\hat{\rho}$ obtained by Fourier transform of (2.7).

From (2.1), (2.4), (2.11), $\hat{u}$, $\hat{p}$ satisfy

$$\frac{\partial}{\partial x}\hat{p} = \frac{i\,\omega}{\varepsilon}\,\rho_o(x)\,[1+\eta\,(x, \frac{x}{\varepsilon^2})]\,\hat{u} \tag{2.13}$$

$$\frac{\partial}{\partial x}\hat{u} = \frac{i\,\omega}{\varepsilon}\,\frac{1}{K_o(x)}\,[1+\nu\,(x, \frac{x}{\varepsilon^2})]\,\hat{p}\,.$$

In the frequency domain a radiation condition as $x \to -\infty$, is required for (2.13). One way to do this is to terminate the random slab at a finite point $x = -L$, and assume the medium is not random for $x > -L$. We can later let $L \to -\infty$ but in any case the reflected signal up to a time t is not affected by how we terminate the slab at a sufficiently distant point $-L$. This is a consequence of the hyperbolicity of (2.1).

We next introduce a right going wave A and a left going wave B, with respect to the macroscopic medium. Let the travel time in the macroscopic medium be given by

$$\tau(x) = \int_x^0 \frac{ds}{c_o(s)}\ ,\quad x<0 \tag{2.14}$$

We define A, B by

$$\hat{u} = \frac{1}{(K_o\rho_o)^{1/4}}\,[\,A\,e^{-i\,\omega\,\tau/\varepsilon} + B\,e^{i\,\omega\,\tau/\varepsilon}\,]$$

$$\hat{p} = (K_o\, \rho_o)^{1/4}\, [\, A\, e^{-i\,\omega\tau/\varepsilon} - B\, e^{i\,\omega\tau/\varepsilon}\,] \tag{2.15}$$

Putting (2.14), (2.15) into (2.13) yields equations for A, B. Define the random functions $m^{\varepsilon}(x)$ and $n^{\varepsilon}(x)$ by

$$m^{\varepsilon}(x) = m(x,\, x/\varepsilon^2) = \frac{1}{2}\, [\eta(x,x/\varepsilon^2) + \nu(x,x/\varepsilon^2)] \tag{2.16}$$

$$n^{\varepsilon}(x) = n(x,\, x/\varepsilon^2) = \frac{1}{2}\, [\eta(x,x/\varepsilon^2) - \nu(x,x/\varepsilon^2)]$$

Then

$$\frac{d}{dx}\begin{bmatrix} A \\ B \end{bmatrix} = \frac{i\,\omega}{\varepsilon}\left(\frac{\rho_o}{K_o}\right)^{1/2}\begin{bmatrix} m^{\varepsilon} & n^{\varepsilon}\, e^{2i\omega\tau/\varepsilon} \\ -n^{\varepsilon}\, e^{-2i\,\omega\tau/\varepsilon} & -m^{\varepsilon} \end{bmatrix}\begin{bmatrix} A \\ B \end{bmatrix}$$

$$+\frac{1}{4}\,\frac{(K_o\,\rho_o)'}{(K_o\,\rho_o)}\begin{bmatrix} 0 & e^{2i\omega\tau/\varepsilon} \\ e^{-2i\omega\tau/\varepsilon} & 0 \end{bmatrix}\begin{bmatrix} A \\ B \end{bmatrix}. \tag{2.17}$$

We take as boundary conditions for (2.17) that there is no right-going wave at $x = -L$, and that there is a unit left-going wave at $x = 0$.

$$A(-L) = 0 \quad B(0) = 1 \tag{2.18}$$

$$B(-L) = T \quad A(0) = R = R^{\varepsilon}(-L,\, \omega)$$

Here T is the transmission coefficient for the slab, and $R^{\varepsilon}(-L,\, \omega)$ is the reflection coefficient. From (2.6), (2.7) we see that

$$\hat{G} = R^{\varepsilon}\,(-L,\, \omega)\,. \tag{2.19}$$

We introduce the fundamental matrix solution of the linear system (2.17). That is, let $Y(x, -L)$ satisfy (2.17) with the initial condition that $Y(-L, -L) = I$ the 2 x 2 identity. From symmetries in (2.17) it is apparent that if $(a,\bar{b})^T$ is a vector solution (bar denotes complex conjugate and T transpose), then so is $(b,\, \bar{a})^T$. Thus

$$Y = \begin{bmatrix} a & b \\ \bar{b} & \bar{a} \end{bmatrix}. \tag{2.20}$$

Furthermore, since the system has trace zero, Y has determinant one. Hence

$$|\, a\,|^2 - |\,b\,|^2 = 1\,. \tag{2.21}$$

Now the reflection coefficient R may be expressed in terms of a, b, by writing (2.18) in terms of propagators, i.e.

$$\begin{bmatrix} a & b \\ \bar{b} & \bar{a} \end{bmatrix} \begin{bmatrix} 0 \\ T \end{bmatrix} = \begin{bmatrix} R \\ 1 \end{bmatrix}$$

and hence

$$R = \frac{b}{\bar{a}}, \qquad T = \frac{1}{\bar{a}} \tag{2.22}$$

Now from (2.17), (2.20) we have that

$$\frac{da}{dx} = \frac{i\,\omega}{\varepsilon}\left[\frac{\rho_o}{K_o}\right]^{1/2}\left[m^\varepsilon\, a + n^\varepsilon\, \bar{b}\, e^{2i\,\omega\tau/\varepsilon}\right] + \frac{1}{4}\,\frac{(\rho_o K_o)'}{\rho_o K_o}\,\bar{b}\, e^{2i\omega\tau/\varepsilon}$$

$$\frac{d\bar{b}}{dx} = -\frac{i\,\omega}{\varepsilon}\left[\frac{\rho_o}{K_o}\right]^{1/2}\left[n^\varepsilon\, a\, e^{-2i\omega\tau/\varepsilon} + m^\varepsilon \bar{b}\right] + \frac{1}{4}\,\frac{(\rho_o K_o)'}{\rho_o K_o}\, a\, e^{-2i\omega\tau/\varepsilon} \tag{2.23}$$

$$a(-L) = 1,\; b(-L) = 0\,.$$

Therefore, from (2.22), (2.23) we can derive the Riccati equation for R

$$\frac{dR^\varepsilon}{dx} = \frac{i\,\omega}{\varepsilon}\left[\frac{\rho_o}{K_o}\right]^{1/2}\left[n^\varepsilon\, e^{2i\,\omega\tau/\varepsilon} + 2m^\varepsilon\, R^\varepsilon + n^\varepsilon (R^\varepsilon)^2\, e^{-2i\omega\tau/\varepsilon}\right]$$

$$+\frac{1}{4}\,\frac{(\rho_o K_o)'}{(\rho_o K_o)}\left[e^{2i\,\omega\tau/\varepsilon} - (R^\varepsilon)^2\, e^{-2i\,\omega\tau/\varepsilon}\right] \tag{2.24}$$

$$R^\varepsilon(-L) = 0\;.$$

The boundary condition at $-L$ in (2.24) is for termination of the random slab by a uniform medium. If the medium is homogeneously random beyond $-L$ ($\rho_o(x)$, $K_o(x)$ constant) then we will have total reflection at $-L$ because the wave cannot penetrate the random medium to infinite depth. In fact in a statistically homogeneous random medium we have that

$$|\,T\,| \to 0 \text{ as } L \to -\infty \tag{2.25}$$

exponentially fast which follows from Furstenberg's theorem [11,12]. Since (2.21), (2.22) imply that $|\,R\,|^2 + |\,T\,|^2 = 1$ we have

$$|\,R\,| \to 1 \text{ as } L \to -\infty\,. \tag{2.26}$$

It is convenient to analyze (2.24) with a **totally reflecting termination**, so that

$$R^\varepsilon = e^{-i\,\psi^\varepsilon}\,. \tag{2.27}$$

and the number of degrees of freedom is reduced by one. This simplification, not possible when we do have transmission, was not made in [8]. Putting (2.27) into (2.24) yields

$$\frac{d}{dx}\psi^{\varepsilon} = -\frac{\omega}{\varepsilon}\left[\frac{\rho_o(x)}{K_o(x)}\right]^{1/2}\left[2\,m^{\varepsilon}(x) + 2n^{\varepsilon}(x)\cos\left(\psi^{\varepsilon} + \frac{2\omega\tau(x)}{\varepsilon}\right)\right. \tag{2.28}$$

$$\left. + \frac{1}{2}\frac{(\rho_o K_o)'}{\rho_o K_o}\sin\left(\psi^{\varepsilon} + \frac{2\omega\tau(x)}{\varepsilon}\right)\right]$$

and we take ψ^{ε} to be asymptotically stationary as $x \to -\infty$.

To recapitulate, the asymptotically stationary solution of (2.28), evaluated at $x=0$ is put into (2.27) to yield the totally reflecting reflection coefficient R^{ε} at frequncy ω. The frequency domain Green's function is then given by (2.19) The result is then transformed back to the time domain by (2.12).

3. Statement of the main results

Let $G^{\varepsilon}_{t,f}(\sigma)$ be the reflection process observed at $x=0$ within the time window centered at t. Then $G^{\varepsilon}_{t,f}(\cdot)$ converges weakly as $\varepsilon \downarrow 0$ to a stationary Gaussian process with mean zero and power spectral density

$$S_t(\omega) = |\,\hat{f}(\omega)\,|^2\,\mu(t,\omega)\,, \tag{3.1}$$

The normalized power spectral density μ is computed as follows.

Let α_{nn} be the integral of the second moment of the medium properties defined by

$$\alpha_{nn}(x) = \int_0^{\infty} E[n(x,y)\,n(x,y+s)]\,ds\,. \tag{3.2}$$

Let $\tau(x)$ be travel time to depth x defined by (2.14), and let $\overline{x}(\tau)$ be its inverse which is depth reached up to time t in the medium without fluctuations. Define

$$\gamma(\tau) = \frac{\alpha_{nn}(\overline{x}(\tau))}{c_o(\overline{x}(\tau))}\,. \tag{3.3}$$

Let $W^{(N)}(\tau, t, \omega)$, $N = 0,1,2...$ be the solution of

$$\frac{\partial W^{(N)}}{\partial \tau} + 2N\frac{\partial W^{(N)}}{\partial t} - 2\omega^2\gamma(\tau)\left\{[N+1]^2 W^{(N+1)}\right.$$

$$\left. - 2N^2 W^{(N)} + [N-1]^2 W^{(N-1)}\right\} = 0 \tag{3.4}$$

for

$$t,\ \tau > 0\,, \quad N = 0,1,2,\cdots$$

with

$$W^{(N)} \equiv 0 \text{ for } t < 0\,,\ N < 0\,.$$

and

$$W^{(N)}(0,t,\omega) = \delta(t)\,\delta_{N,1}\,. \tag{3.5}$$

Then

$$\mu(t,\omega) = \lim_{\tau\to\infty} W^{(0)}(\tau,t,\omega)\,. \tag{3.6}$$

The system (3.4) is hyperbolic so it is not necessary to take a limit in (3.6) because $W^{(0)}$ is constant for $\tau > t/2$. Thus

$$\mu(t,\omega) = W^{(0)}(\frac{t}{2},t,\omega) \tag{3.6a}$$

For the case of a homogeneous medium $[c_o,\ \gamma = const = \ddot{\gamma}]$ the normalized power spectral density can be computed explicitly.

$$(BCI)\quad \mu_1(t,\omega) = \frac{\omega^2\tilde{\gamma}}{[1+\omega^2\tilde{\gamma}t]^2} \tag{3.7}$$

4. Calculation of Power Spectral Density.

We next calculate the power spectrum, as $\varepsilon \downarrow 0$, or the reflection process $G^{\varepsilon}_{t,f}(\sigma)$. From (2.9) we have the correlation function $C^{\varepsilon}_{t,f}$

$$C^{\varepsilon}_{t,f}\,\sigma) \equiv E[G^{\varepsilon}_{t,f}(\sigma)\,G^{\varepsilon}_{t,f}(0)] \tag{4.1}$$

$$= \frac{1}{4\pi^2\varepsilon}\int_{-\infty}^{\infty} d\omega_1 \int_{-\infty}^{\infty} d\omega_2\, e^{-i\,\omega_1 t/\varepsilon}\, e^{-i\omega_1\sigma}\, e^{i\omega_2 t/\varepsilon}$$

$$\cdot \hat{f}(\omega_1)\,\bar{\hat{f}}(\omega_2)\,E[\hat{G}(\omega_1)\,\bar{\hat{G}}(\omega_2)]\,.$$

Let

$$u^{\varepsilon}(\omega,h) = E[\,\hat{G}(\omega - \frac{\varepsilon h}{2})\,\bar{\hat{G}}(\omega + \frac{\varepsilon h}{2})]\,. \tag{4.2}$$

We will show that the limit

$$u(\omega,h) = \lim_{\varepsilon \downarrow 0} u^{\varepsilon}(\omega,h) \tag{4.3}$$

exists, and we will characterize it in this section. Then, after the change of variables $\omega = \frac{1}{2}(\omega_1 + \omega_2)$, $h = (\omega_2 - \omega_1)/\varepsilon$ in (4.1) we obtain in the limit $\varepsilon \to 0$

$$C_{t,f}(\sigma) \equiv \lim_{\varepsilon \downarrow 0} C^{\varepsilon}_{t,f}(\sigma) \tag{4.4}$$

$$= \frac{1}{4\pi^2} \int_{-\infty}^{\infty} \int_{-\infty}^{\infty} e^{-i\omega\sigma}\, e^{iht}\, |\,\hat{f}(\omega)\,|^2\, u(\omega,h)\, dh\,.$$

Let

$$\mu(t,\omega) = \frac{1}{2\pi} \int_{-\infty}^{\infty} e^{iht}\, u(\omega,h)\, dh \tag{4.5}$$

Then from (4.4), (4.5) the power spectral density, $S_t(\omega)$ is given by

$$S_t(\omega) \equiv \int_{-\infty}^{\infty} e^{i\omega\sigma}\, C_{t,f}(\sigma)\, d\sigma \tag{4.6}$$

$$= |\,\hat{f}(\omega)\,|^2\, \mu(t,\omega)\,.$$

In the remainder of this section we characterize $u(\omega,h)$ and its transform $\mu(t,\omega)$. From (4.2) and (2.19) we see that $u^{\varepsilon}(\omega,h)$ can be computed from knowledge of the joint statistics of R^{ε}_1, R^{ε}_2 the reflection coefficients corresponding, respectively, to frequencies $\omega_1 = \omega - \varepsilon h/2$, $\omega_2 = \omega + \varepsilon h/2$. Since we are in the totally reflecting case (2.27)

$$R_1 = e^{-i\psi^{\varepsilon}_1},\; R^{\varepsilon}_2 = e^{-i\psi^{\varepsilon}_2}\,,$$

where ψ^{ε}_1, ψ^{ε}_2 correspond, respectively, to the same frequencies, and each satisfies (2.28). We shall compute the joint distribution of ψ^{ε}_1, ψ^{ε}_2 as ε tends to zero.

Let

$$\psi^{\varepsilon} = \begin{bmatrix} \psi^{\varepsilon}_1 \\ \psi^{\varepsilon}_2 \end{bmatrix}$$

From (2.28) we see that ψ^{ε} satisfies the differential equation

$$\frac{d\psi^\varepsilon}{dx} = \frac{1}{\varepsilon}\,\mathbf{F}(x,\frac{x}{\varepsilon^2},\frac{\tau(x)}{\varepsilon},\psi^\varepsilon) + \mathbf{G}(x,\frac{x}{\varepsilon^2},\frac{\tau(x)}{\varepsilon},\psi^\varepsilon) \tag{4.7}$$

where

$$\mathbf{F}(x,y,h,\psi) = -2\omega\left[\frac{\rho_o(x)}{K_o(x)}\right]^{1/2}\begin{bmatrix} m(x,y)+n(x,y)\cos(\psi_1+2\omega h - h\,\tau(x)) \\ m(x,y)+n(x,y)\cos(\psi_2+2\omega h + h\tau(x)) \end{bmatrix} \tag{4.8}$$

and

$$\mathbf{G}(x,y,h,\psi) = \begin{bmatrix} G(x,y,h,\psi) \\ G_2(x,y,h,\psi) \end{bmatrix}$$

$$G(x,y,h,\psi) = h\left[\frac{\rho_o}{K_o}\right]^{1/2}(m(x,y)+n(x,y)\cos(\psi_1+2\omega h - h\,\tau(x)))$$

$$+\frac{1}{2}\frac{(\rho_o(x)\,K_o(x))'}{\rho_o(x)\,K_o(x)}\sin\,(\psi_1+2\omega h - h\,\tau(x))$$

$$G_2(x,y,h,\psi) = -h\left[\frac{\rho_o}{K_o}\right]^{1/2}(m(x,y)+n(x,y)\cos(\psi_2+2\omega h + \tau(x)))$$

$$+\frac{1}{2}\frac{(\rho_o(x)K_o(x))'}{\rho_o(x)K_o(x)}\sin\,(\psi_2+2\omega h + h\tau(x))$$

We assume, as in Appendix A, that the randomness in equation (4.7) is generated by an ergodic Markov process $q^\varepsilon(x) = q(x,x/\varepsilon^2)$ in Euclidean space R^d of arbitrary dimension d. It is assumed that $q(x,x/\varepsilon^2)$ is a random process on the fast, x/ε^2, spatial scale, but has slowly-varying statistics on the x scale. We express this mathematically by the assumption that $q(x,y)$ is, for fixed x, a stationary ergodic Markov process in y with infinitesimal generator Q_x, depending on x. We then write $m(x,x/\varepsilon^2) = \tilde{m}(x,q(x,\ x/\varepsilon^2))$, etc. (to simplify notation we will drop tildes). A very wide class of processes with small scale randomness but slowly-varying statistics can be generated in this way.

The process $(q^\varepsilon,\ \psi^\varepsilon)\ \varepsilon\, R^{d+2}$, the solution of (4.7) together with its coefficients, is now jointly Markovian, with infinitesimal generator

$$\mathbf{L}_x^\varepsilon = \frac{1}{\varepsilon^2}\,Q_x + \frac{1}{\varepsilon}\,\mathbf{F}(x,\frac{x}{\varepsilon^2},\ \frac{\tau(x)}{\varepsilon},\ \psi)\cdot\nabla_\psi \tag{4.9}$$

$$+\,\mathbf{G}(x,\frac{x}{\varepsilon^2},\ \frac{\tau(x)}{\varepsilon},\psi)\cdot\nabla_\psi$$

From the results of Appendix A, we have that ψ^ε converges (weakly) to a process ψ which is Markovian by itself, without the necessity of including q. The limit process ψ has the x-dependent infinitesimal generator L_x, where

$$L_x = \frac{4\omega^2}{c_o^2(x)} \left\{ \alpha_{mm}(x) \left[\frac{\partial}{\partial\psi_1} + \frac{\partial}{\partial\psi_2}\right]^2 \right. \tag{4.10}$$

$$\left. + \frac{1}{2}\,\alpha_{nn}(x) \left[\frac{\partial^2}{\partial\psi_1^2} + \frac{\partial^2}{\partial\psi_2^2} + 2\cos\left(\psi_2 - \psi_1 + 2h\,\tau(x)\right) \frac{\partial^2}{\partial\psi_1\partial\psi_2}\right] \right\}.$$

In (4.10), the coefficients α_{mm}, α_{nn} are defined by the averaged second moments

$$\alpha_{mm}(x) = \int_0^\infty E[m(x,q(x,y))\, m(x,q(x,y+r))]\, dr \tag{4.11}$$

$$\alpha_{nn}(x) = \int_0^\infty E[n(x,q(x,y))n(x,q(x,y+r))]\, dr\,.$$

In Appendix A we show briefly how these results are obtained.

The generator (4.10) is better expressed in terms of the sum and difference variables

$$\psi = \psi_2 - \psi_1,\; \tilde{\psi} = \frac{1}{2}(\psi_2 + \psi_1)\,.$$

Then

$$L_x = \frac{4\omega^2}{c_o^2(x)} \left\{ \alpha_{mm}(x)\frac{\partial^2}{\partial\tilde{\psi}^2} + \frac{1}{4}\,\alpha_{nn}(x)\left[1 + \cos\left(\psi + 2h\,\tau(x)\right)\right] \frac{\partial^2}{\partial\tilde{\psi}^2} \right. \tag{4.12}$$

$$\left. + \alpha_{nn}(x)\left[1 - \cos\left(\psi + 2h\tau(x)\right)\right] \frac{\partial^2}{\partial\psi^2} \right\}.$$

Using (4.12) we can now formulate the equations for $u(\omega,h)$ and its transform $\mu(t,\omega)$. From (2.19), (2.27) and (4.2), (4.3) we have that $u_1(\omega,h)$ is

$$u(\omega, h) = E[e^{i\psi}]\,. \tag{4.13}$$

Note that the coefficients in (4.12) do not depend on $\tilde{\psi}$, so that ψ is Markovian by itself. The function $u(\omega,h)$ can therefore be calculated from the solution V of the Kolmogorov backward equation

$$\frac{\partial V}{\partial x} + \frac{4\omega^2\alpha_{nn}(x)}{c_o^2(x)}\left[1 - \cos\left(\psi + 2h\,\tau(x)\right)\right] \frac{\partial^2 V}{\partial\psi^2} = 0,\; x < 0\,. \tag{4.14}$$

with the final condition

$$V\,|_{x=0} = e^{i\psi}\,. \tag{4.15}$$

The function u is then

$$u(\omega, h) = \lim_{x \to -\infty} V(x, \psi, \omega, h)\,, \tag{4.16}$$

where the limit in (4.16) exists and is independent of ψ. This follows easily assuming that α_{nn}, τ and c_o are constant for $-x$ large.

Equation (4.14) may be simplified somewhat by the change of variables

$$\hat{\psi} = \psi + 2h\,\tau(x) \tag{4.17}$$

$$\hat{x} = x$$

Then upon dropping hats, it becomes

$$\frac{\partial V}{\partial x} - \frac{2h}{c_o(x)}\frac{\partial V}{\partial \psi} + \frac{4\omega^2 \alpha_{nn}(x)}{c_o^2(x)}[1 - \cos\psi]\frac{\partial^2 V}{\partial \psi^2} = 0\,, \text{ for } x < 0\,. \tag{4.18}$$

To summarize, (4.18) is to be solved for V subject to the final condition (4.15). Then u is obtained from (4.16).

Equation (4.18) can be solved by Fourier series in ψ Let

$$V = \sum_{N=-\infty}^{\infty} V^{(N)} e^{iN\psi}\,, \; \psi \; in \; [-\pi, \pi)\,.$$

Then (4.18) becomes the infinite-dimensional system

$$\frac{\partial V^{(N)}}{\partial x} - \frac{2ihN}{c_o(x)}V^{(N)} + \frac{2\omega^2 \alpha_{nn}(x)}{c_o^2(x)}\Big\{[N+1]^2\, V^{(N+1)} \tag{4.19}$$

$$-2N^2 V^{(N)} + [N-1]^2\, V^{(N-1)}\Big\} = 0 \text{ for } x < 0\,.$$

with boundary condition

$$V^{(N)}\,|_{x=0} = \delta_{N,1} \tag{4.20}$$

We now have that $V^{(N)} \to 0$ as $x \to -\infty$ for $N \neq 0$, and $u(\omega,h)$ is given by $\lim_{x \to -\infty} V^{(o)}$.

Equivalently, we may now formulate an infinite set of coupled linear first order partial differential equations for $\mu(t, \omega)$ the Fourier transform in h of $u(\omega,h)$, equation (4.5). Let $W^{(N)}$ be the Fourier transform in h, of $V^{(N)}$. Then we obtain easily from (4.19), (4.20) that

$$\frac{\partial W^{(N)}}{\partial x} - \frac{2N}{c_o(x)}\frac{\partial W^{(N)}}{\partial t} + \frac{2\omega^2 \alpha_{nn}(x)}{c_o^2(x)}\left\{[N+1]^2 W^{(N+1)} - 2N^2 W^{(N)} + [N-1]^2 W^{(N-1)}\right\} = 0 \tag{4.21}$$

with

$$W^{(N)}|_{x=0} = \delta(t)\,\delta_{N,1} \tag{4.22}$$

The normalized power spectral density $\mu(t,\omega)$ is then given by the limit of $W^{(o)}$ as $x \to -\infty$. The formulation given in section 3 follows from making the change of variables in (4.21) from depth, x, to travel-time $\tau(x)$, equation (2.14).

We will next calculate μ explicitly for the case of a statistically homogeneous medium. That is, we assume that c_o and

$$\tilde{\gamma} = \frac{\alpha_{nn}(x)}{c_o(x)} \tag{4.23}$$

do not depend on x. We will use a forward Kolmogorov equation formulation based upon (4.18) and (4.15). For the case of c_o and $\tilde{\gamma}$ constant, great simplification is achieved by first making the transformation

$$z = \cot\frac{\psi}{2} \tag{4.24}$$

Then $z \in [-\infty, \infty)$ when $\psi \in [-\pi, \pi)$. The Kolmogorov backward equation for z is obtained by change of variables in (4.18)

$$\frac{\partial V}{\partial x} + h\,\frac{(1+z^2)}{c_o}\frac{\partial V}{\partial z} + 2\frac{\omega^2\tilde{\gamma}}{c_o}\frac{\partial}{\partial z}\left\{(1+z_2)\frac{\partial V}{\partial z}\right\} = 0 \tag{4.25}$$

The probability density associated with (4.25), $P(x,z)$, satisfies the Kolmogorov forward equation

$$c_o\frac{\partial P}{\partial x} = -\frac{h}{\partial z}\{(1+z^2)P\} + 2\omega^2\,\tilde{\gamma}\frac{\partial}{\partial z}\left\{(1+z^2)\frac{\partial P}{\partial z}\right\} \tag{4.26}$$

The invariant density $\bar{P}_h(z)$ is obtained by setting $\partial P/\partial x = 0$ in (4.26). For $h > 0$ we have

$$\bar{P}_h(z) = \frac{h}{2\pi\omega^2\tilde{\gamma}}\int_0^\infty \frac{e^{-h\xi/2\omega^2\tilde{\gamma}}}{[1+(\xi+z)^2]}\,d\xi \tag{4.27}$$

For $h < 0$, symmetries in (4.26) imply that

$$\bar{P}_{-h}(-z) = \bar{P}_h(z) . \tag{4.28}$$

Now from (4.24) we have that

$$e^{i\Psi} = (\frac{z+i}{z-i}) . \tag{4.29}$$

Therefore

$$u(\omega, h) = E[e^{i\Psi}] = \int_{-\infty}^{\infty} \bar{P}_h(z) (\frac{z+i}{z-i}) \, dz \tag{4.37}$$

$$= \frac{h}{2\omega^2\tilde{\gamma}} \int_0^{\infty} e^{-h\xi/2\omega^2\tilde{\gamma}} \left[\frac{\xi}{\xi+2i} \right] d\xi$$

$$\text{for} \quad h > 0 .$$

From (4.28), (4.30) it follows that

$$u(\omega, -h) = \overline{u(\omega,h)} . \tag{4.31}$$

Therefore

$$\mu(t, \omega) = \frac{1}{\pi} \mathrm{Re} \int_0^{\infty} e^{iht} u_1(\omega,h) \, dh . \tag{4.32}$$

Substitution of (4.30) into (4.32) then gives, after some elementary integrations

$$\mu_1(t, \omega) = \frac{\omega^2\tilde{\gamma}}{[1+\omega^2\tilde{\gamma}t]^2} . \tag{4.33}$$

which is the result (3.7)

Appendix A. Limit Theorem for a Stochastic Differential Equation.

We consider here the behavior, as $\varepsilon \downarrow 0$, of ψ^{ε} given by (4.7). As discussed in section 4, we assume that the equation is driven by a Markov process with slowly varying parameters. Thus, we let $q(x,y)$ with values in R^d be, for each fixed x, a stationary ergodic Markov process in y, with infinitesimal generator Q_x. Equation (4.7) is then of the form

$$\frac{d}{dx}\psi^{\varepsilon} = \frac{1}{\varepsilon}\,\mathbf{F}(x, q(x, \frac{x}{\varepsilon^2}), \frac{\tau(x)}{\varepsilon}, \psi^{\varepsilon}) \tag{A.1}$$

$$+\, \mathbf{G}(\mathbf{x}, q(x, \frac{x}{\varepsilon^2}), \frac{\tau(x)}{\varepsilon}, \psi^{\varepsilon})$$

By ergodicity $q(x, \cdot)$ has an invariant measure $\tilde{P}_x(dq)$ which satisfies

$$\int Q_x f(q)\tilde{P}_x(dq) = 0 \tag{A.2}$$

for any test function f. We define expectation with respect to $\tilde{P}_x$ by

$$E\{\cdot\} \equiv \int \cdot\, d\,\tilde{P}_x(q) \tag{A.3}$$

Since m, n have mean zero, it is apparent from (4.8) that

$$E\{\mathbf{F}\} = 0 \tag{A.4}$$

Now the infinitesimal generator of the Markov Process $q^{\varepsilon}, \psi^{\varepsilon}$ with $q^{\varepsilon}(x) = q(x, x/\varepsilon^2)$ is given by (4.9), so that the Kolmogorov backward equation for this process may be written as

$$\frac{\partial V^{\varepsilon}}{\partial x} + \frac{1}{\varepsilon^2} Q_x V^{\varepsilon} + \frac{1}{\varepsilon}\mathbf{F}\cdot\nabla_{\psi} V^{\varepsilon} + \mathbf{G}\cdot\nabla_{\psi} V^{\varepsilon} = 0 \quad x < 0\,. \tag{A.5}$$

We consider final conditions at $x = 0$ which do not depend on q i.e.

$$V^{\varepsilon}(x, q, \psi)\,|_{x=0} = H(\psi) \tag{A.6}$$

Let

$$h = \tau/\varepsilon \tag{A.7}$$

so that $\mathbf{F} = \mathbf{F}(x, q, h, \psi)$, etc. in (A.5). We will solve (A.5), (A.6) asymptotically as $\varepsilon \downarrow 0$ by the multiple scale expansion

$$V^{\varepsilon} = \sum_{n=0}^{\infty} \varepsilon^n V^n\,(x, q, h, \psi)\,|_{h=\tau(x)/\varepsilon} \tag{A.8}$$

To expand (A.5) in multiple scales x, h, we replace $\frac{\partial}{\partial x}$ by $\frac{\partial}{\partial x} + \frac{\tau'(x)}{\varepsilon}\frac{\partial}{\partial h}$, and note that $\tau'(x) = -\frac{1}{c_o(x)} < 0$. Thus (A.5) becomes

$$\frac{1}{\varepsilon^2} Q_x V^{\varepsilon} + \frac{1}{\varepsilon}\left\{\mathbf{F}\cdot\nabla_{\psi} V^{\varepsilon} - \frac{1}{c_o(x)}\frac{\partial}{\partial h} V^{\varepsilon}\right\} + \left\{\mathbf{G}\cdot\nabla_{\psi} V^{\varepsilon} + \frac{\partial}{\partial x} V^{\varepsilon}\right\} = 0 \tag{A.9}$$

Now substitution of (A.8) into (A.9) yields a hierachy of equations for $V^{(n)}$ of which the first three are

$$Q_x V^o = 0 \tag{A.10}$$

$$Q_x V^1 + \mathbf{F}\cdot\nabla_\psi V^o - \frac{1}{c_o(x)}\frac{\partial}{\partial h}V^o = 0 \tag{A.11}$$

$$Q_x V^2 + \mathbf{F}\cdot\nabla_\psi V^1 - \frac{1}{c_o(x)}\frac{\partial}{\partial h}V^1 + \mathbf{G}\cdot\nabla_\psi V^o + \frac{\partial}{\partial x}V^o = 0 \tag{A.12}$$

From (A.10) and ergodicity of $q(x,\ \cdot)$ we conclude that V^o does not depend on q.

$$V^o = V^o(x,h,\psi) \tag{A.13}$$

We next take the expectation of (A.11). Since $\mathbf{F}$ has mean zero as noted in (A.4), using (A.2) we see that (A.11) implies that

$$-\frac{1}{c_o(x)}\frac{\partial}{\partial h}V^o = 0$$

whence V^o does not depend on h

$$V^o = V^o(x,\psi) \tag{A.14}$$

Now by ergodicity, Q_x has the one-dimensional null space consisting of functions which do not depend on q. Thus Q_x does not have an inverse. However, by the Fredholm alternative, which we assume to hold for the process q, Q_x has an inverse on the subspace of functions which have mean zero with respect to $\tilde{P}_x$. We defind a particular inverse Q_x^{-1} such that its range consists of functions with vanishing mean

$$-Q_x^{-1} = \int_0^\infty e^{Q_x r}\,dr\,. \tag{A.15}$$

In terms of this Q_x^{-1} we can solve (A.11) for V^1

$$V^1 = -Q_x^{-1}\{\mathbf{F}\cdot\nabla_\psi V^o\} + V^{1,o} \tag{A.16}$$

where $V^{1,o}$ does not depend on q.

We now substitute (A.16) into (A.12) and take expectations. We also average this equation with respect to h

$$<\cdot>_h = \lim_{h_o \to \infty} \frac{1}{h_o} \int_0^{h_o} \cdot \, dh \, . \tag{A.17}$$

Since $E\{\mathbf{F}\} = 0$ and $<\mathbf{G}>_h = 0$, (see (4.8)), we see that V^o must satisfy

$$\frac{\partial}{\partial x} V^o + <E\{\mathbf{F}\cdot\nabla_\psi(-Q_x^{-1})\mathbf{F}\cdot\nabla_\psi\}>_h V^o = 0 \tag{A.18}$$

This is the solvability condition for (A.12).

Equation (A.18) is the limiting backward Kolmogorov equation for ψ. It has the form

$$\frac{\partial}{\partial x} V^o + \mathbf{L}_x V^o = 0 \, , \; x<0 \tag{A.19}$$

$$V^o|_{x=0} = H(\psi)$$

From (A.18) the limit infinitesimal generator $\mathbf{L}_x$ is given by

$$\mathbf{L}_x = \int_0^\infty dr <E\{\mathbf{F}\cdot\nabla_\psi e^{rQ_x}\mathbf{F}\cdot\nabla_\psi\}>_h \tag{A.20}$$

Using the probabilistic interpretation of the semigroup e^{rQ_x} and expectation $E\{\cdot\}$ with respect to the invariant measure $P_x(dq)$ and the averaging $<\cdot>$, we can write (A.20) in the form

$$\mathbf{L}_x = \int_0^\infty dr E\{<\mathbf{F}(x,q(x,y),h,\psi)\cdot\nabla_\psi(\mathbf{F}(x,q(x,y+r),h,\psi)\cdot\nabla_\psi \cdot)>_h\} \tag{A.21}$$

In the application of this result in section 4, the explicit form of $\mathbf{F}$ in (4.7) is used in (A.21) to obtain (4.11).

Acknowledgement

The work of George Papanicolaou was supported by the National Science Foundation and the Office of Naval Research.

References

[1] Kohler W. and Papanicolaou G., Power statistics for wave propagation in one dimension and comparison with transport theory, J. Math. Phys. 14 (1973), 1733-1745 and 15 (1974), 2186-2197.

[2] Morrison J. A., J. Math. Anal. Appl., 39 (1972), 13

[3] Papanicolaou G., Asymptotic analysis of stochastic equations, in MAA Studies in Mathematics vol 18, M. Rosenblatt ed., MAA 1978.

[4] Klyatskin V. I., Stochastic equations and waves in random media, Nauka Moscow, 1980.

[5] Klyatskin V. I. Method of imbedding in the theory of random waves, Nauka, Moscow, 1986.

[6] Lifschitz I. M., Gradetski C. A., and Pastur L. A., Introduction to the theory of disordered systems, Nauka, Moscow, 1982.

[7] P. Richards and W. Menke, The apparent attenuation of a scattering medium, Bull. Seismol. Soc. Amer., 73 (1983), 1005-1021

[8] R. Burridge, G. Papanicolaou and B. White, Statistics for pulse reflection from a randomly layered medium, Siam J. Appl. Math., 47 (1987), 146-168

[9] P. Sheng, Z.-Q. Zhang, B. White and G. Papanicolaou, Multiple scattering noise in one dimension: universality through localization lenght scales, Phys. Rev. Letters 57, number 8, 1986, 1000-1003.

[10] R. Burridge, G. Papanicolaou, P. Sheng and B. White, Direct and inverse problems for pulse reflection from inhomogeneously random halfspaces, to appear

[11] Furstenberg H., Noncommuting random products, Trans. Am. Soc., 108 (1963), 377-428.

[12] Carmona R., Ecole d'ete de Saint-Flour 1984, Springer Lecture Notes in Mathematics, P. -L. Hennequin, editor, 1985.

SHAPE MEMORY ALLOYS-PHENOMENOLOGY AND SIMULATION

Ingo Müller
Hermann-Föttinger-Institut, TU Berlin

Abstract. Shape memory and pseudoelasticity provide a paradigma of thermomechanical interaction in a solid. In a schematic manner we introduce some observed properties of shape memory alloys and we proceed to simulate them by a structural model. The exploitation of the properties of the model uses simple ideas from statistical mechanics and from the theory of activated processes.

1 PHENOMENOLOGY

Shape memory and pseudoelasticity provide a paradigma of thermomechanical interaction. This observation is put in evidence by Figure 1 which shows schematic local-deformation diagrams for increasingly higher temperatures. These diagrams clearly imply shape memory, i.e. the ability of a deformed sample to recover the natural state of zero load and deformation upon heating. The behaviour shown in the last two diagrams is called pseudoelastic, because a loading-unloading cycle shows a hysteresis but it does return the sample to the natural state. The temperature range in Figure 1 is about 30 K around room temperature. A typical recoverable deformation is 6%.

Figure 1: Load-deformation diagrams at different temperatures

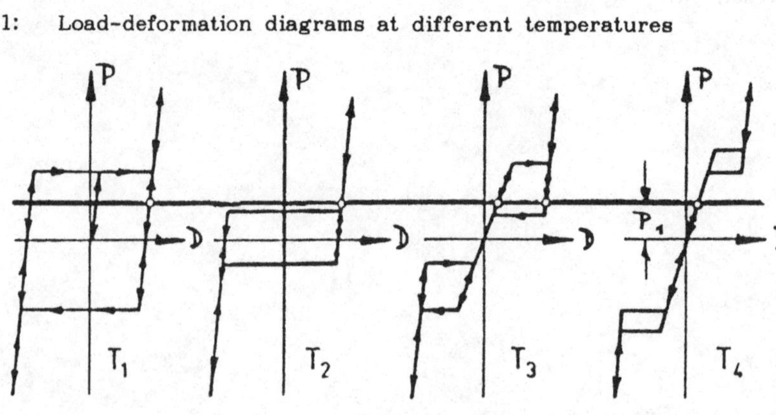

Figure 2: Deformation-temperature diagram of constant load

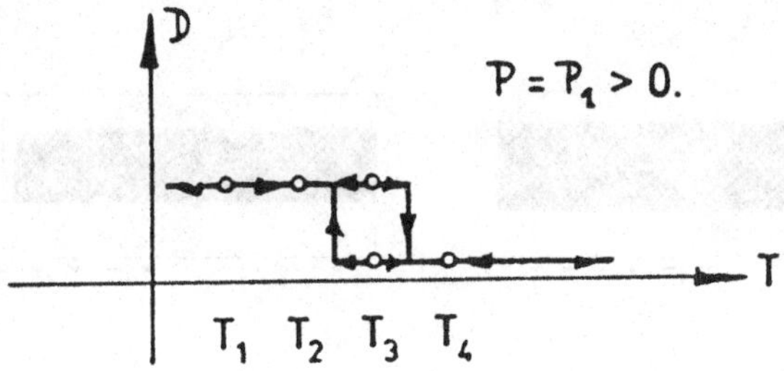

Figure 2 shows the schematic deformation-temperature diagram at a constant load P_1. This curve is implied by the sequence of load-deformation diagrams of Figure 1.

If a tensile specimen of NiTi (50 at %) is subject to a slowly oscillating tensile load and to an increase and subsequent decrease in temperature, the deformation will develop as shown in Figure 3.

In the lower plot on the left hand side of Figure 3 the time has been eliminated between the D(t) and T(t) curve on top. Thus a deformation-temperature diagram has appeared which must be compared to Figure 2. Obviously the D(t) curves on the left and right hand sides of Figure 3 are qualitatively different (apart from their difference in scale). This is due, as we shall see, to a difference in initial conditions.

The purpose of this research is to unterstand the phenomena described above and to simulate them. The key to the understanding is the observation, made by the metallurgists, that a phase transition occurs in the body. At high temperature the lattice structure is highly symmetric and we say that the body is austenitic. At low temperature the structure is less symmetric, the body is said to be in the martensitic phase and martensite tends to form twins.

Figure 3: **Deformation as a result of an oscillating tensile load with a changing temperature**

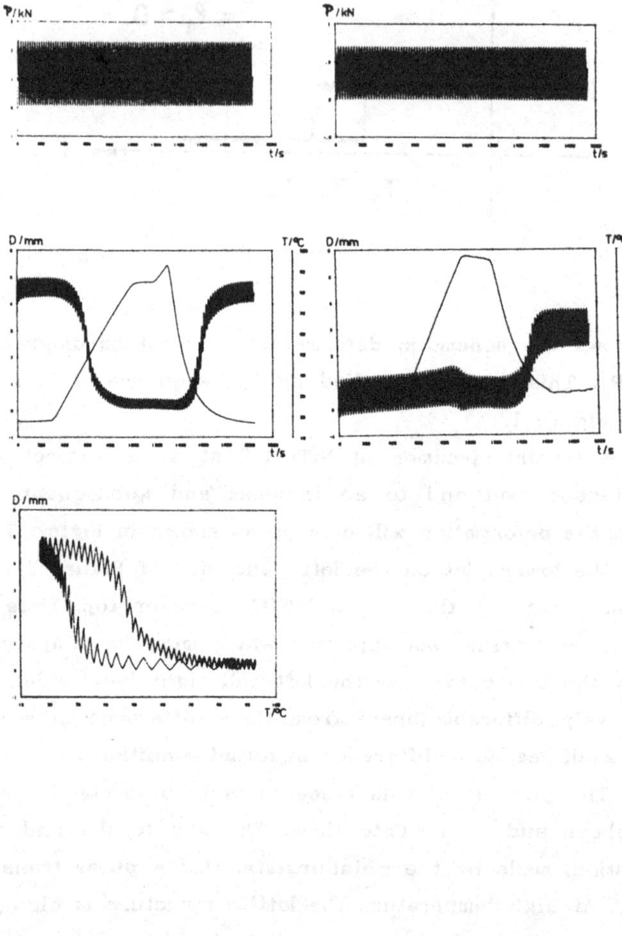

2 THE MODEL

Figure 4 in its upper part shows three lattice particles, i.e. small parts of the metallic lattice, which are denoted by A for austenitic and $M_\pm$ for the martensitic twins. We may think of these particles as sheared versions of one of them. To each shear length Δ we assign a potential energy whose postulated form is also shown in Figure 4. There are two stable minima corresponding to the martensitic twins and a metastable one for the austenitic particle.

Figure 4: Top: Lattice particles and their potential energy
Bottom: Deformation of the model body

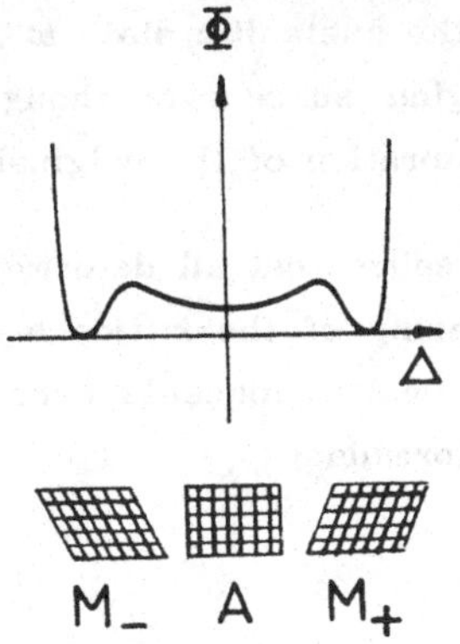

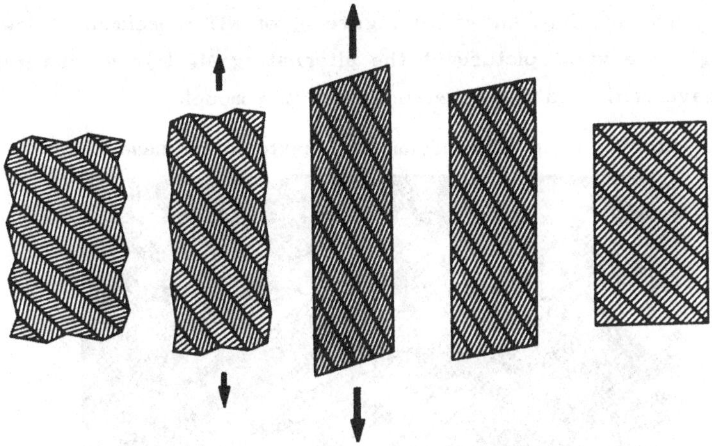

The model for the body is constructed by joining lattice particles in a layer and then stack the layers on top of each other as shown in the lower part of Figure 4 on the left. The sequence of model bodies shown in Figure 4 under different loads and at different temperatures is supposed to give a qualitative understanding of

i) the initial elastic deformation at low temperature that is due to a slight shearing of the martensitic layers under a small load. Removal of the load will restore the original shape of the model.

ii) the yielding of the body, which is due to the flipping of the M_- layers into the M_+ state,

iii) the residual deformation after unloading, which comes about, because all layers now settle into the equilibrium state M_+,

iv) the creation of the austenitic state at high temperature, which restores the original shape even though only macroscopically,

v) the complete restoration of the original state after cooling.

It is important to realize that all deformations depicted in Figure 4 come about solely by shearing of the lattice layers. The deformation is equal to the sum of the vertical components over the shear length of all layers as described by the formula

$$D-D_0 = \frac{1}{\sqrt{2}} \sum_{i=1}^{N} \Delta_i . \quad (1)$$

The etching, shown in Figure 5, of NiTi specimen at low temperature gives a vivid picture of the alternating $M_\pm$ layers. Pictures like this one have motivated the construction of the model.

Figure 5: Etching of a NiTi specimen in martensitic phase

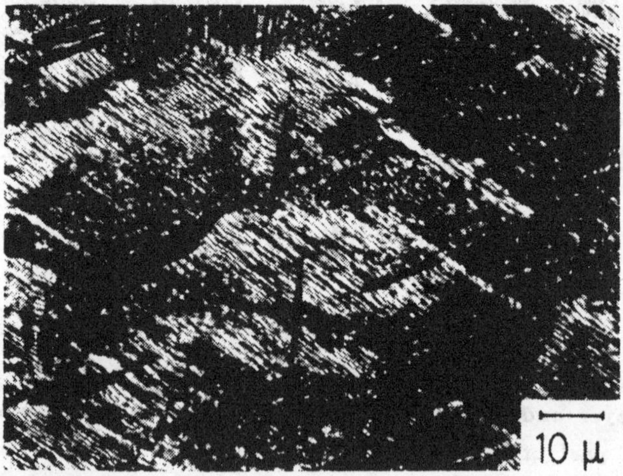

Sometimes it helps to think about the energetic aspects of the model starting with the potential energy curve $\Phi(\Delta)$ of the postulated form shown in Figure 4. If the body is loaded, the potential energy of the load must be taken into account and this is a linear function of the shear length. This must be added to $\Phi(\Delta)$ and thus we obtain the deformed potential energies shown in Figure 6. We see that the load affects the minima and the barriers. At a low temperature, where all particles lie still in their potential wells, the yielding from the M_- to the M_+ phase will occur when the force is big enough to eliminate the left barrier. This is shown on the left hand side of Figure 6. The right hand side refers to high temperature where the particles fluctuate about their minima with a mean kinetic energy that is proportional to temperature. The height of the pools of particles in the potential wells of Figure 6 indicates the strength of the fluctuation and the figure indicates that at a high temperature the yielding from the M_- to M_+ phase will occur at a lower load that at low temperature.

If there is fluctuation, we shall characterize the state of the body by the distribution function N_Δ, giving the number of layers at a certain shear length. In that case an alternative to equation (1) reads

$$D-D_o = \frac{1}{\sqrt{2}} \sum_\Delta \Delta N_\Delta \qquad (2)$$

where the summation is now over all shear lengths. Yet another form of equation reads

$$D-D_o = \frac{1}{\sqrt{2}} \left\{ N_{M_-} \bar{\Delta}_{M_-} + N_A \bar{\Delta}_A + N_{M_+} \bar{\Delta}_{M_+} \right\}, \qquad (3)$$

where $\bar{\Delta}_{M_\pm}$, $\bar{\Delta}_{M_A}$ denote the expectation values of shear length in $M_\pm$- and A-range respectively.

Figure 6: Energetic view of yielding at low and high temperature

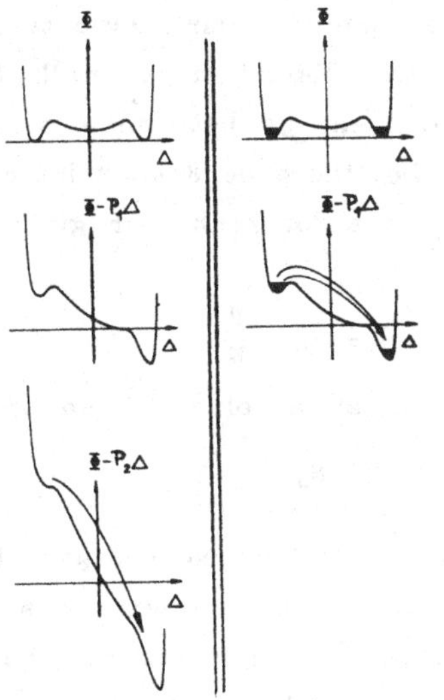

3 Static Theory (Statistical Mechanics)

While it was very easy to visualize the behaviour of the body at low temperature, this is difficult at higher temperatures. The reason is that the body is then subject to two conflicting tendencies: The energy E triesto be minimal by pulling all particles into the depths of the potential wells and the entropy H attempts to be maximal by distributing the particles evenly over the available range of shear lengths. In this competition it is actually the free energy

$$\Psi = E - T H \tag{4}$$

that achieves a minimum. But it is not easy to appreciate what the outcome of the competition will be at a given temperature.

Therefore we turn to statistical mechanics which allows us to set up a formula for the free energy in a state characterized by the distribution function N_Δ. We may write

$$Y = \left[\sum_\Delta \Phi(\Delta)\, N_\Delta + eK \right] - T \left[k \ln \frac{N!}{\pi_\Delta N_\Delta!} \right] \tag{5}$$

and we proceed to interpret the three terms involved:

i) $\sum_\Delta \Phi(\Delta) N_\Delta$ is the potential energy of the layers, if they are independent which we have tacitly assumed sofar.

ii) But in reality the orientation, i.e. shear length, of the layers is not independent. Indeed, whereever two layers of different orientation meet, there is a lattice distortion and an interfacial energy as a result. Between two martensitic twins we ignore that energy, because the distortion is small. But we take account of the interfacial energy between a martensitic layer and an austenitic one. Let there be K such interfaces and let the interfacial energy be e for each. For given phase fractions

$$x_M = \frac{N_M}{N} \qquad x_A = 1 - x_M = \frac{N_A}{N} \tag{6}$$

we can calculate the expectation value of K and come up with

$$eK = 2eNx_A x_M = \frac{2e}{N} \sum_{[A]} N_\Delta \sum_{[M]} N_\Delta\,. \tag{7}$$

for the expectation value of interfacial energy. The derivation of (7) makes use of a statistical argument whose validity requires temperatures that are high enough that fluctuation between the phases can actually seek out the most probable value of K.

iii) The last term in (5) gives the entropy in its usual form of k.ln W where W is the number of possibilities to realize a given distribution N_Δ.

We minimize Ψ under the constraints of constant number of layers and constant deformation, i.e.

$$\sum_\Delta N_\Delta = N \quad \text{and} \quad \frac{1}{\sqrt{2}} \sum_\Delta \Delta N_\Delta = D - D_o \, , \tag{8}$$

and obtain

$$N_\Delta = N\, e^{\frac{\alpha+P\Delta}{kT} - 1}\, e^{-\frac{\Phi(\Delta)+2ex_A}{kT}} \quad \text{for} \quad \Delta \in [M]$$

$$N_\Delta = N\, e^{\frac{\alpha+P\Delta}{kT} - 1}\, e^{-\frac{\Phi(\Delta)+2ex_M}{kT}} \quad \text{for} \quad \Delta \in [A] \tag{9}$$

for the distribution in the M and A ranges. α and P are Lagrange multipliers taking care of the constraints. α can readily be calculated from $(8)_1$ and P can be shown - by use of some thermodynamics - to be equal to the load necessary to maintain the given deformation.

Insertion of (9) into (5), $(8)_2$ and into $x_M = \frac{1}{N} \sum_{[M]} N_\Delta$ gives after slight rearrangement of terms

$$\Psi = -NkT \ln\left\{ e^{-\frac{2ex_A}{kT}} \sum_{]\,[} e^{\frac{P\Delta-\Phi(\Delta)}{kT}} + e^{-\frac{2ex_M}{kT}} \sum_{[\,]} e^{\frac{P\Delta-\Phi(\Delta)}{kT}} \right\} + \tag{10}$$
$$+\sqrt{2}\, P(D-D_o) - 2Ne\, x_A\, x_M,$$

$$D - D_o = \frac{N}{\sqrt{2}} \frac{e^{-\frac{2ex_A}{kT}} \sum_{]\,[} \Delta e^{\frac{P\Delta-\Phi(\Delta)}{kT}} + e^{-\frac{2ex_M}{kT}} \sum_{[\,]} \Delta e^{\frac{P\Delta-\Phi(\Delta)}{kT}}}{e^{-\frac{2ex_A}{kT}} \sum_{]\,[} e^{\frac{P\Delta-\Phi(\Delta)}{kT}} + e^{-\frac{2ex_M}{kT}} \sum_{[\,]} e^{\frac{P\Delta-\Phi(\Delta)}{kT}}} \, , \tag{11}$$

$$x_M = \frac{e^{-\frac{2ex_A}{kT}} \sum_{]\,[} e^{\frac{P\Delta-\Phi(\Delta)}{kT}}}{e^{-\frac{2ex_A}{kT}} \sum_{]\,[} e^{\frac{P\Delta-\Phi(\Delta)}{kT}} + e^{-\frac{2ex_M}{kT}} \sum_{[\,]} e^{\frac{P\Delta-\Phi(\Delta)}{kT}}} \, . \tag{12}$$

The last equation may serve to calculate $x_M = x_M(P,T)$. Insertion of that function into (11) gives $D = D(P,T)$. Or, by inversion $P = P(D,T)$ and if this and $x_M(P,T)$ are inserted into Ψ we get $\Psi = \Psi(D,T)$.

None of these calculations can be done analytically, but they have been carried out graphically and numerically and Figure 7 shows the result: temperature increases from left to right. We proceed to discuss these curves.

The second and third load-deformation curves are non-monotone which in a load-controlled experiment will imply break throughs along the dotted horizontal lines. Thus we see that the model can simulate a pseudoelastic hysteresis of the type shown schematically in Figure 1. The left (P,D)-curve must be ignored, because here temperature is too low to permit statistical arguments. Also the right (P,D)-curve, which suggests purely elastic behaviour at high temperature, is never observed, because here temperature is so high that true plastic deformation governs the mechanical behaviour of the body and this is not provided for by the model.

Figure 7: Free energy-deformation curves and load-deformation curves in their dependence upon temperature.

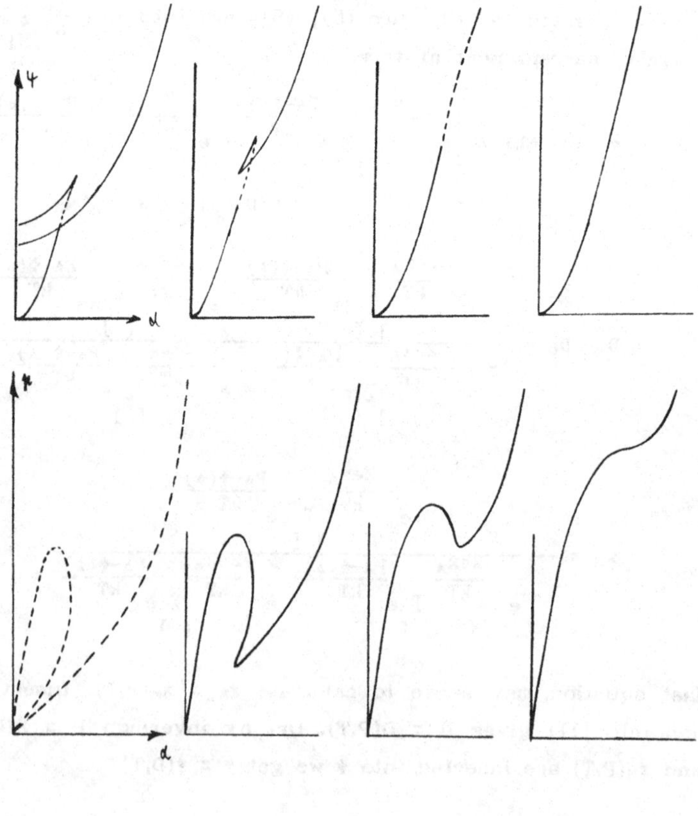

The free energy-deformation curves on the top of Figure 7 are the integrals of the (P,D)-curves below them. It seems worthwhile to draw attention to their complexity at intermediate temperature. There is more here than mere non-convexity of the free energy.

Statistical arguments of the above kind are useful to provide a static theory of thermodynamic behaviour. Correspondingly they have given us isothermes in the (P,D)-diagram. Also the validity of such arguments requires elevated temperature and therefore does not permit the simulation of low-temperature behaviour which is characterized by "frozen equilibria".

We should also like, however, to predict deformation and loads under time-dependent temperatures and certainly we are also interested in the low temperature behaviour. Therefore we now turn to a kinetic theory of the model.

4 Kinetic Theory (Activated Processes)

The kinetic theory envisages changes in deformation as an activated process. What is activated here by either an increase in temperature or by an increase in load is the jump of lattice layers across a potential barrier. The rate at which this process occurs is governed by rate laws for the phase fractions $x_{M\pm}$ and x_A which are postulated to have the forms

$$\begin{aligned} x_{M_-} &= - p\, x_{M_-} + p\, x_A \\ \dot{x}_A &= + \overset{-o}{p}\, x_{M_-} - \overset{o-}{p}\, x_A - \overset{o-}{p}\, x_A + \overset{+o}{p}\, x_{M_+} \\ \dot{x}_{M_+} &= \overset{o+}{p}\, x_A - \overset{+o}{p}\, x_{M_+} . \end{aligned} \tag{13}$$

$\overset{-o}{p}$ etc. are the transition probabilities between the potential minima. The form of $\overset{-o}{p}$ is given in (14)

$$\overset{o-}{p} = \sqrt{\frac{kT}{2\pi m}}\, e^{-\frac{\Phi(m_L)-Pm_L}{kT}}\, e^{-\frac{2e(1-2x_M)}{kT}} . \tag{14}$$

There are three contributions

i) $\overset{o-}{p}$ is proportional to the probability of finding a layer from M_- on top of the left barrier, because then it will presumably cross the barrier. This probability is given by the Boltzmann factor $e^{-\frac{E}{kT}}$ where E is the height of the left barrier which depends on P:

ii) When a layer flips from the M_- to the A phase there is also a change in the number of (M,A)-interfaces and consequently of interfacial energy. $\overset{o-}{p}$ in (14) takes care of this fact by the last factor which is again a Boltzmann factor $e^{-\frac{E}{kT}}$ with E now being the change of interfacial energy in a jump.

iii) The transition probability increases with the frequency with which a layer runs against the confining barrier. This frequency is proportional to $\sqrt{T}$ and the first factor in (14) takes care of this effect.

The other transition probabilities are formed in an analogous manner.

Figure 8: Energies $H_\pm$ associated with jumps.

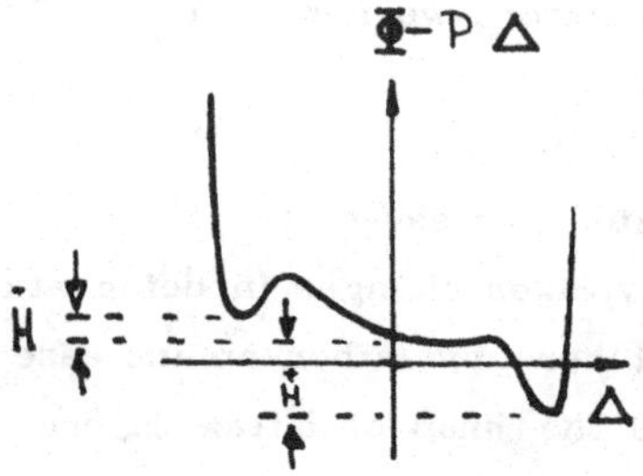

There is also a rate law for the temperature of the body which is infact a truncated form of the energy balance. It reads

$$C\,\dot{T} = -\,\alpha(T-T_E) - \left(\dot{x}_{M_-}H_-(P) + \dot{x}_{M_+}H_+(P)\right) \tag{15}$$

where C is the heat capacity. The first term on the right is due to an efflux of heat, if the temperature of the body differs from that of the exterior. α is the heat transfer coefficient. The second term on the right hand side of (15) takes care of the fact that there is a conversion of the potential energy H_- into kinetic energy (i.e.heat) when there is a jump from the left minimum into the middle. Also heat is converted into the potential energy H_+ if a layer jumps from the right minimum into the middle; see Figure 8.

Inspection of (13) with (14) and (15) shows that we have a set of ordinary differential equations which will be able to predict $x_{M\pm}(t)$, $x_A(t)$ and T(t), if only P(t) and $T_E(t)$ are given - and initial values of course. Once $x_{M\pm}(t)$, $x_A(t)$ and T(t) are thus calculated, we can determine D(t) from (3) which reads more explicitly

$$D-D_o = \frac{N}{\sqrt{2}}\left\{x_{M_-}\frac{\sum\limits_{M_-}\Delta e^{-\frac{\Phi(\Delta)-P\Delta}{kT}}}{\sum\limits_{M_-} e^{-\frac{\Phi-P\Delta}{kT}}} + x_A\frac{\sum\limits_{A}\Delta e^{-\frac{\Phi(\Delta)-P\Delta}{kT}}}{\sum\limits_{A} e^{-\frac{\Phi(\Delta)-P\Delta}{kT}}} + x_{M_+}\frac{\sum\limits_{M_+}\Delta e^{-\frac{\Phi(\Delta)-P\Delta}{kT}}}{\sum\limits_{A_+} e^{-\frac{\Phi(\Delta)-P\Delta}{kT}}}\right\} \tag{16}$$

Here we have used the expectation that the probability of an M_- layer to have shear length Δ is given by the Boltzmann factor

$$\frac{e^{-\frac{\Phi-P\Delta}{kT}}}{\sum\limits_{M_-} e^{-\frac{\Phi-P\Delta}{kT}}} \tag{17}$$

and, of course, similar expressions hold for the A and M_+ layers.

The solutions of the equations (13), (15) and the evaluation of (16) must proceed numerically, because of the strong non-linearity of these equations. For a suitable choice of the potential $\Phi(\Delta)$ and for suitable choices of the parameters of the model we obtain the curves of Figure 9. On the left hand side the plots show the "input" consisting of a triangular tensile and compressive load and of a constant temperature and the "output" which are $x_{M\pm}(t)$, $x_A(t)$ and D(t). The temperature grows from top to bottom. It is instructive to eliminate the time between the given function P(t) and the calculated function D(t) and obtain the (P,D)-diagrams on the right hand side of Figure 9. These must be compared with the schematic curves of Figure 1 and we see that there is good qualitative agreement.

In the numerical evaluation of the equations (13, (15) it makes not much difference whether the external temperature T_E is constant or changing and it is possible to simulate the load and temperature input that was remarked on in Figure 3 where D(t) was the output. The result of that simulation is shown in Figure 10. We observe qualitatively similar D(t) curves as in Figure 3. And, of course, now we can also appreciate what happens inside the body as Figure 10 gives us an idea how the phase fractions change and, by their change, dictate the deformation. Actually we can even see in Figure 10, if we look closely, how the prescribed T_E differs from the calculated T, particulary in the periods of transition. We see that the essential difference between the D(t)-curves on the left and right of Figure 10 is caused by different initial conditions for $x_{M\pm}(0)$.

Figures like 10 serve to identify the parameters of the model, if we compare them to observations as those in Figure 3.

Figure 9: Response of the model to a triangular tensile and compressive load at constant temperature.

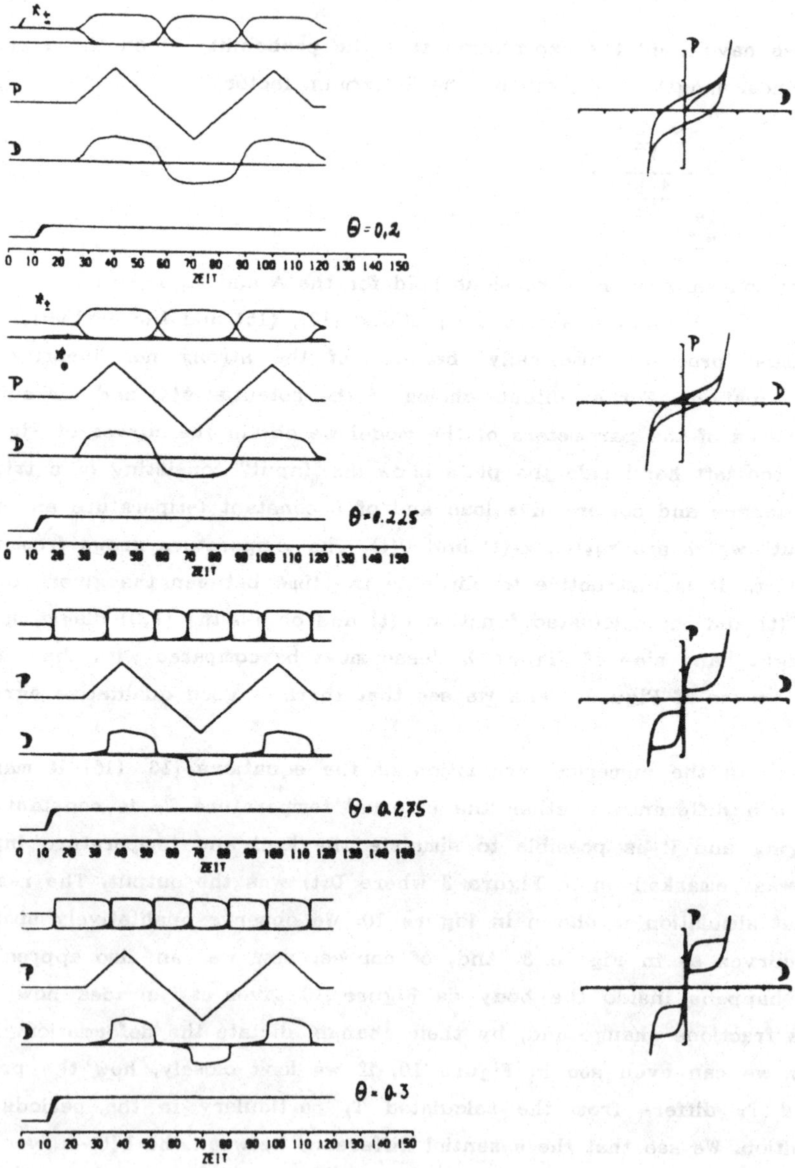

Figure 10: Response of the model under an oscillating tensile load and variable temperature T_E.

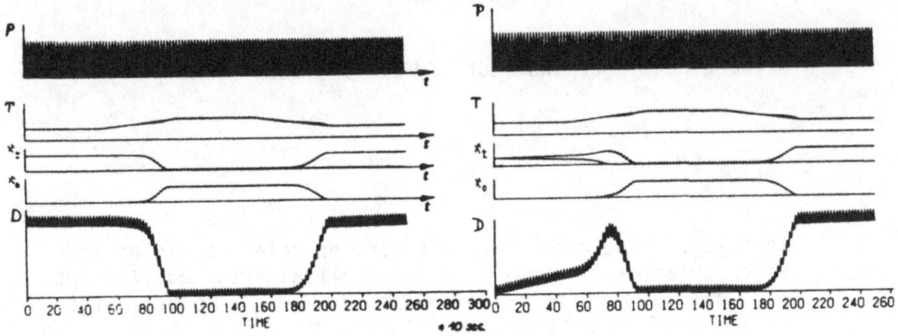

Acknowledgements and References

The material presented in this survey is mostly taken from research papers that were published either by myself or by my co-workers Dr. Achenbach and Dr. Ehrenstein. The following list of publications may help the interested reader to better understand the phenomena and their simulation by a model.

On phenomenology:

J. Perkins (ed.) Shape Memory Effects in Alloys. (Plenum Press, N.Y. London, 1976)

Delaey, L., Chandrasekharan, L. (eds.) Proc. Int. Conf. on Martensitic Transformation Leuven (1982). J. de Physique 43 (1982)

Ehrenstein, H., Formerinerungsvermögen in NiTi, Dissertation, Technische Universität Berlin.

Müller, I., Pseudoelasticity in Shape Memory Alloys - An extreme Case of Thermoelasticity. IMA preprint No. 169, July 1985. Also: Proc. Convegno Termoelasticitá, Rome, May 1985.

Achenbach, M., Ein Modell zur Simulation des Last-Verformungs-Temperaturverhaltens von Legierungen mit Formerinnerungsvermögen. Dissertation, Technische Universität Berlin (1986).

RELATIVISTIC EXTENDED THERMODYNAMICS

Ingo Müller
Hermann-Föttinger-Institut, TU Berlin, West Germany

Abstract. The constitutive theory of relativistic extended thermodynamics of gases determines all material coefficients in terms of the thermal equation of state. Specific forms of the coefficients can be obtained for the limiting cases where the gas is either non-relativistic or ultrarelativistic or where it is not degenerate or strongly degenerate.

1. INTRODUCTION

Relativistic extended thermodynamics is a field theory with the principal objective of determining the 14 fields

A^A - particle flux vector

A^{AB} - energy-momentum tensor.

The necessary field equations are based upon

conservation law of particle number

conservation law of energy-momentum

balance of fluxes.

Constitutive equations must be given for the flux tensor and the flux productions. The constitutive functions are restricted in generality by the

principle of relativity

entropy principle

requirement of hyperbolicity.

It turns out that, in a linear theory, all constitutive coefficients are determined by the thermal equation of state in equilibrium to within two functions of a single variable.

It is indicated how the thermal equation of state can be derived from statistical mechanics.

The notation is the usual index notation. Capital indices run between 0 and 3 and lower case indices run between 1 and 3.

The metric tensor of space time is denoted by g_{AB}.

2. THERMODYNAMIC PROCESSES

The principal objective of relativistic extended thermodynamics is the determination of the 14 fields

$$A^A \text{ - particle flux vector,} \qquad A^{AB} \text{ - energy-momentum tensor,} \tag{2.1}$$

in all events x^D of a body. The energy-momentum tensor is assumed symmetric so that it has 10 independent components.

The components of (1) have the following physical interpretations

$$\begin{aligned} &A^o \text{ - c·number density,} \\ &A^a \text{ - particle flux,} \\ &A^{00} \text{ - energy density,} \\ &A^{oa} \text{ - c·energy flux,} \\ &A^{ao} \text{ - c·momentum density} \\ &A^{ab} \text{ - momentum flux.} \end{aligned} \tag{2.2}$$

The fields of ordinary thermodynamics are the first 5 among the 14 fields (2).

For the determination of the fields (1) we need field equations and these are formed by the conservation laws of particle number and energy-momentum, viz

$$A^A{}_{,A} = 0 \qquad (2.3)_1$$

$$A^{AB}{}_{,B} = 0 \qquad (2.3)_2$$

and by the equations of balance of fluxes.

$$A^{ABC}{}_{,C} = I^{AB} \qquad (2.3)_3$$

A^{ABC} is the completely symmetric tensor of fluxes and I^{AB} is its production density. The set (3) consists of 15 equations for only 14 fields. Therefore we assume

$$I^A{}_A = 0 \quad \text{and} \quad A^A{}_A{}^C \sim A^C \tag{2.4}$$

so that the trace of $(3)_3$ reduces to $(3)_1$. We are then left with only 14 independent equations (3).

The motivation for the choice of equations (3), and in particular for the equation $(3)_3$, stems from the kinetic theory of monatomic gases (see [1], [2], [3]). Indeed A^A and A^{AB} are the first two moments of the distribution function in the relativistic kinetic theory and $A^A{}_{,A} = 0$, $A^{AB}{}_{,B} = 0$ are the first two equations of transfer. It is then reasonable to take the further equations from the equation of transfer for the third moment A^{ABC} and these have the form $(3)_3$. In the kinetic theory the two conditions (4) are satisfied and the factor of proportionality in $(4)_2$ turns out to be m^2c^2, where m is the rest mass of the atoms. Therefore we rewrite (4) in the form

$$I^A{}_A = 0 \quad \text{and} \quad A^A{}_A{}^C = m^2c^2A^C. \tag{2.5}$$

The kinetic theory makes it also clear that the assumed symmetry of A^{ABC} is characteristic for a monatomic gas. The present theory is therefore restricted to that case. This is not much of a restriction, however, because all bodies are monatomic at temperatures high enough for relativistic effects to be important.

Of course, the set of equations (3) is not a set of field equations for the basic fields (1), because two new quantities appear in them: the flux tensor A^{ABC} and the flux production I^{AB}. In this situation we let ourselves be guided by the arguments of non-relativistic continuum mechanics and thermodynamics and assume that A^{ABC} and I^{AB} are constitutive quantities. Such quantities are related to the basic fields in a materially dependent manner by constitutive functions. In particular, when the constitutive relations are of the general form

$$\begin{aligned} A^{ABC} &= \hat{A}^{ABC}\,(A^M, A^{MN}) \\ I^{AB} &= \hat{I}^{AB}\,(A^M, A^{MN}) \end{aligned} \tag{2.6}$$

we say that they characterize a viscous heat-conducting gas.

If the constitutive functions $\hat{A}^{ABC}$ and $\hat{I}^{AB}$ were known, we could eliminate A^{ABC} and I^{AB} between (3) and (6) and obtain an explicit set of field equations for the 14 fields A^A and A^{AB}. Each solution of that set of field equations is called a <u>thermodynamic process</u>.

3. PRINCIPLES OF THE CONSTITUTIVE THEORY

In reality, of course, the constitutive functions $\hat{A}^{ABC}$ and $\hat{I}^{AB}$ are not known and therefore the thermodynamicist tries to determine them or at least to restrict their generality. In these efforts the thermodynamicist is guided by universal principles to which the constitutive functions must conform. The most important such principles are

i.) the entropy principle,

ii.) the requirement of hyperbolicity,

iii.) the principle of relativity.

The entropy principle postulates the existence of the entropy flux vector h^A which is a constitutive quantity so that in a viscous, heat-conducting gas we have

$$h^A = \hat{h}^A(A^M, A^{MN}). \tag{3.1}$$

Moreover, the principle requires that h^A satisfy the entropy inequality

$$h^A{}_{,A} \geqslant 0 \tag{3.2}$$

for all thermodynamic processes.

Hyperbolicity is a condition on the character of the field equations. It requires that these equations are symmetric hyperbolic. This property guarantees that Cauchy's initial value problem is well-posed and that all wave speeds are finite.

The principle of relativity assumes that the field equations and the entropy inequality are invariant under space-time transformations

$$\overset{*}{x}{}^A = \overset{*}{x}{}^A(x^B) \tag{3.3}$$

Since the balance equations (2.3) are tensor equations which are naturally invariant, the principle requires that the constitutive functions be invariant. If G and G^* stand for A^{ABC}, I^{AB} and h^A in the two frames connected by the transformation (3), the principle of relativity can be stated as follows

$$G = \hat{G}(A^M, A^{MN}) \quad \text{and} \quad G^* = \hat{G}(\overset{*}{A}{}^M, \overset{*}{A}{}^{MN}). \tag{3.4}$$

Note that $\hat{G}$ is the same function in both equations.

The exploitation of the above restrictive principles is the subject of the thermodynamic constitutive theory. The details of this theory are explained in [4] and in the forthcoming book [5]. Here I shall only list the results and indicate certain special cases of interest.

4. RESULTS

4.1. Flux tensor and entropy flux vector

It has proved convenient to introduce the 4-velocity U^A, $U^A U_A = c^2$ and the projector $h_{AB} = \frac{1}{c^2} U_A U_B - g_{AB}$. We use these to decompose the particle flux vector, the energy-momentum tensor and the entropy flux vector by the equations

$$\begin{aligned} A^A &= nU^A \\ A^{AB} &= t^{\langle AB\rangle} + (p(n,e) + \pi)h^{AB} + \frac{1}{c^2}(U^A q^B + U^B q^A) + \frac{e}{c^2} U^A U^B \qquad (4.1) \\ h^A &= hU^A + \phi^A. \end{aligned}$$

The advantage of this decomposition is that the different parts have the following suggestive meanings:

$$\begin{aligned} &n && \text{- number density} \\ &t^{\langle AB\rangle} && \text{- stress deviator} \\ &p+\pi && \text{- pressure} \\ &q^A && \text{- heat flux} \qquad (4.2) \\ &e && \text{- energy density} \\ &h && \text{- entropy density} \\ &\phi^A && \text{- entropy flux.} \end{aligned}$$

The stress deviator $t^{\langle AB\rangle}$, the heat flux q^A and the dynamic pressure π vanish in equilibrium.

We present here the results of the constitutive theory whose derivation can be found in [4] or [5].

To within second order terms in $t^{<AB>}$, q^A and π the flux tensor A^{ABC} and the flux production I^{AB} have the forms

$$A^{ABC} = \left(\frac{1}{2c^2}\frac{\Gamma_1'}{T} + C_1^\pi \pi\right) U^A U^B U^C \tag{4.3}$$

$$- \left(\frac{1}{2}\frac{\Gamma_1}{6} + \frac{c^2}{6} C_1^\pi \pi\right)(g^{AB}U^C + g^{BC}U^A + g^{CA}U^B) + C_3 (g^{AB}q^C + g^{BC}g^A + g^{CA}q^B)$$

$$- \frac{6}{c^2} C_3 (U^A U^B q^C + U^B U^C q^A + U^C U^A q^B) + C_5 (t^{<AB>}U^C + t^{<BC>}U^A + t^{<CA>}U^B),$$

$$I^{AB} = B_1^\pi \pi g^{AB} - \frac{4}{c^2} B_1^\pi \, U^A U^B + B_3 t^{<AB>} + B_4 (q^A U^B + q^B U^A). \tag{4.4}$$

Also the entropy flux vector h^A to within third order terms in $t^{<AB>}$, q^A and π assumes the form

$$h^A = (h|_E + A_1^{\pi^2}\pi^2 + A_1^q q_E q^E + A_1^t t_{<EF>} t^{<EF>}) U^A$$

$$+ (A_2^o + A_2^\pi \pi) q^A + A_3^o t^{<AE>} q_E. \tag{4.5}$$

The equilibrium entropy density $h|_E$ satisfies the Gibbs equation

$$d\left(\frac{h|_E}{n}\right) = \frac{1}{T}\left(d\left(\frac{e}{n}\right) + p\, d\left(\frac{1}{n}\right)\right), \tag{4.6}$$

where T is the absolute temperature.

The coefficients C in the flux tensor (3) and A in the entropy flux vector (5) are related to the three functions p, Γ_1 and Γ_2 of the two variables

fugacity $\alpha = -\frac{1}{T}\frac{1}{n}(e - Th|_E + p)$ and absolute temperature T

in the following manner:

$$C_1^{\pi} = -\frac{3}{c^2}\frac{1}{T}\frac{\begin{vmatrix} -\ddot{p} & \dot{p}-\dot{p}' & \dot{\Gamma}_1 \\ \dot{p}-\dot{p}' & p'-p'' & \Gamma_1'-\Gamma_1 \\ \dot{\Gamma}_1 & \Gamma_1'-\Gamma_1 & \frac{5}{3}\Gamma_2 \end{vmatrix}}{\begin{vmatrix} -\ddot{p} & \dot{p}-\dot{p}' & \dot{\Gamma}_1 \\ \dot{p}-\dot{p}' & p'-p'' & \Gamma_1'-\Gamma_1 \\ -\dot{p} & -p' & \frac{5}{3}\Gamma_1 \end{vmatrix}} \qquad (4.7)_1$$

$$C_3 = -\frac{1}{2}\frac{1}{T}\frac{\begin{vmatrix} \dot{p} & -\dot{\Gamma}_1 \\ \Gamma_1 & \Gamma_2 \end{vmatrix}}{\begin{vmatrix} \dot{p} & -\dot{\Gamma}_1 \\ p' & \Gamma_1-\Gamma_1' \end{vmatrix}} \qquad (4.7)_2$$

$$C_5 = -\frac{1}{2}\frac{1}{T}\frac{\Gamma_2}{\Gamma_1} \qquad (4.7)_3$$

$$A_1^{\pi^2} = -\frac{c^2}{6}C_1^{\pi}\frac{\begin{vmatrix} \ddot{p} & \dot{p}-\dot{p}' \\ \dot{p}-\dot{p}' & p'-p'' \end{vmatrix}}{\begin{vmatrix} -\ddot{p} & \dot{p}-\dot{p}' & \dot{\Gamma}_1 \\ \dot{p}-\dot{p}' & p'-p'' & \Gamma_1'-\Gamma_1 \\ -\dot{p} & -p' & \frac{5}{3}\Gamma_1 \end{vmatrix}} \qquad (4.8)_1$$

$$A_1^{q} = -\frac{1}{2c^2}\frac{-\frac{1}{T}\dot{\Gamma}_1 - 10C_3\dot{p}}{\begin{vmatrix} \dot{p} & -\dot{\Gamma}_1 \\ p' & \Gamma_1-\Gamma_1' \end{vmatrix}} \qquad (4.8)_2$$

$$A_1^{t} = -\frac{1}{T}C_5\frac{1}{\Gamma_1} \qquad (4.8)_3$$

$$A_2^0 = \frac{1}{T} \tag{4.8$_4$}$$

$$A_2^\pi = -\frac{\frac{1}{T}\begin{vmatrix} -\ddot{p} & \dot{\Gamma}_1 \\ \dot{p}-\dot{p}' & \Gamma_1'-\Gamma_1 \end{vmatrix} - \frac{10}{3}C_5\begin{vmatrix} -\ddot{p} & \dot{p}-\dot{p}' \\ \dot{p}-\dot{p}' & p'-p'' \end{vmatrix}}{\begin{vmatrix} -\ddot{p} & \dot{p}-\dot{p}' & \dot{\Gamma}_1 \\ \dot{p}-\dot{p}' & p'-p'' & \Gamma_1'-\Gamma_1 \\ \dot{p} & -p' & \frac{5}{3}\Gamma_1 \end{vmatrix}} = -\frac{\frac{1}{T}\dot{\Gamma}_1 - \frac{c^2}{3}C_1^\pi\dot{p}}{\begin{vmatrix} \dot{p} & -\dot{\Gamma}_1 \\ p' & \Gamma_1-\Gamma_1' \end{vmatrix}} \tag{4.8$_5$}$$

$$A_3^0 = \frac{\frac{1}{T}\dot{\Gamma}_1 - 2C_5\dot{p}}{\begin{vmatrix} \dot{p} & -\dot{\Gamma}_1 \\ p' & \Gamma_1-\Gamma_1' \end{vmatrix}} = -C_3\frac{2}{\Gamma_1}\,. \tag{4.8$_6$}$$

Dots and primes denote derivatives with respect to α and $\ln T$ respectively.

Fugacity and absolute temperature are the natural variables of statistical mechanics of a gas as we shall see later. Also they are easier to measure than the original variables n and e. A switch between the sets of variables (α, T) and (n, e) can be done by use of the formulae

$$n = -\frac{1}{T}\dot{p} \quad \text{and} \quad e = p' - p, \tag{4.9}$$

provided that the thermal equation of state $p = p(\alpha, T)$ is known.

The results (7) and (8) seem to indicate that the three functions $p(\alpha,T)$, $\Gamma_1(\alpha,T)$ and $\Gamma_2(\alpha,T)$ determine the flux tensor A^{ABC} and the entropy flux vector h^A. However, the functions $\Gamma_1(\alpha,T)$, $\Gamma_2(\alpha,T)$ are not independent of $p(\alpha,T)$, because we have

$$-\frac{1}{2}\Gamma_1' + 3\Gamma_1 = m^2c^2p, \quad -\frac{1}{2}\Gamma_2' + 4\Gamma_2 = -m^2c^2\Gamma_1 \tag{4.10}$$

so that Γ_1 and Γ_2 can be related to p by integration. We obtain from (10)

$$\Gamma_1 = T^6\{-2m^2c^2 \int \frac{\dot{p}}{T^7}dT + A_1(\alpha)\}$$

$$\Gamma_2 = T^8\{2m^2c^2 \int \frac{1}{T^3}[-2m^2c^2 \int \frac{\ddot{p}}{T^7}dT + A_1(\alpha)]dT + A_2(\alpha)\}, \tag{4.11}$$

and we conclude that Γ_1 and Γ_2 each may be determined from $p(\alpha,T)$ to within a function of the single variable α, the fugacity.

It follows from (7), (8) that all coefficients of the flux tensor A^{ABC} and of the entropy flux vector h^A can be determined to within two functions of α, if only the thermal equation of state $p = p(\alpha,T)$ is known.

The requirement of hyperbolicity furnishes conditions on the coefficients of A^{ABC} and h^A which in effect restrict the thermal equation of state $p = p(\alpha,T)$ and the functions $A_1(\alpha), A_2(\alpha)$. Most of these restrictions are not easy to interpret but we mention some which are:

$$\left|\begin{matrix} \frac{1}{T}\ddot{p} & -\frac{1}{2T}(\dot{p}-\dot{p}') \\ -\frac{1}{2T}(\dot{p}-\dot{p}') & \frac{1}{4T}(p''-p') \end{matrix}\right| \text{ - positive definite,}$$

$$A_1^{\pi^2} < 0, \quad A_1^q > 0, \quad A_1^t < 0. \tag{4.12}$$

Inspection of (5) shows that the inequalities $(12)_{2,3,4}$ are necessary for the entropy density $h = \frac{1}{c^2} U_A h^A$ to have a maximum in equilibrium. The positive definiteness of the matrix $(12)_1$ ensures that the compressibility $\frac{1}{n} (\frac{\partial n}{\partial p})_T$ and the specific heat $(\frac{\partial e}{\partial T})_n$ are positive. These are two well-known "thermodynamic stability conditions", i.e. requirements on the constitutive functions which guarantee that the entropy of a body in homogeneous equilibrium is bigger than in non-homogeneous equilibrium (e.g. see [6], p. 22 ff.)

4.2 Viscosity, heat conductivity and bulk viscosity

There are three dissipative "mechanisms" in the gas, due to dynamic pressure, heat flux and stress deviator, because the entropy production is a quadratic form in these variables. Necessary and sufficient conditions for the entropy production to be non-negative are given by the inequalities

$$B_1^{\pi} \frac{\begin{vmatrix} -\ddot{p} & \dot{p}-\dot{p}' \\ \dot{p}-\dot{p}' & p'-p'' \end{vmatrix}}{\begin{vmatrix} \ddot{p} & \dot{p}-\dot{p}' & \dot{\Gamma}_1 \\ \dot{p}-\dot{p}' & p'-p'' & \Gamma_1'-\Gamma_1 \\ -\dot{p} & -p' & \frac{5}{3}\Gamma_1 \end{vmatrix}} \geqslant 0, \quad B_4 \frac{\dot{p}}{\begin{vmatrix} \dot{p} & -\dot{\Gamma}_1 \\ p' & \Gamma_1-\Gamma_1' \end{vmatrix}} \geqslant 0, \quad B_3\frac{1}{\Gamma_1} \leqslant 0, \qquad (4.13)$$

which provide restrictions for the coefficients B in the flux production.

It is possible to identify the coefficients B_1^{π}, B_3, B_4 in the flux production by relating them to the bulk viscosity, the shear viscosity and the heat conductivity. These are the transport coefficients in ordinary thermodynamics which can in principle be measured. In order to find that relation - and for that purpose only - we give a brief account of ordinary thermodynamics.

First of all we recall that the objective of <u>extended</u> thermodynamics is the determination of the 14 fields

$$A^A,\ A^{AB} \quad \text{or of} \quad n,\ U^A,\ e,\ \pi,\ q^A,\ t^{\langle AB\rangle}. \qquad (4.14)$$

The field equations are based upon the balance equations (2.3) which, by (1) and the constitutive relation (4) for I^{AB}, we may write in the forms

$$(nU^A)_{,A} = 0,$$

$$\left(t^{\langle AB\rangle} + (p+\pi)h^{AB} + \frac{1}{c^2}(q^AU^B + q^BU^A) + \frac{e}{c^2}U^AU^B\right)_{,A} = 0, \qquad (4.15)$$

$$A^{ABC}{}_{,C} = B_1^{\pi}\pi g^{AB} - \frac{4}{c^2}B_1^{\pi}\pi U^AU^B + B_3 t^{\langle AB\rangle} + B_4(q^AU^B + q^BU^A).$$

Of course, A^{ABC} in $(15)_3$ must be replaced by the constitutive expression on the right hand side of (3) but we omit that for brevity.

The objective of <u>ordinary</u> thermodynamics is the determination of the 5 fields

$$n,\ U^A,\ e. \tag{4.16}$$

The necesary field equations are based upon the conservation laws $(2.3)_{1,2}$ or $(15)_{1,2}$, and constitutive equations must be found for π, $t^{\langle AB\rangle}$ and q^A which relate these quantities to the fields (16). For viscous, heat conducting fluids the quantities π, $t^{\langle AB\rangle}$ and q^A at one event x^D are related algebraically to the values n, U^A, e. Note that the equation $(15)_3$ as yet plays no role in ordinary thermodynamics; but we shall now use that equation to motivate constitutive equations for π, $t^{\langle AB\rangle}$ and q^A in an approximate manner.

The approximation proceeds by setting π, q^A and $t^{\langle AB\rangle}$ equal to zero on the left hand sides of the equations (15). In particular this means setting

$$A^{ABC}|_E = \frac{1}{2c^2}\frac{\Gamma_1'}{T}U^AU^BU^C - \frac{1}{2}\frac{\Gamma_1}{T}(g^{AB}U^C + g^{BC}U^A + g^{CA}U^B) \tag{4.17}$$

into the left hand side of $(15)_3$. We decompose the energy-momentum balance and the flux balance into temporal and spatial parts and obtain for the balance equations of particle number, energy-momentum and fluxes

$$0 = -\ddot{p}\frac{d\alpha}{d\tau} - (\dot{p}' - \dot{p})\frac{d\ln T}{d\tau} - \dot{p}\,U^A{}_{,A} \tag{4.18$_1$}$$

$$0 = -\frac{\dot{\Gamma}_1}{2T}\frac{d\alpha}{d\tau} + (p'' - p')\frac{d\ln T}{d\tau} - p'U^A{}_{,A} \tag{4.18$_2$}$$

$$0 = -\frac{\Gamma_1}{2T}h^A_M\alpha_{,A} + p'h^A_M(\ln T)_{,A} - \frac{p'}{c^2}\frac{dU_M}{d\tau} \tag{4.18$_3$}$$

$$-3B_1^\pi\pi = \frac{\dot{\Gamma}_1' - 3\dot{\Gamma}_1}{2T}\frac{d\alpha}{d\tau} + \frac{\Gamma_1'' - 4\Gamma_1' + 3\Gamma_1}{2T}\frac{d\ln T}{d\tau} + \frac{\Gamma_1' - \Gamma_1}{2T}U^A{}_{,A} \tag{4.18$_4$}$$

$$-B_4q_M = -\frac{\dot{\Gamma}_1}{2T}h^A_M\alpha_{,A} - \frac{\Gamma_1' - \Gamma_1}{2T}h^A_M(\ln T)_{,A} + \frac{1}{c^2}\frac{\Gamma_1' - \Gamma_1}{2T}\frac{dU_M}{d\tau} \tag{4.18$_5$}$$

$$B_3t_{\langle MN\rangle} = -\frac{\Gamma_1}{2T}h^B_Mh^C_N\,U_{\langle B,C\rangle}. \tag{4.18$_6$}$$

The equations (9) have been used here to replace $\dot{n}, \dot{e}, n'$ and e' by derivatives of p. The symbol $\frac{dR}{d\tau}$ stands for the material time derivative $U^A R_{,A}$. Elimination of the derivatives $\frac{d\alpha}{d\tau}, \frac{d\ln T}{d\tau}$ between $(18)_{1,2,4}$ and of the derivatives $h^A{}_M \alpha_{,A}$ between $(18)_{3,5}$ allows us to determine π, q_M and $t_{\langle MN\rangle}$ from $(18)_{4-6}$ as follows

$$\Pi = -\frac{1}{2T}\frac{1}{B_1^{\pi}} \frac{\begin{vmatrix} \ddot{p} & \dot{p}-\dot{p}' & \dot{\Gamma}_1 \\ \dot{p}-\dot{p}' & p'-p'' & \Gamma_1'-\Gamma_1 \\ -\dot{p} & -p' & \frac{5}{3}\Gamma_1 \end{vmatrix}}{\begin{vmatrix} -\ddot{p} & \dot{p}-\dot{p}' \\ \dot{p}-\dot{p}' & p'-p'' \end{vmatrix}} \left[U^A{}_{,A}\right] \tag{4.19$_1$}$$

$$q_M = -\frac{1}{2T}\frac{1}{B_4} \frac{\begin{vmatrix} \dot{p} & -\dot{\Gamma}_1 \\ \dot{p}' & \Gamma_1-\Gamma_1' \end{vmatrix}}{\dot{p}} \left[h^A_M\{(\ln T)_{,A} - \frac{1}{c^2}\frac{dU_A}{d\tau}\}\right] \tag{4.19$_2$}$$

$$t_{\langle MN\rangle} = -\frac{1}{2T}\frac{1}{B_3}\Gamma_1 \left[h^B_M h^C_N U_{\langle B,C\rangle}\right]. \tag{4.19$_3$}$$

The equations $(19)_{2,3}$ are the relativistic analogues to the phenomenological equations of Fourier and Navier-Stokes. Equation $(19)_1$ gives a linear relation between the dynamic pressure and the divergence of velocity. In analogy to the nomenclature of ordinary irreversible thermodynamics we may call

$$\pi,\ q_M,\ t_{\langle MN\rangle}$$

and

$$[U^A{}_{,A}],\quad \left[h^A{}_M T_{,A} - \frac{T}{c^2}\frac{dU_M}{d\tau}\right],\quad [h^B_M h^C_N U_{\langle B,C\rangle}] \tag{4.20}$$

the thermodynamic forces and fluxes respectively. The factors of proportionality between these forces and fluxes, called transport coefficients, are known as the bulk viscosity ν, the heat conductivity κ and the shear viscosity μ. We have

$$\nu + \frac{2}{3}\mu = \frac{1}{2T}\frac{1}{B_1^{\pi}} \frac{\begin{vmatrix} -\ddot{p} & \dot{p}-\dot{p}' & \dot{\Gamma}_1 \\ \dot{p}-\dot{p}' & p'-p'' & \Gamma_1'-\Gamma_1 \\ -\dot{p} & -p' & \frac{5}{3}\Gamma_1 \end{vmatrix}}{\begin{vmatrix} -\ddot{p} & \dot{p}-\dot{p}' \\ \dot{p}-\dot{p}' & p'-p'' \end{vmatrix}} \geq 0, \tag{4.21$_1$}$$

$$\kappa = \frac{1}{2T^2}\frac{1}{B_4} \frac{\begin{vmatrix} \dot{p} & -\dot{\Gamma}_1 \\ p' & \Gamma_1 - \Gamma_1' \end{vmatrix}}{\dot{p}} \geq 0, \tag{4.21$_2$}$$

$$\mu = -\frac{1}{4T}\frac{1}{B_3}\Gamma_1 \geq 0. \tag{4.21$_3$}$$

By comparison with the inequalities (13) we conclude that the requirement of the non-negative entropy production determines the sign of the transport coefficients ν, μ and κ as indicated in (21).

Since in principle viscosity, heat conductivity and bulk viscosity can be measured, we may use the equations (21) to identify the coefficients B_1^{π}, B_3 and B_4.

The equations (19) may be regarded as first iterates for π, q_M and $t_{<MN>}$ in an iteration scheme: if these first iterates were inserted on the left hand sides of (15), we could use those equations to calculate second iterates for π, q_M, $t_{<MN>}$, etc. Such a scheme is akin to the Maxwellian iteration in the kinetic theory of gases (e.g. see [7]). Purely formal as this iterative scheme is, it achieves some plausibility by producing Fourier's law and the Navier-Stokes equation as first iterates. Also the corresponding scheme in the kinetic theory of gases provides successively higher powers of the mean free path in higher iterates.

Note that by $(19)_2$ the "thermodynamic force" driving the heat flux is not just equal to the temperature gradient. Rather there is an additional acceleration term which was first derived by Eckart [8]. This term has the interesting consequence that a gas in equilibrium within a gravitational field has a non-uniform temperature.

A discussion of this effect and related ones may be found in [6], p. 427 ff, in [2], or elsewhere in the literature.

5. THE SPECTRUM OF PROPERTIES OF AN IDEAL GAS

We recall that the coefficients in A^{ABC} and h^A have all been related to the thermal equation of state $p = p(\alpha, T)$ and to two functions $A_1(\alpha)$ and $A_2(\alpha)$ (see (7), (8), and (11)). Specific expressions for the coefficients require an exact knowledge of the thermal equation of state. In principle that equation may be determined experimentally, but in practice it is difficult to produce an analytical expression in this manner, particularly for a relativistic gas. Instead we calculate the function $p(\alpha, T)$ from equilibrium statistical mechanics.

In statistical mechanics the particle flux vector A^A, the energy-momentum tensor A^{AB} and the flux tensor A^{ABC} are defined as moments of the distribution function $f(x^A, p^a)$

$$
\begin{aligned}
A^A &= c\int p^A f dP, \\
A^{AB} &= c\int p^A p^B f dP, \qquad (5.1)\\
A^{ABC} &= c\int p^A p^B p^C f dP,
\end{aligned}
$$

where p^A is the 4-momentum of a particle and $dP = (\sqrt{-g}) dp^0\, dp^1\, dp^2\, dp^3$ is the invariant element of momentum space. In particular, in equilibrium the equations (1) read by virtue of (4.1) and (4.3)

$$
\begin{aligned}
nU^A &= c\int p^A f_E dP, \\
ph^{AB} + \frac{e}{c^2} U^A U^B &= c\int p^A p^B f_E dP, \qquad (5.2)\\
\frac{1}{c^2}\frac{\Gamma_1'}{T} U^A U^B U^C - \frac{1}{2}\frac{\Gamma_1}{T}(g^{AB}U^C + g^{BC}U^A + g^{CA}U^B) &= c\int p^A p^B p^C f_E dP,
\end{aligned}
$$

where f_E is the equilibrium distribution function. From (2) we obtain

$$n = \frac{1}{c} U_A \int p^A f_E dP,$$

$$e = \frac{1}{c} U_A U_B \int p^A p^B f_E dP,$$

$$p = \frac{c}{3} h_{AB} \int p^A p^B f_E dP, \tag{5.3}$$

$$\frac{\Gamma_1}{T} = \frac{2}{3c} h_{AB} U_C \int p^A p^B p^C f_E dP.$$

The general form of f_E is the Jüttner distribution (see [9], [10])

$$f_E = y \left[\exp\left(\frac{\alpha}{k} + \frac{U_A p^A}{kT}\right) \mp 1\right]^{-1} \tag{5.4}$$

where k is the Boltzmann constant and y is equal to w/h^3. Here, h is Planck's constant and w is equal to $2s+1$ for particles with spin $s\frac{h}{2\pi}$. Thus $\frac{1}{y}$ is the smallest element of phase space that can accommodate a particle. The $\mp$ sign refers to Bosons and Fermions.

The equilibrium distribution (4) assumes different forms depending on the degree of degeneracy of the gas and on the degree to which relativistic effects are important. An appreciation of the whole spectrum of the gas properties is gleaned from an inspection of the Table - reproduced on the following page - which characterizes various limiting cases and gives the appropriate equilibrium distribution function in the rest Lorentz frame. In the upper left corner we have the classical Maxwellian distribution function and in the lower right corner we have the Planck distribution appropriate to a gas of mass-less Bosons. The general Jüttner distribution appears in the framed field in the centre while γ in the Table, defined as $\frac{mc^2}{kT}$, may be called the relativistic coldness. For γ much larger than 1 we have a non-relativistic gas, and for γ much smaller, we have an ultrarelativistic gas.

In [4], [5] and [11] specific expressions for the constitutive coefficients (7) and (8) are derived for particular cases.

	Non-relativistic	relativistic	ultrarelativistic
non-degenerate	$y e^{-(\frac{\alpha}{k}+\gamma)} e^{-\frac{p^2}{2mkT}}$	$y e^{-\frac{\alpha}{k}} e^{-p\sqrt{1+\frac{p^2}{m^2c^2}}}$	$y e^{-\frac{\alpha}{k}} e^{-\frac{cp}{kT}}$
degenerate	$\dfrac{y}{e^{(\frac{\alpha}{k}+\gamma)} e^{\frac{p^2}{2mkT}} \mp 1}$	$\boxed{\dfrac{y}{\exp\left\{\frac{\alpha}{k}+\gamma\sqrt{1+\frac{p^2}{m^2c^2}}\right\} \mp 1}}$	$\dfrac{y}{e^{\frac{\alpha}{k}+\frac{cp}{kT}} \mp 1}$
strongly degenerate FERMI	$\begin{cases} y & 0 \leq p < \sqrt{-2mkT(\frac{\alpha}{k}+\gamma)} \\ 0 & p > \sqrt{-2mkT(\frac{\alpha}{k}+\gamma)} \end{cases}$	$\begin{cases} y & 0 \leq p < mc\sqrt{(\frac{\alpha/k}{\gamma})^2-1} \\ 0 & p > mc\sqrt{(\frac{\alpha/k}{\gamma})^2-1} \end{cases}$	$\begin{cases} y & 0 \leq p < -\alpha\frac{T}{c} \\ 0 & p > -\alpha\frac{T}{c} \end{cases}$
strongly degenerate BOSE	$\dfrac{y}{e^{\frac{p^2}{2mkT}}-1}$ $p \neq 0$	$\dfrac{y}{e^{\gamma\left(\sqrt{1+\frac{p^2}{m^2c^2}}-1\right)}-1}$ $p \neq 0$	$\dfrac{y}{e^{\frac{cp}{kT}}-1}$ $p \neq 0$

Table 1. Characterization of Limiting Cases
Equilibrium Distribution Functions in Rest Lorentz Frame

REFERENCES

[1] Chernikov, N.A. (1963). The relativistic gas in the gravitational field. Acta Phys. Polonica 23.

[2] Chernikov, N.A. (1964). Equilibrium distribution of the relativistic gas. Acta Phys. Polonica 25.

[3] Chernikov, N.A. (1964). Microscopic foundation of relativistic hydrodynamics. Acta Phys. Polonica 27.

[4] Liu, I-Shih, Müller, I. & Ruggeri, T. (1986). Relativistic thermodynamics of gases. Annals of Physics 169.

[5] Müller, I. & Dreyer, W. (in preparation) Extended Thermodynamics, Relativistic and Classical. Springer Tracts of Natural Philosophy.

[6] Müller, I. (1985). Interaction of Mechanics and Mathematics Series. London, Pitman.

[7] Ikenberg, E. & Truesdell, C. (1956). On the pressures and the flux of energy in a gas according to Maxwell's kinetic theory. J. Rational Mech. Anal. 5.

[8] Eckart, C. (1940). The thermodynamics of irreversible processes III: Relativistic theory of the simple fluid. Phys. Rev. 58.

[9] Jüttner, F. (1911). Das Maxwell'sche Gesetz der Geschwindigkeits-verteilung in der Relativitätstheorie. Annalen der Physik 34.

[10] Jüttner, F. (1928). Die relativistiche Quantentheorie des idealen Gases. Zeitschrift für Physik 47.

[11] Dreyer, W. Statistical thermodynamics of relativistic and degenerate gases in non-equilibrium. Submitted to Annals of Physics.

HOMOGENIZATION PROBLEMS IN ELASTICITY. SPECTRA OF SINGULARLY PERTURBED OPERATORS.

O.A. Oleinik
Moscow State University
Moscow
U S S R

ABSTRACT

The following homogenization problems are considered. 1) Asymptotic expansions of eigenvalues and eigenfunctions of the Sturm-Liouville problem for differential equations with rapidly oscillating coefficients. 2) Homogenisation problems for stratified elastic composites: a necessary and sufficient condition for the G-convergence, formulas for the coefficients of the homogenized operator, estimates for the difference between displacements, strain and stress tensors, energy integrals, eigenvalues and eigenfunctions of a stratified body and the corresponding effective characteristics. 3) Estimates for the difference of eigenvalues and eigenfunctions for a perforated elastic body and the eigenvalues and eigenfunctions of the homogenized boundary value problem. Moreover the behaviour of frequencies of free vibrations of bodies with concentrated masses is considered.

In recent years various problems in mechanics, physics and technology have promoted the development of a new branch of mathematical physics called the homogenization theory of partial differential equations (see [1] - [38]). We consider here several problems of this theory.

Let us consider first the spectral properties of linear differential operators connected with boundary value problems for differential equations with rapidly oscillating coefficients. Such problems arise in particular in the theory of elastic composites. The simplest case is the Sturm-Liouville problem.

1. ASYMPTOTIC EXPANSIONS FOR EIGENVALUES AND EIGENFUNCTIONS OF THE STURM-LIOUVILLE PROBLEM WITH RAPIDLY OSCILLATING PERIOD COEFFICIENTS

Consider the following eigenvalue problem

$$\frac{d}{dx}\left[a\left(\frac{x}{\epsilon}\right)\frac{du^{\epsilon,k}}{dx}\right] + b\left(\frac{x}{\epsilon}\right)u^{\epsilon,k} + \lambda_k^\epsilon \rho\left(\frac{x}{\epsilon}\right)u^{\epsilon,k} = 0, \quad x \in (0,1), \tag{1}$$

$$u^{\epsilon,k}(0) = 0, \quad u^{\epsilon,k}(1) = 0, \quad \int_0^1 (u^{\epsilon,k})^2 \rho\left(\frac{x}{\epsilon}\right)dx = 1, \tag{2}$$

$$0 < \lambda_1^\epsilon < \lambda_2^\epsilon < \lambda_3^\epsilon < \ldots < \lambda_k^\epsilon < \ldots$$

$$a\left(\frac{x}{\epsilon}\right) \geq a_0 = \text{const} > 0, \ \rho\left(\frac{x}{\epsilon}\right) \geq \rho_0 = \text{const}, \ \epsilon = \text{const}, \ b \leq 0.$$

We suppose for simplicity that $a(\xi)$, $b(\xi)$, $\rho(\epsilon)$ are smooth periodic functions with period 1 (1-periodic functions in ξ). A more general case is considered in [24].

For a fixed k we look for asymptotic expansions of λ_k^ϵ and $u^{\epsilon,k}$ in the form (the index k is omitted)

$$\lambda_\epsilon^{(M)} = \lambda_0 + \epsilon\lambda_1(\epsilon) + \ldots + \epsilon^M\lambda_M(\epsilon), \tag{3}$$

$$u_\epsilon^{(M)} = \sum_{i=0}^{M} \epsilon^i \left[\sum_{s=0}^{i} N^{(i,s)}(\xi)\,\frac{d^s v_\epsilon(x)}{dx^2}\right], \quad \xi = \frac{x}{\epsilon}, \tag{4}$$

where $M \geq 2$ is an integer, $N^{(i,s)}(\xi), v_\epsilon(x)$ are the functions to be determined, $\lambda_i(\epsilon)$ are bounded in ϵ; $i = 1,\ldots,M$. Set

$$\langle f \rangle = \lim_{t\to\infty} \frac{1}{2T}\int_{-T}^{T} f(t)dt.$$

For a function $\varphi(\xi)$ which is 1-periodic in ξ we have $\langle\varphi\rangle = \int_0^1 \varphi(\xi)d\xi$. Let us replace λ_k^ϵ and $u^{\epsilon,k}$ in the left-hand side of (1) by λ_ϵ and $u_\epsilon^{(M)}$ respectively and consider the coefficients by the different powers of ϵ and by the different derivatives of v_ϵ. Setting them equal to zero we get a system of equations for $N^{(i,s)}(\xi)$. Set

$$N^{(0,0)} = 1, \ N^{(1,0)} = 0, \ N^{(2,1)} = 0$$

and define $N^{(1,1)}$ as a solution of the problem:

$$\frac{d}{d\xi}\left[a(\xi)\,\frac{dN^{(1,1)}}{d\xi}\right] = -\,\frac{da(\xi)}{d\xi},\ \ \xi \in R^1, \qquad < N^{(1,1)}(\xi)\rho(\xi) > = 0.$$

$N^{(1,1)}(\xi)$ is a 1-periodic function in ξ.

The function $N^{(2,2)}(\xi)$, with $\xi \in \mathbb{R}^1$, we define as a solution of the problem:

$$\frac{d}{d\xi}\left[a(\xi)\frac{dN^{(2,2)}}{d\xi}\right] = -\,\frac{d}{d\xi}\left[a(\xi)N^{(1,1)}\right] - a(\xi)\frac{dN^{(1,1)}}{d\xi} - a(\xi) + h^{(2,2)},$$

$N^{(2,2)}(\xi)$ is 1-periodic in ξ and $< N^{(2,2)}(\xi)\rho(\xi) > = 0$,

$$h^{(2,2)} = < a(\xi) + a(\xi)\,\frac{dN^{(1,1)}}{d\xi} >.$$

We take $N^{(2,0)}$ to be a solution of the problem:

$$\frac{d}{d\xi}\left[a(\xi)\,\frac{dN^{(2,0)}}{d\xi}\right] = -\,\rho(\xi)\lambda_0 - b(\xi) + h^{(2,0)},\quad \xi \in \mathbb{R}^1,$$

$N^{(2,0)}(\xi)$ is a 1-periodic in ξ and $< N^{(2,0)}(\xi)\rho(\xi) > = 0$,

$$h^{(2,0)} = \lambda_0\langle\rho\rangle + \langle b\rangle.$$

For $i > 2$ we define $N^{(i,s)}(\xi)$, $s \leq i$, as solutions of the problems:

$$\frac{d}{d\xi}\left[a(\xi)\,\frac{dN^{(i+2,s)}}{d\xi}\right] = -\,a(\xi)^{(i,s-2)} - a(\xi)\frac{dN^{(i+1,s-1)}}{d\xi}$$

$$-\,\frac{d}{d\xi}\left[a(\xi)N^{(i+1,s-1)}\right] - \sum_{j=0}^{i}\lambda_j N^{(i-j,s)}\rho(\xi) - b(\xi)N^{(i,s)} + h^{(i+2,s)}, \tag{5}$$

$N^{(i+2,s)}$ is 1-periodic in ξ, $< N^{(1+2,s)}(\xi)\rho(\xi) > = 0$,

where $i = 1,\ldots, M-2$, and

$$h^{(i+2,s)} = \langle bN^{(i,s)}\rangle + \lambda_i\langle\rho\rangle \equiv \tilde{h}^{(i+2,s)} + \lambda_i\langle\rho\rangle,\quad s = 0;$$

$$h^{(i+2,s)} = \langle aN^{(i,s-2)} - bN^{(i,s)} + a\frac{dN^{(i+1,s-1)}}{d\xi}\rangle \equiv \tilde{h}^{(i+2,s)}, \quad s > 0.$$

It is easy to prove by induction in i,s, $s \leqslant i$ that $N^{(i,s)}$ exist, provided that $N^{(0,0)}, N^{(1,0)}, N^{(1,1)}, N^{(2,0)}, N^{(2,1)}, N^{(2,2)}$ are as defined above. We thus have

$$L_\epsilon^{(M)}(u_\epsilon^M) \equiv \frac{d}{dx}\left[a\left(\frac{x}{\epsilon}\right)\frac{du_\epsilon^{(M)}}{dx}\right] + b\left(\frac{x}{\epsilon}\right)u_\epsilon^{(M)} + \lambda_\epsilon^{(M)}\rho\left(\frac{x}{\epsilon}\right)u_\epsilon^{(M)}$$

$$= \sum_{i=0}^{M-2}\epsilon^i\left[\sum_{s=0}^{i+2}\tilde{h}^{(i+2,s)}\frac{d^s v_\epsilon}{dx^s} + \lambda_i\,\langle\rho\rangle v_\epsilon\right] + \epsilon^{M-1}F_1(x.\epsilon), \quad (6)$$

where $F_1(x,\epsilon)$ is a sum of terms having the form $\epsilon^t\varphi(\xi)d^l v_\epsilon/dx^l$, $l \leqslant M + 2$, $t \geqslant 0$, φ is a bounded function. We look for v_ϵ in the form $v_\epsilon\,(x) = v_0(x) + \epsilon v_1(x) + \ldots + \epsilon^{M-2}v_{M-2}$. It then follows from (6) that

$$L_\epsilon^{(M)}\left[u_\epsilon^{(M)}\right] = \sum_{i=0}^{M-2}\epsilon^i\left[\sum_{p=0}^{i}\sum_{s=0}^{i+p+2}\tilde{h}^{(i-p+2,s)}\frac{d^s v_p}{dx^s}\right.$$

$$\left. + \sum_{p=0}^{i}\lambda_{i-p}\,\langle\rho\rangle v_p(x)\right] + \epsilon^{M-1}F_2(x,\epsilon).$$

We now define $v_p(x)$, $p = 0,\ldots,M-2$, as functions which satisfy the equations

$$\sum_{p=0}^{i}\sum_{s=0}^{i-p+2}\tilde{h}^{(i-p+2,s)}\frac{d^s v_p}{dx^s} + \sum_{p=0}^{i}\lambda_{i-p}\langle\rho\rangle v_p = 0, \; i = 0,1,\ldots,M-2, \quad (7)$$

and the boundary conditions

$$\sum_{p=0}^{i}\sum_{s=0}^{i-p}N^{(i-p,s)}(0)\frac{d^s v_p}{dx^s}(0) = \sum_{p=0}^{i}\sum_{s=0}^{i-p}N^{(i-p,s)}(\epsilon^{-1})\frac{d^s v_p}{dx^s}(1) = 0. \quad (8)$$

Let us rewrite (7), (8) leaving in the left-hand side only those terms which contain v_i and its derivatives. For $i = 0$ we have

$$h^{(2,2)} \frac{d^2 v_o}{dx^2} + \langle b \rangle v_o + \lambda_o \langle \rho \rangle v_o = 0, \tag{9}$$

$$v_o(0) = v_o(1) = 0.$$

For $i = 1, 2, \ldots, M-2$ we get the following boundary value problems

$$h^{(2,2)} \frac{d^2 v_i}{dx^2} + \langle b \rangle v_i + \lambda_o \langle \rho \rangle v_i$$

$$= - \sum_{p=0}^{i-1} \sum_{s=o}^{i-p+2} \tilde{h}^{(i-p+2,s)} \frac{d^s v_p}{dx^s} - \sum_{p=1}^{i-1} \lambda_{i-p} \langle \rho \rangle v_p - \lambda_i \langle \rho \rangle v_o, \tag{10}$$

$$v_i(0) = - \sum_{p=0}^{i-1} \sum_{s=0}^{i-p} N^{(i-p,s)}(0) \frac{d^s v_p}{dx^s}(0),$$

$$v_i(1) = - \sum_{p=0}^{i-1} \sum_{s=0}^{i-p} N^{(i-p,s)}(\epsilon^{-1}) \frac{d^s v_p}{dx^s}(1) . \tag{11}$$

We take for λ_o the k-th eigenvalue of the homogenized problem (9). By induction, assuming $\lambda_o, \ldots \lambda_{i-1}(\epsilon), v_o, \ldots, v_i$, $N^{(0,0)}, \ldots, N^{(i+1,s)}$ to be determined, we find $\lambda_i(\epsilon)$ from the condition of solvability of problem (10),(11) and then we take for $v_i(x)$ the solution of (10),(11) with this $\lambda_i(\epsilon)$ and define $N^{(i+2,s)}$ from equation (5). In this way we obtain

$$L_\epsilon^{(M)}(u_\epsilon^{(M)}) = \epsilon^{M-1} F_2(x,\epsilon), \quad u_\epsilon^{(M)}(0) = \epsilon^{M-1} \psi_o(\epsilon), \quad u_\epsilon^{(M)}(1) = \epsilon^{M-1} \psi_1(\epsilon),$$

where F_2, ψ_o, ψ_1 are bounded uniformly in x and ϵ. Thus we get a formal asymptotic expansion for λ_k^ϵ and $u^{\epsilon,k}(x)$. A complete proof of the following theorem is given in [24].

Theorem 1. Let λ_k^ϵ be an eigenvalue of problem (1),(2) and $u^{\epsilon,k}(x)$ be the corresponding eigenfunction. Then

$$|\lambda_k^\epsilon - \lambda_\epsilon^{(M)}| \leq C_1(k)\epsilon^{M-1}, \tag{12}$$

$$\|u^{\epsilon,k} - u_\epsilon^{(M)}\|_{L^2(0,1)} \leq C_2(k)\epsilon^{M-1}, \tag{13}$$

where constants $C_1(l)$ and $C_2(k)$ do not depend on ϵ. In formula (3) for $\lambda_\epsilon^{(M)}$ we have $\lambda_1(\epsilon) = 0$ for any k.

Remark 1. In a similar way one can obtain asymptotic expansions for eigenvalues and eigenfunctions in the case when the coefficients of the equation (1) have the form

$$a = a(x,\xi), \quad b = b(x,\xi), \quad \rho = \rho(x,\xi), \quad \xi = \frac{x}{\epsilon}.$$

where a,b,ρ are smooth functions of x and are either 1-periodic in ξ or belong to the class A which is described in [24]. The same method can be applied to get asymptotic expansions for simple eigenvalues and eigenfunctions in the case of higher order ordinary differential equations with boundary conditions of a wide class.

The proof of Theorem 1 is based on the following lemmas.

Lemma 1. Let $A: H \to H$ be a self-adjoint positive compact operator in a Hilbert space H. Suppose that there exist a $\mu \in R^1$ and a vector $u \in H$, $\|u\|_H = 1$ such that

$$\|Au - \mu u\|_H \leq \alpha, \quad \alpha = \text{const} > 0. \tag{14}$$

Then there is an eigenvalue μ_i of the operator A such that $|\mu_i - \mu| \leq \alpha$. Moreover for any $d > \alpha$ there exists a vector $\bar{u}$ such that

$$\|u - \bar{u}\|_H \leq 2\alpha d^{-1}, \quad \|\bar{u}\|_H = 1, \tag{15}$$

and $\bar{u}$ belongs to the linear space formed by eigenvectors of A corresponding to eigenvalues from the interval $(\mu - d, \mu + d)$.

The proof of Lemma 1 is given in [39].

Lemma 2. Let λ_k^ϵ be eigenvalues of problem (1), (2) and

$\lambda_{k(t)}^{\epsilon_t} \to \lambda^*$ for a sequence $\epsilon_t \to 0$. Then $u^{\epsilon,k(t)} \to u^*$ in $L^2(0,1)$ as $\epsilon_t \to 0$, where u^* is an eigenfunction of the problem

$$h^{(2,2)} \frac{d^2u^*}{dx^2} + \langle b\rangle u^* + \lambda^*\langle\rho\rangle u^* = 0, \quad u^*(0) = u^*(1) = 0.$$

Lemma 2 can be proved using Tartar's method described in [1].

2. HOMOGENIZATION OF STRATIFIED STRUCTURES

Of great importance for applications in technology is the study of processes in bodies made of stratified elastic materials. The elastic properties of such bodies vary rapidly in only one direction. In this case one can obtain explicit formulae which express the coefficients of the homogenized elasticity system through the coefficients of the initial system for a stratified elastic material. Thus it becomes possible to find in explicit form many effective characteristics of a stratified elastic body. Stratified composites form the subject of a vast literature in mechanics (see e.g. [11]). Composites with plane layers are considered in [36]. We study here stratified structures without any assumptions on their periodicity (see [37]).

Consider a sequence $\mathfrak{L}_\epsilon$ of differential operators of linear elasticity

$$\mathfrak{L}_\epsilon(u) \equiv \frac{\partial}{\partial x_i}\left[C_\epsilon^{ij}(\varphi(x),x_1,\dots,x_n)\frac{\partial u}{\partial x_j}\right] = f, \tag{16}$$

where $C_\epsilon^{ij}(\varphi(x),x_1,\dots,x_n)$ are $(n \times n)$ - matrices whose elements $C_{\epsilon\,lh}^{ij}(t,y)$, $y = (y_1,\dots,y_n)$ are bounded (uniformly in ϵ) measurable functions of $t \in R^1$, $y \in R^n$ with bounded (uniformly in ϵ) first derivatives in $y_1,\dots,y_n$; $u = (u_1,\dots u_n)$ is a column-vector with components $u_1,\dots,u_n$; $\varphi(x)$ is a scalar function in $C^3(\bar{\Omega})$, Ω is a smooth domain in R_n, and $0 \leqslant \varphi(x) \leqslant 1$, $\varphi_{x_i}\varphi_{x_i} \geqslant \text{const} > 0$.

Here and in what follows we assume summation over repeated indices from 1 to n.

The elements of matrices $C_\epsilon^{ij}(t,y)$ are required to satisfy the following conditions

$$\kappa_1 \eta_l^i \eta_l^i \leq C_{\epsilon,hl}^{ij}(t,y)\eta_h^i \eta_l^j \leq \kappa_2 \eta_l^i \eta_l^i,$$

$$C_{\epsilon,lh}^{ij}(t,y) = C_{\epsilon,hl}^{ji}(t,y) = C_{\epsilon,ih}^{lj}(t,y), \tag{17}$$

where $\eta = \|\eta_l^i\|$ is an arbitrary symmetric matrix and κ_1,κ_2 are positive constants which do not depend on ϵ,t,y,η.

Let us also consider another system of equations of linear elasticity

$$\hat{\mathfrak{L}}(u) \equiv \frac{\partial}{\partial x_i}\left[\hat{C}^{ij}(\varphi(x),x_1,\ldots,x_n)\frac{\partial u}{\partial x_j}\right] = f,$$

whose coefficients satisfy conditions similar to (17) but with constants $\hat{\kappa}_1,\hat{\kappa}_2$ and are smooth functions of t and y. We obtain an estimate for the difference between a solution of the Dirichlet problem for operator $\mathfrak{L}_\epsilon$ in a smooth domain Ω and a solution of the corresponding Dirichlet problem for operator $\hat{\mathfrak{L}}$. This result is used here to estimate the difference between the respective energies contained in a subdomain ω of a stratified elastic body Ω and also between the respective eigenvalues and eigenfunctions of the operators $\mathfrak{L}_\epsilon$ and $\hat{\mathfrak{L}}$ for the Dirichlet problem. We also obtain necessary and sufficient conditions for the strong G-convergence of the sequence $\mathfrak{L}_\epsilon$ to the operator $\hat{\mathfrak{L}}$ as $\epsilon \to 0$ and give explicit formulae for the coefficients of $\hat{\mathfrak{L}}$.

By $H^m(\Omega)$, $(H_0^m(\Omega))$ we denote the completion in the norm

$$\|v\|_m = \left[\sum_{|\gamma|\leq m}\int_\Omega |D^\gamma v|^2 dx\right]^{1/2}, \ \gamma = (\gamma_1,\ldots,\gamma_n), |\gamma| = \gamma_1+\ldots+\gamma_n,$$

of the space of vector-valued functions $v = (v_1,\ldots,v_n)$ with components in $C^\infty(\bar{\Omega})$ $(C^\infty{}_0(\bar{\Omega}))$. The scalar product in $H^m(\Omega)$ is denoted by $(u,v)_m$ while the Hilbert space dual to $H_0^1(\Omega)$ is denoted by $H^{-1}(\Omega)$. We assume that Ω is a bounded smooth domain in $\mathbb{R}^n$.

Consider the following Dirichlet problems

$$\mathfrak{L}_\epsilon(u^\epsilon) = f, \quad u^\epsilon \in H_0^1(\Omega), \quad f \in L^2(\Omega), \tag{18}$$

$$\hat{\mathfrak{L}}(u) = f, \quad u \in H_0^1(\Omega), \quad f \in L^2(\Omega). \tag{19}$$

Define matrices $N_s^\epsilon(t,y)$ and $M_{is}^\epsilon(t,y)$ by the formulae

$$N_s^\epsilon(t,y) = \int_0^t \Big[\varphi_l(y)\varphi_k(y)C_\epsilon^{kl}(\tau,y)\Big]^{-1} \varphi_p(y)\Big[\hat{C}^{ps}(\tau,y) - C_\epsilon^{ps}(\tau,y)\Big]d\tau,$$

$$M_{is}^\epsilon(t,y) = \int_0^t \Big\{\varphi_j(y)C_\epsilon^{ij}(\tau,y)\Big[\varphi_l(y)\varphi_k(y)C_\epsilon^{kl}(\tau,y)\Big]^{-1} \varphi_p(y)\Big[\hat{C}^{ps}(\tau,y) - C_\epsilon^{ps}(\tau,y)\Big] + C_\epsilon^{is}(\tau,y) - \hat{C}^{is}(\tau,y)\Big\}d\tau, \quad i,s = 1,\dots,n, \tag{20}$$

where $(\varphi_1(y),\dots,\varphi_n(y)) = \Big(\frac{\partial\varphi}{\partial x_1},\dots,\frac{\partial\varphi}{\partial x_n}\Big) = \operatorname{grad}\varphi$ and A^{-1} is the inverse matrix to A. It can be proved that the matrix $\big[\varphi_l(y)\varphi_k(y)C_\epsilon^{kl}(\tau,y)\big]^{-1}$ exists and its elements are functions bounded uniformly in ϵ. Set

$$\delta_\epsilon = \max_{x\in\bar{\Omega}} \Big\{\Big|M_{ij}^\epsilon(\varphi(x),x)\Big|, \Big|N_j^\epsilon(\varphi(x),x)\Big|, \Big|\frac{\partial}{\partial y_l}M_{ij}(\varphi(x),x)\Big|, \Big|\frac{\partial}{\partial y_l}N_j^\epsilon(\varphi(x),x\Big|\Big\}.$$

$l,i,j = 1,\dots,n$

For a matrix $A = \|a^{kl}\|$ we denote $a^{kl}a^{kl}$ by $|A|$. Solutions of problems (18) and (19) are taken to be vector valued functions from $H_0^1(\Omega)$ which satisfy (18), (19) in the sense of distributions.

<u>Theorem 2</u>. Let $f \in L^2(\Omega)$. Then for the solutions u^ϵ and u of problems (18) and (19) respectively, the following estimates hold

$$\|u^\epsilon - u - N_s^\epsilon(\varphi(x),x)\frac{\partial u}{\partial x_s}\|_1 \le C_0\delta_\epsilon^{1/2}\|f\|_0, \tag{22}$$

$$\|\gamma_\epsilon^i - \hat{\gamma}^i - \frac{\partial M_{is}^\epsilon}{\partial t}(\varphi(x),x)\frac{\partial u}{\partial x_s}\|_0 \le C_1\delta_\epsilon^{1/2}\|f\|_0, \tag{23}$$

$$|E_\omega^\epsilon(u^\epsilon) - E_\omega(u)| \le C_2(\omega)\delta_\epsilon^{1/2}\|f\|_0^2, \tag{24}$$

where $\gamma_\epsilon^i \equiv C_\epsilon^{ij} \frac{\partial u^\epsilon}{\partial x_j}$, $\gamma^i \equiv \hat{C}^{ij} \frac{\partial u}{\partial x_j}$ and the constants C_0, C_1, C_2 do not depend on ϵ; ω is a smooth subdomain of Ω, $(u,v) = u_j v_j$ for any vector-valued functions $v = (v_1,\ldots,v_n)$, $u = (u_1,\ldots,u_n)$ and

$$E_\omega^\epsilon(u^\epsilon) \equiv \int_\omega \left[C_\epsilon^{ij} \frac{\partial u^\epsilon}{\partial x_j}, \frac{\partial u}{\partial x_j}\right] dx, \quad E_\omega(u) \equiv \int_\omega \left[\hat{C}^{ij} \frac{\partial u}{\partial x_j}, \frac{\partial u}{\partial x_i}\right] dx.$$

We describe briefly the main ideas of the proof of Theorem 2.

Define a function $v_\epsilon(x)$ as a solution of the problem

$$\mathfrak{L}_\epsilon(v_\epsilon) = 0 \text{ in } \Omega, \quad v_\epsilon = N_j^\epsilon(\varphi(x),x)\frac{\partial u}{\partial x_j} \text{ on } \partial\Omega, \quad v_\epsilon \in H^1(\Omega).$$

It is easy to prove that

$$\mathfrak{L}_\epsilon\left(u_\epsilon - u - N_s^\epsilon(\varphi(x),x)\frac{\partial u}{\partial x_s} + v_\epsilon\right)$$

$$= \frac{\partial}{\partial x_i}\left[\left[\hat{C}^{is} - C_\epsilon^{is} - C_\epsilon^{ij} \frac{\partial N_s^\epsilon}{\partial x_j}\right]\frac{\partial u}{\partial x_s}\right] - \frac{\partial}{\partial x_i}\left[C_\epsilon^{ij} N_s^\epsilon \frac{\partial^2 u}{\partial x_s \partial x_j}\right] \equiv F_\epsilon . \quad (25)$$

This equation is satisfied in the sense of distributions. One can easily get that

$$\frac{\partial}{\partial t}\left[M_{is}^\epsilon(\varphi(x),x)\right] = \varphi_k |\text{grad } \varphi|^{-2} \frac{\partial}{\partial x_k}\left[M_{is}^\epsilon(\varphi(x),x)\right] + \beta_{is}(x,\epsilon), \quad (26)$$

$$\frac{\partial}{\partial x_i}\left[M_{is}^\epsilon(\varphi(x),x)\right] = \alpha_s(x,\epsilon),$$

$$\hat{C}^{is}(\varphi(x),x) - C_\epsilon^{is}(\varphi(x),x) - C_\epsilon^{ij}(\varphi(x),x)\frac{\partial}{\partial x_j} N_s^\epsilon(\varphi(x),x) = -\frac{\partial M_{is}}{\partial t} + \alpha_{is}(x,\epsilon),$$

where

$$|\beta_{is}(x,\epsilon)| < C_3\delta_\epsilon, \quad |\alpha_s(x,\epsilon)| \leq C_4\delta_\epsilon, \quad |\alpha_{is}(x,\epsilon))| \leq C_5\delta_\epsilon,$$

and the constants C_3, C_4, C_5 do not depend on ϵ. Let us estimate the right-hand side of (25) in the norm of $H^{-1}(\Omega)$. Using (25) and (26) we obtain

$$\|F_\epsilon\|_{H^{-1}(\Omega)} \leq C_6\delta_\epsilon\|u\|_2, \tag{27}$$

where the constant C_6 does not depend on ϵ and u. Since the coefficients of $\hat{\ell}$ and the boundary of Ω are smooth we have

$$\|u\|_2 \leq C_7\|f\|_0, \qquad C_7 = \text{const.}$$

From (27) and (25) we get

$$\|u_\epsilon - u - N_s^\epsilon(\varphi(x),x)\frac{\partial u}{\partial x_s} + v_\epsilon\|_1 \leq C_8\delta_\epsilon\|f\|, \tag{28}$$

where the constant C_8 does not depend on ϵ and f.

Now we have to estimate $\|v_\epsilon\|_1$. Set $w_\epsilon = v_\epsilon - \theta_\epsilon$, where $\theta_\epsilon = N_j^\epsilon(\varphi(x),x)\frac{\partial u}{\partial x_j}\psi_\epsilon$, $\psi_\epsilon = 1$ in the δ_ϵ – neighbourhood of $\partial\Omega$, $\psi_\epsilon = 0$ outside the $2\delta_\epsilon$ – neighbourhood of $\partial\Omega$, $\psi_\epsilon \in C^\infty(\bar{\Omega})$, $0 \leq \psi_\epsilon \leq 1$. Using the integral identity for v_ϵ, conditions (17) and Korn's inequality, we obtain

$$\|w_\epsilon\|_1 \leq C_9\|\theta_\epsilon\|_1.$$

To estimate $\|\theta_\epsilon\|_1$ we consider

$$\frac{\partial\theta_\epsilon}{\partial x_k} = \frac{\partial N_j^\epsilon}{\partial x_k}\frac{\partial u}{\partial x_j}\psi_\epsilon + N_j^\epsilon\frac{\partial^2 u}{\partial x_j\partial x_k}\psi_\epsilon + N_j^\epsilon\frac{\partial u}{\partial x_j}\frac{\partial\psi_\epsilon}{\partial x_k}. \tag{29}$$

It follows from (29) that

$$\|\theta_\epsilon\|_1^2 \leq C_{10}\delta_\epsilon^2\|u\|_2^2 + C_{11}\|\text{grad } u\|^2_{L^2(\omega_1)},$$

where ω_1 is the $2\delta_\epsilon$ – neighbourhood of $\partial\Omega$. It is easy to prove that

$$\|\frac{\partial u}{\partial x_j}\|_{L^2(\omega_1)} \leq C_{12}\delta_\epsilon\|u\|_2^2,$$

since according to the imbedding theorem we have $\|\frac{\partial u}{\partial x_j}\|_{L^2(\Gamma)} \leq C\|u\|_2^2$ for any $u \in H^2(\Omega)$ and any smooth surface Γ belonging to a neighbourhood of $\partial\Omega$. Therefore

$$\|v_\epsilon\|_1 \leq C_{13}\delta^{1/2}\|f\|_0. \tag{30}$$

Estimates (30) and (28) imply (22).

From (22) it follows that

$$\frac{\partial u_\epsilon}{\partial x_j} = \frac{\partial u}{\partial x} + \frac{\partial N_s^\epsilon}{\partial x_j}\frac{\partial u}{\partial x_s} + q_j^\epsilon, \tag{31}$$

where $\|q_j^\epsilon\|_0 \leq C_{14}\delta_\epsilon^{1/2}\|f\|_0$. From (26) and (31) we get

$$C_\epsilon^{ij}\frac{\partial u}{\partial x_j} - \hat{C}^{ij}\frac{\partial u}{\partial x_j} = \frac{\partial M_{is}^\epsilon}{\partial t}\frac{\partial u}{\partial x_s} - \alpha_{is}\frac{\partial u}{\partial x_s} + C_\epsilon^{ij}q_j^\epsilon. \tag{32}$$

The estimate (23) follows easily from (32). The estimate (24) can be proved using estimates (22) and (23).

Now we consider the strong G-convergence of operators $\mathfrak{L}_\epsilon$ to $\hat{\mathfrak{L}}$ as $\epsilon \to 0$.

A sequence of operators $\mathfrak{L}_\epsilon$ is called strongly G-convergent to operator $\hat{\mathfrak{L}}$ if for any $f \in H^{-1}(\Omega)$ the sequence u^ϵ of solutions of problem (18) converges to a solution u of problem (19) weakly in $H_0^1(\Omega)$ as $\epsilon \to 0$ and $\gamma_\epsilon^i \equiv C_\epsilon^{ij}\frac{\partial u_\epsilon}{\partial x_j}$ converges to $\hat{\gamma}^i \equiv \hat{C}^{ij}\frac{\partial u}{\partial x_j}$ weakly in $L^2(\Omega)$ as $\epsilon \to 0$, $i = 1,\dots,n$ (see [4],[9]).

Denote by $C^{0,\beta}$ the space of functions $g(t,y)$, $t \in [0,1]$, $y \in \bar{\Omega}$ with the norm

$$\|g\|_{0,\beta} = \sup_{t,y}|g(t,y)| + \sup_{t,y'y''}\frac{|g(t,y')-g(t,y'')|}{|y'-y''|^\beta},$$

$$t \in [0,1],\quad y,y',y'' \in \bar{\Omega},\quad y' \neq y'',\quad 0 < \beta < 1.$$

It is easy to prove the following lemmas.

<u>Lemma 3</u>. Suppose that a family of functions $\psi_\epsilon(t,y)$ has norms in $C^{0,\beta}$ uniformly bounded in ϵ. Then there is a subsequence $\epsilon' \to 0$ and a function $\phi \in C^{0,\beta}$ such that $\psi_{\epsilon'}(t,y) \to \phi(t,y)$ weakly in $L^2(0,1)$ for every $y \in \bar{\Omega}$. If $\psi_\epsilon(t,y) \to 0$ as $\epsilon \to 0$ weakly in

$L^2(0,1)$ for every $y \in \bar{\Omega}$, then $\phi_\epsilon(t,y) = \int_0^t \psi_\epsilon(\tau,y)d\tau \to 0$ in the norm $C([0,1] \times \bar{\Omega})$, $\phi_\epsilon(\varphi(x),x) \to 0$ weakly in $H^1(\Omega)$ as $\epsilon \to 0$.

Lemma 4. Suppose that there exist matrices $N_{\epsilon,s}(x)$, $x \in \Omega$, such that for $\epsilon \to 0$ and $\tilde{C}_\epsilon^{ij} = C_\epsilon^{il} \frac{\partial N_{\epsilon,j}}{\partial x_l} + C_\epsilon^{ij}$ we have

1) $N_{\epsilon,s} \in H^1(\Omega)$, $N_{\epsilon,s} \to 0$ weakly in $H^1(\Omega)$, $s = 1\ldots,n$, (33)

2) $\tilde{C}_\epsilon^{ij} \to \tilde{C}^{ij}$ weakly in $L^2(\Omega)$, $i,j = 1,\ldots,n$, (34)

3) $\frac{\partial}{\partial x_j}(\tilde{C}_\epsilon^{ij} - \tilde{C}^{ij}) \to 0$ in the norm of $H^{-1}(\Omega)$, $i = 1,\ldots,n$. (35)

Then the sequence $\mathfrak{L}_\epsilon$ is strongly G-convergent to an operator $\hat{\mathfrak{L}}$ with the matrix of coefficients $\tilde{C}^{ij}(x)$ satisfying conditions similar to (17).

Conditions (33) - (35) are known as the N-condition (see [9],[30]). The proof of Lemma 4 is given in [30].

We introduce the following notation

$$B_\epsilon^o(t,y) \equiv \left[\varphi_l(y)\varphi_k(y)C_\epsilon^{kl}(t,y)\right]^{-1},$$

$$B_\epsilon^s(t,y) \equiv \left[\varphi_l(y)\varphi_k(y)C_\epsilon^{kl}(t,y)\right]^{-1}\varphi_j(y)C_\epsilon^{ps}(t,y),$$

$$B_\epsilon^{is}(t,y) \equiv \varphi_j(y)C_\epsilon^{ij}(t,y)\left[\varphi_l(y)\varphi_k(y)C_\epsilon^{kl}(t,y)\right]^{-1}\varphi_p(y)C_\epsilon^{ps}(t,y) - C_\epsilon^{is}(t,y),$$

$$\hat{B}^o(t,y) \equiv \left[\varphi_l(y)\varphi_k(y)\hat{C}^{kl}(t,y)\right]^{-1},$$

$$\hat{B}^s(t,y) \equiv \left[\varphi_l(y)\varphi_k(y)\hat{C}^{kl}(t,y)\right]^{-1}\varphi_p(y)\hat{C}^{ps}(t,y),$$

$$B^{is}(t,y) \equiv \varphi_j(y)\hat{C}^{ij}(t,y)\left[\varphi_l(y)\varphi_k(y)\hat{C}^{kl}(t,y)\right]^{-1}\varphi_p(y)\hat{C}_\epsilon^{ps}(t,y) - \hat{C}_\epsilon^{is}(t,y).$$

The following theorem gives the necessary and sufficient conditions for the strong G-convergence of operators $\mathfrak{L}_\epsilon$ to $\hat{\mathfrak{L}}$ as $\epsilon \to 0$.

Theorem 3. Suppose that the elements of matrices $C_\epsilon^{ij}(t,y)$ have norms in $C^{0,\beta}$ uniformly bounded in ϵ. Then the sequence $\mathfrak{L}_\epsilon$ is strongly G-convergent to $\hat{\mathfrak{L}}$ if and only if the following conditions are satisfied

$$B_\epsilon^0(t,y) \to \hat{B}^0(t,y), \quad B_\epsilon^s(t,y) \to \hat{B}^s(t,y),$$

$$B_\epsilon^{is}(t,y) \to \hat{B}^{is}(t,y), \tag{36}$$

$$\frac{\partial}{\partial y_l} B_\epsilon^0(t,y) \to \frac{\partial}{\partial y_l} \hat{B}^0(t,y), \quad \frac{\partial}{\partial y_l} B_\epsilon^s(t,y) \to \frac{\partial}{\partial y_l} \hat{B}^s(t,y),$$

$$\frac{\partial}{\partial y_l} B_\epsilon^{is}(t,y) \to \frac{\partial}{\partial y_l} \hat{B}^{is}(t,y), \quad i,s,l = 1,\dots,n,$$

weakly in $L^2(0,1)$ as $\epsilon \to 0$ for any $y \in \bar{\Omega}$.

We outline here the proof of Theorem 3. Let us prove first that conditions (36) are sufficient for $\mathfrak{L}_\epsilon$ to be strongly G-convergent to $\hat{\mathfrak{L}}$ as $\epsilon \to 0$. Let us show that in this case $\delta_\epsilon \to 0$ as $\epsilon \to 0$. One can easily check that

$$N_s^\epsilon(t,y) = \int_0^t \Big[B_\epsilon^0(\tau,y)(\hat{B}^0(\tau,y))^{-1}\hat{B}(t,y) - B_\epsilon^s(\tau,y) \Big] d\tau, \tag{37}$$

$$M_{is}^\epsilon(t,y) = \int_0^t \Big[(B_\epsilon^i(\tau,y))^*(\hat{B}^0(\tau,y))^{-1}\hat{B}^s(\tau,y) - B_\epsilon^{is}(\tau,y)$$

$$- (\hat{B}^i(\tau,y))^*(\hat{B}^0(\tau,y))^{-1}\hat{B}^s(\tau,y) + \hat{B}^{is}(\tau,y) \Big] d\tau.$$

Therefore by virtue of Lemma 3 and conditions (36) we get that $\delta_\epsilon \to 0$ as $\epsilon \to 0$. From Theorem 2 (estimate (22)) it follows that for $f \in L^2(\Omega)$, $u^\epsilon \to u$ weakly in $H_0^1(\Omega)$ as $\epsilon \to 0$ since

$N_s^\epsilon(\varphi(x),x) \dfrac{\partial u}{\partial x_s} \to 0$ weakly in $H^1(\Omega)$ as $\epsilon \to 0$ according to Lemma 3.

The estimate (23) implies that $\gamma_\epsilon^i \to \hat{\gamma}^i$, $i = 1,\dots,n$, weakly in $L^2(\Omega)$

as $\epsilon \to 0$ since $\frac{\partial}{\partial t} M^{\epsilon}_{is}(\varphi(x),x) \to 0$ weakly in $L^2(\Omega)$ due to (36), (37). Approximating any $f \in H^{-1}(\Omega)$ by functions from $L^2(\Omega)$ one can easily obtain that $u^{\epsilon} \to u$ weakly in $H^1_0(\Omega)$, for $\gamma^i_{\epsilon} \to \hat{\gamma}^i$ weakly in $L^2(\Omega)$ as $\epsilon \to 0$ and for any $f \in H^{-1}(\Omega)$.

The proof of the necessity of conditions (36) for the strong G-convergence of $\mathfrak{L}_{\epsilon}$ to $\hat{\mathfrak{L}}$ as $\epsilon \to 0$ is based on Lemmas 3,4 and the theorem about the uniqueness of the strong G-limit (see [9],[30]). It is easy to prove that if conditions (36) are valid for some matrices $\tilde{B}^o(t,y)$, $\tilde{B}^s(t,y)$, $\tilde{B}^{ij}(t,y)$, then for these matrices the condition N is satisfied.

Theorem 4. Let the elements of matrices $C^{ij}_{\epsilon}(t,y)$ have norms in $C^{o,\beta}$ uniformly bounded in ϵ. Suppose that there are matrices $\hat{B}^o(t,y)$, $\hat{B}^s(t,y)$, $\hat{B}^{is}(t,y)$, $s,i, = 1,\dots,n$, such that conditions are satisfied. Then $\mathfrak{L}_{\epsilon}$ is strongly G-convergent to $\hat{\mathfrak{L}}$ as $\epsilon \to 0$ and the coefficients of $\hat{\mathfrak{L}}$ are given by the formulae

$$\hat{C}^{is} = (\hat{B}^i)^*(\hat{B}^o)^{-1}\hat{B}^s - \hat{B}^{is}. \tag{38}$$

Let us consider some examples when conditions (36) are satisfied.

Theorem 5. Suppose that the elements of matrices C^{ij}_{ϵ} have the form $C^{ij}_{hl}(\epsilon\ x)$ and $C^{ij}_{hl}(\xi)$ are almost periodic functions of $\xi \in \mathbb{R}^1$ satisfying conditions (17). Then the sequence $\mathfrak{L}_{\epsilon}$ is strongly G-convergent to the operator $\hat{\mathfrak{L}}$ whose matrices of coefficients are given by the formulae

$$\hat{C}^{ij} = \langle C^{ij}\rangle - \langle C^{i1}(C^{11})^{-1}C^{1j}\rangle + \langle C^{i1}(C^{11})^{-1}\rangle \langle (C^{11})^{-1}\rangle^{-1}\langle (C^{11})^{-1}C^{1j}\rangle,$$

where $\langle C^{ij}\rangle$ are matrices with elements $\langle C^{ij}_{hl}\rangle \equiv \lim_{T\to\infty} \int_{-T}^{T} C^{ij}_{hl}(\xi)d\xi.$

Moreover estimates (22), (23), (24) are valid with $\delta_{\epsilon} \to 0$ as $\epsilon \to 0$. If $C^{ij}_{hl}(\xi)$ are 1-periodic functions in $\xi, \xi \in \mathbb{R}^1$, then $\delta_{\epsilon} \leq C\epsilon$, where the constant C does not depend on ϵ.

We introduce a class A_σ consisting of functions $f(t,y)$ such that for some $C_f(y)$, $g(t,y)$ one has

$$\int_0^t f(s,y)ds - C_f(y)t = g(t,y).$$

The functions $f, \frac{\partial f}{\partial y_l}, C_f, g, \frac{\partial g}{\partial y_l}$ are supposed to be Hölder continuous in $y \in \bar{\Omega}$ uniformly in t; $l = 1, \ldots, n$,

$$|g(t,y)| + |\frac{\partial}{\partial y_l} g(t,y)| \leq C_0(1 + |t|)^{1-\sigma},$$

where the constants C_0, σ do not depend on t and $\sigma \in (0,1]$. Set

$$<f(.,y)> = \lim_{T\to\infty} \frac{1}{2T}\int_{-T}^{T} f(s,y)ds.$$

Obviously, for $f \in A_\sigma$ we have $<f(.,y)> = C_f(y)$.

<u>THEOREM 6</u>. In (16) let

$$C_\epsilon^{ij}(t,y) \equiv \| C_{\epsilon,lh}^{ij}(t,y)\| = \| C_{lh}^{ij}(\tfrac{t}{\epsilon},y)\|, \quad t = \varphi(x), \ y = (y_1,\ldots,y_n).$$

Suppose that the elemeents of matrices

$$B_\epsilon^o(t,y) \equiv B^o(\tfrac{t}{\epsilon},y) \equiv B^o(\tau,y), \quad B_\epsilon^s(t,y) \equiv B^o(\tfrac{t}{\epsilon},y) \equiv B^o(\tau,y),$$
$$B_\epsilon^{is}(t,y) \equiv B^{is}(\tfrac{t}{\epsilon},y) \equiv B^{is}(\tau,y), \ i,s = 1,\ldots,n,$$

defined above are functions of class A_σ, $\sigma\in(0,1]$, in τ and y and $<B^o(.,y)>$, $<B^s(.,y)>$, $<B^{is}(.,y)>$ are matrices whose elements are smooth functions. Then the sequence of operators $\mathfrak{L}_\epsilon$ corrresponding to the matrices $C_\epsilon^{ij} = \| C_{lh}^{ij}\left[\frac{\varphi(x)}{\epsilon},x\right]\|$ is strongly G-convergent to the operator $\hat{\mathfrak{L}}$ with the coefficients

$$\hat{C}^{is}(x) = <B^i(.,x)>^*<B^o(.,x)>^{-1}<B^s(.,x)> - <B^{is}(.,x)>.$$

Moreover, δ_ϵ in Theorem 1 satisfies the inequality $\delta_\epsilon \leq C\epsilon^\sigma$, where the constant C does not depend on ϵ.

Theorems 5 and 6 are obtained as a consequence of Theorems 2 and 3.

Now we consider eigenvalue problems related to the operators $\mathfrak{L}_\epsilon$ and $\hat{\mathfrak{L}}$:

$$\mathfrak{L}_\epsilon(u^{\epsilon,k}) + \lambda_k^\epsilon \rho_\epsilon(x) u^{\epsilon,k} = 0 \quad \text{in } \Omega,$$
$$u^{\epsilon,k} \in H_0^1(\Omega), \quad (u^{\epsilon,k}, u^{\epsilon,m}\rho_\epsilon)_0 = \delta^{km}, \tag{39}$$
$$0 < \lambda_1^\epsilon \leq \lambda_2^\epsilon \leq \dots \leq \lambda_k^\epsilon \leq \dots,$$

$$\hat{\mathfrak{L}}(u^k) + \lambda_k \hat{\rho}(x) u^k = 0 \quad \text{in } \Omega,$$
$$u^k \in H_0^1(\Omega), \quad (u^k, u^m\hat{\rho})_0 = \delta^{km}, \tag{40}$$
$$0 < \lambda_1 \leq \lambda_2 \leq \dots \leq \lambda_k \leq \dots,$$

where δ^{km} is the Kronecker symbol, each eigenvalue is counted as many times as its multiplicity, and ρ_ϵ, $\hat{\rho}$ are bounded measurable functions such that $\rho_\epsilon \geq \text{const} > 0$, $\hat{\rho} \geq \text{const} > 0$.

We also consider an auxiliary eigenvalue problem

$$\mathfrak{L}_\epsilon(v^{\epsilon,k}) + \tilde{\lambda}_k^\epsilon \hat{\rho}(x) v^{\epsilon,k} = 0 \quad \text{in } \Omega,$$
$$v^{\epsilon,k} \in H_0^1(\Omega), \quad (v^{\epsilon,k}, v^{\epsilon,k}\hat{\rho})_0 = \delta^{km}, \tag{41}$$
$$0 < \tilde{\lambda}_1^\epsilon \leq \tilde{\lambda}_2^\epsilon \leq \dots \leq \tilde{\lambda}_k^\epsilon \leq \dots .$$

Let $\rho(x)$ be a bounded measurable function which can be represented in the form

$$\rho = f_0 + \frac{\partial f_j}{\partial x_j}, \quad f_0,\ f_j \in L^p(\Omega), \quad p \geq n. \tag{42}$$

Set

$$\|\rho\|_{-1,p} = \inf_{f_0, f_1} \left\{ \|f_0\|_{L^p(\Omega)} + \sum_{i=1} \|f_i\|_{L^p(\Omega)} \right\},$$

where the infimum is taken over all representations of ρ in the form (42).

<u>Lemma 5</u>. Let λ_k^ϵ and $\tilde{\lambda}_k^\epsilon$ be eigenvalues of problems (39) and (41) respectively. Then

$$|(\lambda_k^\epsilon)^{-1} - (\tilde{\lambda}_k^\epsilon)^{-1}| \leq C\|\rho_\epsilon - \hat{\rho}\|_{-1,p}, \quad p \geq n,$$

where C is a constant which does not depend on ϵ and k.

The proof of this lemma is based on the variational theory

of eigenvalues, Korn's inequality and the imbedding theorem:

$$\|u\|_{L^p(\Omega)} \leq C_0\|u\|_1, \quad \text{if} \quad 1 - \frac{n}{2} + \frac{n}{p} \geq 0.$$

<u>Theorem 7</u>. Suppose that $\hat{\rho} \in C^1(\bar{\Omega})$ and matrices C_ϵ^{ij} in (16) satisfy the conditions of Theorem 2. Then

$$|(\lambda_k^\epsilon)^{-1} - (\lambda_k)^{-1}| \leq C_1(\|\rho_\epsilon - \hat{\rho}\|_{-1,p} + \delta_\epsilon^{1/2}), \quad p \geq n, \tag{43}$$

where λ_k^ϵ, λ_k are eigenvalues of problems (39), (40) respectively and C_1 is a constant which does not depend on ϵ.

The inequality (43) is proved on the basis of H. Weyl's theorem (see [40]).

<u>Theorem 8</u>. Let all the conditions of Theorem 7 be satisfied. Then

$$r(u^{\epsilon,k}, N(\lambda_k, \hat{\ell})) \leq C_k(\delta_\epsilon^{1/2} + \|\rho_\epsilon - \hat{\rho}\|_{-1,p}),$$

where the constant C_k does not depend on ϵ; $u^{\epsilon,k}$ is the k-th eigenfunction of (39) corresponding to λ_k^ϵ, $N(\lambda_k, \hat{\ell})$ is the linear space formed by all eigenfunctions of (40) corresponding to the eigenvalue λ_k and $r(u^{\epsilon,k}, N(\lambda_k, \hat{\ell}))$ is the distance between $u^{\epsilon,k}$ and $N(\lambda_k, \hat{\ell})$ in the space $L^2(\Omega)$ with scalar product $(u,v) = (u, v\hat{\rho})_0$.

Note that in the case $\rho_\epsilon(x) = \rho\left[\frac{\varphi(x)}{\epsilon}, x\right]$, where $\rho(t,y) \in A_\sigma$, we have

$$\left\|\rho\left[\frac{\varphi(x)}{\epsilon}, x\right] - \hat{\rho}(x)\right\|_{-1,\infty} \leq C\,\epsilon^\sigma,$$

when $\hat{\rho}(x) = \langle\rho(.,x)\rangle$. The constant C does not depend on ϵ.

Proofs of Theorems 7 and 8 are given in [37].

3 <u>HOMOGENIZATION OF EIGENVALUES AND EIGENFUNCTIONS OF BOUNDARY VALUE PROBLEMS IN PERFORATED DOMAINS FOR ELLIPTIC EQUATIONS AND THE ELASTICITY SYSTEM WITH NON-UNIFORMLY OSCILLATING COEFFICIENTS</u>

Boundary value problems for partial differential equations describing processes in strongly non-homogeneous media are a subject of intensive research (see [1] - [38]). In many applications it is important to estimate various characteristics of these processes in elastic composites such as displacements, stres tensors, energy, eigenvalues, etc. First we consider a mixed boundary value problem for second order elliptic equations with non-uniformly oscillating coefficients in a perforated domain with a periodic structure of period ϵ and then a similar problem for the system of linear elasticity. We obtain estimates for the difference between eigenvalues and eigenfunctions of this problem and the eigenvalues and the eigenfunctions of the related homogenized problem. In the case of ellpitic second order equations these differences are of order ϵ (see [17]) and in the case of the elasticity system these differences are estimated by $C\sqrt{\epsilon}$, $C = \text{const}$, [25].

Let G^0 be a set which consists of a finite number of non-intersecting smooth domains belonging to a cube $Q = \{x \in \mathbb{R}^n, 0 < x_j < 1, j = 1,\dots,n\}$. Denote by $X + z$ a shift of the set $X \in \mathbb{R}^n$ by a vector z and denote by ϵX the set $\{x \in \mathbb{R}^n, \epsilon^{-1}x \in X\}$. Let $G_1 = \bigcup_{z\in\mathbb{Z}^n} (G^0 + z)$, where $\mathbb{Z}$ is the set of all vectors $z = (z_1,\dots,z_n)$ with integer components,

$$G_\epsilon = \epsilon G_1, \quad \Omega^\epsilon = \Omega\setminus\bar{G}_\epsilon, \quad S_\epsilon = \partial\Omega^\epsilon\setminus\partial\Omega, \quad \Gamma_\epsilon = \partial\Omega \cap \partial\Omega^\epsilon,$$

where Ω is a smooth domain in $\mathbb{R}^n$. We assume that Ω^ϵ and $\mathbb{R}^n\setminus\bar{G}_1$ are connected open sets in $\mathbb{R}^n$. The domain Ω^ϵ is called a perforated domain.

In Ω_ϵ we consider second order elliptic differential operators

$$L_\epsilon \equiv \frac{\partial}{\partial x_i}\left[a_{ij}(x,\frac{x}{\epsilon})\frac{\partial}{\partial x_j}\right],$$

where $a_{ij}(x,\xi)$ are functions of class C^∞ in (x,ξ), periodic in $\xi_1,\dots\xi_n$ with period 1 (1-periodic in ξ),

$$a_{ij}(x,\xi) = a_{ji}(\xi,x),$$

$$\varkappa_1|\eta|^2 \leq a_{ij}(x,\xi)\eta_i\eta_j \leq \varkappa_2|\eta|^2, \quad \varkappa_1,\varkappa_2 = \text{const} > 0, \tag{44}$$

for any $\eta \in \mathbb{R}^n$, $\xi \in \mathbb{R}^n$, $x \in \Omega$. In Ω^ϵ we consider the following eigenvalue problem

$$L_\epsilon(u^{\epsilon,k}) + \lambda_k^\epsilon \rho(x,\frac{x}{\epsilon})u^{\epsilon,k} = 0 \quad \text{in } \Omega^\epsilon,$$
$$\sigma_\epsilon(u^{\epsilon,k}) = 0 \text{ on } S_\epsilon, \quad u^{\epsilon,k} = 0 \text{ on } \Gamma_\epsilon, \quad k = 1,\ldots, \tag{45}$$

where $\sigma_\epsilon(u^{\epsilon,k}) \equiv a_{ij}(x,\frac{x}{\epsilon}) \frac{\partial u^{\epsilon,k}}{\partial x_j} \nu_i$, $\nu = (\nu_1,\ldots,\nu_n)$ is a unit outward normal to $\partial\Omega^\epsilon$. We assume that $\rho(x,\xi)$ is a function of class C^∞ in $(x,\xi) \in \bar{\Omega} \times \mathbb{R}^n$, $\rho(x,\xi) \geq \rho_0 = \text{const} > 0$,

$$0 < \lambda_1^\epsilon \leq \ldots \leq \lambda_k^\epsilon \leq \ldots$$

and each eigenvalue is counted as many times as its multiplicity. We consider such functions $u^{\epsilon,k}$ which belong to $H^1(\Omega^\epsilon)$, $u^{\epsilon,k} = 0$ on Γ_ϵ and satisfy the integral identity

$$\int_{\Omega^\epsilon} (-a_{ij}(x,\frac{x}{\epsilon}) \frac{\partial u^{\epsilon,k}}{\partial x_j} \frac{\partial v}{\partial x_i} + \lambda_k^\epsilon \rho(x,\frac{x}{\epsilon}) u^{\epsilon,k} v)dx = 0$$

for any function $v \in H^1(\Omega^\epsilon)$ and $v = 0$ on Γ_ϵ, and

$$\int_{\Omega^\epsilon} u^{\epsilon,k} u^{\epsilon,m} \rho(x,\frac{x}{\epsilon})\, dx = \delta^{km}.$$

Let us define functions $N_s(x,\xi)$, $s = 1,\ldots,n$, as 1-periodic in ξ solutions of the following boundary value problems

$$\frac{\partial}{\partial \xi_i}\left[a_{ij}(x,\xi) \frac{\partial N_s(x,\xi)}{\partial \xi_j}\right] = -\frac{\partial a_{is}(x,\xi)}{\partial \xi_i} \quad \text{in } \mathbb{R}^n \setminus G_1 \tag{46}$$

$$\sigma(N_s) = -a_{is}\nu_i \quad \text{on } \partial G_1,$$

$$\int_{Q\setminus G^0} N_s(x,\xi)d\xi = 0,$$

where $(\nu_1,\ldots,\nu_n)$ is the unit outward normal to the domain $\mathbb{R}^n \setminus \bar{G}_1$, $\sigma(N) \equiv a_{ij}(x,\xi) \frac{\partial N}{\partial \xi_j} \nu_i$; $x \in \Omega$ is considered as a parameter. Set

$$h^{pq}(x) = \left[\text{meas } (Q\setminus G^0)\right]^{-1} \int_{Q\setminus G^0} \left[a_{pq}(x,\xi) + a_{jq}(x,\xi) \frac{\partial N_p}{\partial \xi_j}\right] d\xi \tag{47}$$

and consider the differential operator

$$\hat{L} \equiv \frac{\partial}{\partial x_p}\left[h^{pq}(x)\frac{\partial}{\partial x_q}\right],$$

which is called a homogenized operator for L_ϵ.

It is proved in [1] that $\hat{L}$ satisfies conditions (44) with some positive constants $\hat{\chi}_1$, $\hat{\chi}_2$.

Consider the eigenvalue problem for the homogenized operator

$$\hat{L}(u^k) + \lambda_k\rho(x)u^k = 0 \quad \text{in } \Omega, \qquad (48)$$
$$u^k = 0 \quad \text{on } \partial\Omega,$$

where

$$\hat{\rho}(x) = \left[\text{meas } (Q\backslash G^o)\right]^{-1}\int_{Q\backslash G^o}\rho(x,\xi)d\xi$$

and λ_k are also enumerated according to their multiplicity:

$$0 < \lambda_1 \leqslant \lambda_2 \leqslant \ldots \leqslant \lambda_k \leqslant \ldots,$$

$$\int_\Omega u^k(x)u^m(x)\hat{\rho}(x)dx = \delta^{km}.$$

Theorem 9. Let λ_k^ϵ and λ_k be eigenvalues of problems (45), (48) respectively. Then

$$|\lambda_k^\epsilon - \lambda_k| \leqslant C(k)\epsilon, \quad k = 1,2,\ldots,$$

where the constant $C(k)$ does not depend on ϵ. Assume that for some $j > 0$

$$\lambda_{j-1} < \lambda_j = \ldots = \lambda_{j+r-1} < \lambda_{j+r}, \; r \geqslant 1, \; \lambda_0 = 0.$$

Then for sufficiently small ϵ

$$\|u^{j+t} - v_\epsilon^{j+t}\|_{L^2(\Omega^\epsilon)} \leqslant C_{j+t}\epsilon, \quad t = 0,1,\ldots, r-1,$$

where u^{j+t} is an eigenfunction of problem (48), v_ϵ^{j+t} is a function from the linear space formed by eigenfunctions $u^{\epsilon,j},\ldots,u^{\epsilon,j+r-1}$ of problem (45) and C_{j+t} is a constant independent of ϵ.

To prove Theorem 9 we use Lemma 1 and a generalization of

the maximum principle for solutions of the mixed boundary value problem for elliptic second order operators.

Let ω be a bounded domain in $\mathbb{R}^n$. Assume that the boundary $\partial\omega$ of ω has the form $\partial\omega = \Gamma_0 \cup \Gamma_1$, where Γ_0 is such that for any function $v \in \mathrm{Lip}(\omega)$ and $v = 0$ on Γ_0 the Friedrichs inequality

$$\|v\|_{L^2(\omega)} \leq C \sum_{j=1}^{n} \left\|\frac{\partial v}{\partial x_j}\right\|_{L^2(\omega)}$$

holds with constant C independent of v. By $\mathrm{Lip}(\omega)$ we denote a space of continuous functions in ω with the finite norm

$$\|v\| = \sup_{x \in \bar{\omega}} |v(x) + \sup_{x,y \in \bar{\omega}} \frac{|v(x) - v(y)|}{|x - y|} .$$

By $\mathcal{H}^1(\omega)$ we denote the completion of $\mathrm{Lip}(\omega)$ in the norm

$$H^1(\omega): \|v\|_1 = \left[\|v\|^2_{L^2(\omega)} + \sum_{j=1}^{n} \left\|\frac{\partial v}{\partial x_j}\right\|^2_{L^2(\omega)}\right]^{1/2} . \tag{49}$$

By $\mathcal{H}^1(\omega,\Gamma_0)$ we denote the completion in the norm (49) of a subspace in $\mathrm{Lip}(\omega)$ consisting of the functions equal to zero on Γ_0. Consider the boundary value problem

$$\begin{aligned} &\frac{\partial}{\partial x_i}\left[A_{ij}(x) \frac{\partial u}{\partial x_j}\right] = 0 \quad \text{in } \omega, \\ &A_{ij} \frac{\partial u}{\partial x_j} \nu_i = 0 \quad \text{on } \Gamma_1, \quad u = \varphi \quad \text{on } \Gamma_0, \quad \Gamma_0 \neq \emptyset. \end{aligned} \tag{50}$$

Here $A_{ij}(x)$ are bounded measurable functions, $A_{ij}(x) = A_{ji}(x)$ and the condition similar to (44) is satisfied. We set

$$a(v,w) \equiv \int_\omega A_{ij}(x) \frac{\partial v}{\partial x_j} \frac{\partial w}{\partial x_i} dx,$$

and let $\varphi \in \mathcal{H}^1(\omega)$.

The function $u(x)$ is called a weak solution of problem (50), if $u \in \mathcal{H}^1(\omega)$, $u - \varphi \in \mathcal{H}^1(\omega,\Gamma_0)$ and the integral identity $a(u,v) = 0$ is satisfied for any $v \in \mathcal{H}^1(\omega,\Gamma_0)$.

The existence and uniqueness of a weak solution of problem (50) can be proved using the F.Riesz theorem.

<u>Lemma 6</u>. (Maximum principle). Let $\varphi \in C^1(\bar{\omega})$. Then any weak solution $u(x)$ of problem (50) satisfies the inequality

$$\min_{\overline{\Gamma}_0} \varphi \leqslant u(x) \leqslant \max_{\overline{\Gamma}_0} \varphi \tag{51}$$

for almost all $x \in \omega$.

To prove estimates (51) we use the variational method for the solution of problem (50).

In order to prove Theorem 9 we use the following lemmas.

Lemma 7. Let $\psi \in L^2(S_\epsilon)$, $v(x)$ be a weak solution of the problem

$$L_\epsilon(v) = 0 \ \text{ in } \Omega^\epsilon, \quad v = 0 \text{ on } \Gamma_\epsilon, \ \sigma_\epsilon(v) = \psi \ \text{ on } S_\epsilon.$$

Then

$$\|v\|_{H^1(\Omega^\epsilon)} \leqslant C_1 \epsilon^{-1/2} \|\psi\|_{L^2(S_\epsilon)}; \quad C_1 - \text{const.}$$

Lemma 8. Let $f(x,\xi)$ be a measurable, 1-periodic in ξ function defined in $\overline{\Omega} \times \mathbb{R}^n$ and such that

$$|f(x,\xi) - f(x^0,\xi)| \leqslant C|x - x^0| \quad \text{for all } x,x^0 \in \overline{\Omega},\ \xi \in \mathbb{R}^n,$$

where the constant C does not depend on x, x^0, ξ. Then for any $\varphi \in L^1(\Omega)$, we have

$$\int_\Omega \varphi(x) f(x,\frac{x}{\epsilon})dx \longrightarrow \int_\Omega \varphi(x)\hat{f}(x)dx \quad \text{as } \epsilon \to 0,$$

where

$$\hat{f}(x) = \int_Q f(x,\xi)d\xi.$$

We construct an approximate solution of the problem (45) in order to use Lemma 1 and to estimate the eigenvalues of the problem (45). Let us define functions $N_{pq}(x,\xi)$, $p,q = 1,\ldots,n$, as 1-periodic in ξ solutions of the following boundary value problems

$$\frac{\partial}{\partial \xi_i}(a_{ij}(x,\xi)\frac{\partial Npq(x,\xi)}{\partial \xi_j}) = h^{pq}(x) - apq(x,\xi)$$

$$- \frac{\partial aiq(x,\xi)}{\partial \xi_i} N_p(x,\xi) - 2a_{iq}(x,\xi)\frac{\partial Np(x,\xi)}{\partial \xi_i} \quad \text{in } \mathbb{R}^n \backslash G_1, \tag{52}$$

$$\sigma(N_{pq}) = - a_{iq}(x,\xi)Np(x,\xi)\nu_i \quad \text{on } \partial G_1,$$

$$\int_{Q\setminus G^o} N_{pq}(x,\xi)d\xi = 0.$$

Set

$$h_s^1(x) = \left[\text{meas } (Q\setminus G^o)\right]^{-1} \int_{Q\setminus G^o} \left[\frac{\partial}{\partial x_i}\left(a_{ij}(x,\xi)\frac{\partial N_s}{\partial \xi_j}\right) + \frac{\partial a_{is}(x,\xi)}{\partial x_i}\right]d\xi$$

and define N_s^1, N as 1-periodic in ξ solutions of the problems

$$\frac{\partial}{\partial \xi_i}\left[a_{ij}(x,\xi)\,\frac{\partial N(x,\xi)}{\partial \xi_j}\right] = \hat{\rho}(x) - \rho(x,\xi) \quad \text{in } \mathbb{R}^n\setminus G_1,$$

$$\sigma(N) = 0 \quad \text{on } \partial G_1, \quad \int_{Q\setminus G^o} N(x,\xi)d\xi = 0, \tag{53}$$

$$\frac{\partial}{\partial \xi_i}\left[a_{ij}(x,\xi)\,\frac{\partial N_s^1(x,\xi)}{\partial \xi_j}\right] = h_s^1(x) - \frac{\partial}{\partial x_i}\left[a_{ij}(x,\xi)\,\frac{\partial N_s(x,\xi)}{\partial \xi_j}\right]$$

$$- a_{ij}(x,\xi)\,\frac{\partial^2 N_s(x,\xi)}{\partial \xi_i \partial x_j} - \frac{\partial a_{is}(x,\xi}{\partial x_i} - \frac{\partial a_{ij}(x,\xi)}{\partial \xi_i}\,\frac{\partial N_s(x,\xi)}{\partial x_j}, \tag{54}$$

$$\text{in } \mathbb{R}^n\setminus G_1$$

$$\sigma(N_s^1) = a_{ij}(x,\xi)\,\frac{\partial N_s(x,\xi)}{\partial x_j}\,\nu_i \quad \text{on } \partial G_1, \quad \int_{Q\setminus G^o} N_s^1(x,\xi)d\xi = 0.$$

Existence of solutions of problems (46), (52), (53), (54) can be proved by the methods of functional analysis.

In these problems $x_1,\ldots,x_n$ are considered as parameters, and due to the smoothness of the boundary and of the coefficients of the equations the functions N_s, N_{pq}, N, N_s^1 are smooth with respect to (x,ξ) in $\bar{\Omega} \times (\mathbb{R}^n\setminus G_1)$.

Let us define $\tilde{u}_\epsilon(x)$ by the formula

$$\tilde{u}_\epsilon(x) \equiv u^k(x) + \epsilon N_p(x,\frac{x}{\epsilon}) \frac{\partial u^k}{\partial x_p} + \epsilon^2\Big[N_{pq}(x,\frac{x}{\epsilon}) \frac{\partial^2 u^k}{\partial x_p \partial x_q} + N_p^1(x,\frac{x}{\epsilon}) \frac{\partial u^k}{\partial x_p} + \lambda_k N(x,\xi)u^k\Big],$$

where $u^k(x)$ is an eigenfunction of the problem (48) corresponding to the eigenvalue λ_k. We define $v_\epsilon(x)$ as a solution of the boundary value problem

$$L_\epsilon(v_\epsilon) = 0 \quad \text{in } \Omega^\epsilon, \tag{55}$$

$$v_\epsilon = \tilde{u}_\epsilon \quad \text{on } \Gamma_\epsilon, \quad \sigma_\epsilon(v_\epsilon) = \sigma_\epsilon(\tilde{u}_\epsilon) \quad \text{on } S_\epsilon.$$

Denote by $\mathcal{H}_\epsilon$ the space $L^2(\Omega^\epsilon)$ with scalar product

$$(u,v)_{\mathcal{H}_\epsilon} = \int_{\Omega^\epsilon} u(x)v(x)\rho(x,\epsilon^{-1}x)dx.$$

We introduce an operator $A_\epsilon:\mathcal{H}_\epsilon \to \mathcal{H}_\epsilon$ which maps a function $f \in \mathcal{H}_\epsilon$ into a solution $u_\epsilon(x)$ of the problem

$$L_\epsilon(u_\epsilon) = f \quad \text{in } \Omega^\epsilon,$$

$$\sigma_\epsilon(u_\epsilon) = 0 \quad \text{on } S_\epsilon, \quad u_\epsilon = 0 \quad \text{on } \Gamma_\epsilon.$$

Define an operator $\tilde{A}_\epsilon:\mathcal{H}_\epsilon \to \mathcal{H}_\epsilon$ by the formula

$$\tilde{A}_\epsilon w \equiv A_\epsilon \rho(x,\frac{x}{\epsilon})w.$$

The operator $-\bar{A}_\epsilon$ is positive, self-adjoint and compact in $\mathcal{H}_\epsilon$.

The proof of Theorem 9 is based on the following lemma.

<u>Lemma 9</u>. Let $w_\epsilon = \tilde{u}_\epsilon - v_\epsilon$. Then for sufficiently small ϵ we have

$$\|w_\epsilon + \lambda^k \tilde{A}_\epsilon w_\epsilon\|_{\mathcal{H}_\epsilon} \leq C\epsilon, \tag{56}$$

$$w_\epsilon(x) = u^k(x) + \epsilon W_\epsilon, \quad \|W_\epsilon\|_{\mathcal{H}_\epsilon} \leq K, \tag{57}$$

$$K_2 \geq \|w_\epsilon\|_{\mathcal{H}_\epsilon} \geq K_1 > 0, \tag{58}$$

where constants C, K, K_1, K_2 do not depend on ϵ.

The proof of Theorem 9 is given in [17].

A theorem similar to Theorem 9 is also proved for the elasticity system with rapidly oscillating coefficients in a perforated domain.

Consider the system of elasticity given by

$$\mathfrak{L}_\epsilon(u) \equiv \frac{\partial}{\partial x_h}\left[C^{hk}(x,\epsilon^{-1}x)\,\frac{\partial u}{\partial x_k}\right] = f,$$

where $C^{h,k}(x,\xi)$ are $(n \times n)$ matrices with elements $C^{hk}_{ij}(x,\xi)$ which are C^∞ - functions in $\bar{\Omega} \times \mathbb{R}^n$, 1-periodic in $\xi = (\xi_1,\dots,\xi_n)$, u is a column-vector with components $u_1,\dots,u_n$. Assume that

$$C^{hk}_{ij}(x,\xi) = C^{kh}_{ji}(x,\xi) = C^{ik}_{hj}(x,\xi) \tag{59}$$

for any $(x,\xi) \in \bar{\Omega} \times \mathbb{R}^n$, and for any symmetric matrix $\eta = \|\eta^h_i\|$,

$$\varkappa_1\eta^h_i\eta^h_i \leq C^{hk}_{ij}(x,\xi)\eta^h_i\eta^k_j \leq \varkappa_2\eta^h_i\eta^h_i, \qquad \varkappa_1,\varkappa_2 = \text{const} > 0. \tag{60}$$

Denote by $H^1(\omega)$ the space of vector valued functions $u = (u_1,\dots,u_n)$ with norm

$$\|u\|_{H^1(\omega)} = \left[\int_\omega\left(u_iu_i + \frac{\partial u_i}{\partial x_j}\frac{\partial u_i}{\partial x_j}\right)dx\right]^{1/2}.$$

In the domain Ω^ϵ we consider the following eigenvalue problem

$$\mathfrak{L}_\epsilon(u^{\epsilon,k}) + \lambda^\epsilon_k\rho(x,\epsilon^{-1}x)u^{\epsilon,k} = 0 \quad \text{in } \Omega^\epsilon, \tag{61}$$

$$\sigma_\epsilon(u^{\epsilon,k}) = 0 \text{ on } S_\epsilon, \quad u^{\epsilon,k} = 0 \text{ on } \Gamma_\epsilon, \ k = 1,2,\dots,$$

$$0 < \lambda^\epsilon_1 \leq \lambda^\epsilon_2 \leq \dots \leq \lambda^\epsilon_k \leq \dots, \tag{62}$$

where $\sigma_\epsilon(u) \equiv C^{hk}\dfrac{\partial u}{\partial x_k}\nu_h$, $(\nu_1,\dots,\nu_n)$ is a unit outward normal to $\partial\Omega^\epsilon$.

We assume that $\rho(x,\xi)$ is a function of class $C^\infty(\bar{\Omega} \times \mathbb{R}^n)$, $\rho(x,\xi) \geq \rho_0 = \text{const} > 0$ and each eigenvalue in (62) is counted as many times as its multiplicity. We consider eigenfunctions of problem (61) such that $u^{\epsilon,k} \in H^1(\Omega^\epsilon)$, $u^{\epsilon,k} = 0$ on Γ_ϵ and $u^{\epsilon,k}$ satisfies the integral identity

$$-\int_{\Omega^\epsilon}\left[C^{kh}(x,\epsilon^{-1}x)\frac{\partial u^{\epsilon,k}}{\partial x_k},\frac{\partial v}{\partial x_h}\right]dx+\lambda_k^\epsilon\int_{\Omega^\epsilon}\rho(x,\epsilon^{-1}x)(u^{\epsilon,k},v)dx=0$$

for any vector valued function $v \in H^1(\Omega^\epsilon)$, $v = 0$ on Γ_ϵ, $(v,w) \equiv v_i w_i$. We assume that

$$\int_{\Omega^\epsilon}(u^{\epsilon,k},u^{\epsilon,m})\rho(x,\epsilon^{-1}x)dx=\delta^{km}.$$

We introduce matrices $N_s(x,\xi)$, $s = 1,\ldots,n$, which are solutions of the boundary value problems

$$\frac{\partial}{\partial\xi_h}\left[C^{hk}(x,\xi)\frac{\partial N_s(x,\xi)}{\partial\xi_k}\right]=-\frac{\partial}{\partial\xi_k}C^{ks}(x,\xi)\quad\text{in } \mathbb{R}^n\backslash G_1,$$

$$\sigma(N_s)\equiv C^{hk}(x,\xi)\frac{\partial N_s}{\partial\xi_k}\nu_h=-C^{ks}(x,\xi)\nu_k\quad\text{on } \partial G_1,$$

$$\int_{Q\backslash G^0}N_s(x,\xi)d\xi=0, \tag{63}$$

N_s is 1-periodic in $\xi = (\xi_1,\ldots,\xi_n)$.

Variables $x_1,\ldots,x_n$ are considered in problems (63) as parameters.

We set

$$h^{pq}=\left[\text{meas }(Q\backslash G^0)\right]^{-1}\int_{Q\backslash G^0}\left[C^{pq}(x,\xi)+C^{pj}(x,\xi)\frac{\partial N_p}{\partial\xi_j}\right]d\xi.$$

The operator

$$\hat{\mathfrak{L}}(u)\equiv\frac{\partial}{\partial x_p}\left[h^{pq}(x)\frac{\partial u}{\partial x_q}\right],\quad u=(u_1,\ldots,u_n)^*,$$

is called a homogenized operator for $\mathfrak{L}_\epsilon$. The matrices $h^{pq}(x) = \|h^{pq}_{hl}(x)\|$ satisfy conditions similar to (59), (60). (See [2], [12]).

Consider the eigenvalue problem for the homogenized operator

$$\hat{\mathfrak{L}}(u^k) + \lambda_k \hat{\rho}(x) u^k = 0 \quad \text{in} \quad \Omega, \quad u^k = 0 \quad \text{on} \quad \partial\Omega, \tag{64}$$

$$\hat{\rho}(x) = \left[\text{meas } (Q\backslash G^o)\right] \int_{Q\backslash G^o} \rho(x,\xi)d\xi,$$

$$0 < \lambda_1 \leq \lambda_2 \leq \ldots \leq \lambda_k \text{ H } \ldots \quad . \tag{65}$$

Each eigenvalue is counted as many times as its multiplicity. We assume that

$$\int_\Omega (u^k, u^m)\hat{\rho}(x)dx = \delta^{km}.$$

Theorem 10. Let λ_k^ϵ and λ_k be eigenvalues of problems (61) and (64) respectively. Then

$$|\lambda_k^\epsilon - \lambda_k| \leq C(k)\epsilon^{1/2}, \quad k = 1,2,\ldots, \tag{66}$$

where the constant $C(k)$ does not depend on ϵ. Suppose that for some $j > 0$

$$\lambda_{j-1} < \lambda_j = \ldots = \lambda_{j+r-1} < \lambda_{j+r}, \quad r \geq 1, \ \lambda_o = 0.$$

Then for sufficiently small ϵ

$$\| u^{j+t} - v_\epsilon^{j+t} \|_{L^2(\Omega^\epsilon)} \leq C_{j+t}\, \epsilon^{1/2}, \ t = 0,1,\ldots,r-1, \tag{67}$$

where u^{j+t} is the eigenfunction of the problem (64), v_ϵ^{j+t} is a linear combination of the eigenfunctions $u^{\epsilon,j},\ldots,u^{\epsilon,j+r-1}$ of problem (61), and C_{j+t} is a constant which does not depend on ϵ. Moreover,

$$\| u^{j+t} - v_\epsilon^{j+t} + \epsilon N_s(x,\xi^{-1}x)\frac{\partial u^{j+t}}{\partial x_s} \|_{H^1(\Omega^\epsilon)} \leq C'_{j+t}\, \epsilon^{1/2},$$

where the constant C'_{j+t} does not depend on ϵ and N_s are solutions of problems (63).

The proof of Theorem 10 follows the same scheme as the proof of Theorem 10. However Lemma 6 used in the proof of Theorem 9 is not known for the system of elasticity. Instead of Lemma 6 we use here the following lemma which gives integral estimates for solutions of the elasticity system. That is why we have $\epsilon^{1/2}$ in (66), (67) instead of ϵ.

Lemma 10. Let $f \in L^2(\Omega^\epsilon)$, $\varphi \in L^2(S_\epsilon)$, $\phi \in H^{1/2}(\partial\Omega)$ and suppose that w_ϵ belongs to $H^1(\Omega^\epsilon)$ and is a weak solution of the problem

$$\mathfrak{L}_\epsilon(w_\epsilon) = f \quad \text{in} \quad \Omega^\epsilon,$$

$$w_\epsilon = \phi \quad \text{on} \quad \Gamma_\epsilon, \quad \sigma_\epsilon(w_\epsilon) = \varphi \quad \text{on} \quad S_\epsilon.$$

Then

$$\|w_\epsilon\|_{H^1(\Omega^\epsilon)} \leq C\Big[\|f\|_{L^2(\Omega^\epsilon)} + \|\phi\|_{H^{1/2}(\partial\Omega)} + \epsilon^{-1/2}\|\varphi\|_{L^2(S_\epsilon)}\Big],$$

where C = const does not depend on ϵ.

The proof of Theorem 10 is given in [25].

4 ON VIBRATION OF SYSTEMS WITH CONCENTRATED MASSES.

We consider an eigenvalue problem in a bounded domain Ω for the Laplace operator with Dirichlet boundary conditions. The density is assumed to be constant everywhere except in the ϵ-neighbourhood of the origin in $0 \in \Omega$. The problem is given by

$$\Delta u^{\epsilon,k} + \lambda_k^\epsilon \rho_\epsilon(x) u^{\epsilon,k} = 0 \quad \text{in} \quad \Omega, \tag{1}$$

$$u^{\epsilon,k} = 0 \quad \text{on} \quad \partial\Omega, \tag{2}$$

where

$$\rho_\epsilon = 1 \ \text{for}\ |x| > \epsilon, \quad \rho_\epsilon(x) = 1 + \epsilon^{-m}\hbar_\epsilon(x) \ \text{for}\ |x| \leq \epsilon,\ \hbar_\epsilon(x) = \hbar(\tfrac{x}{\epsilon}),$$

$$\hbar(\xi) = 0 \ \text{for}\ |\xi| > 1,\ \hbar(\xi) \geq 0 \ \text{for}\ |\xi| \leq 1,$$

$$\int_{|\xi|<1} \hbar(\xi)d\xi = M, \quad M = \text{const} > 0,$$

and $\hbar(\xi)$ is a bounded measurable function in $\mathbb{R}^n$, $x = (x_1,\dots,x_n)$, M is a positive constant and $\epsilon > 0$ is a small parameter.

In this section we number the formulae, theorems and lemmas starting again from 1.

As is known, the problem (1), (2), for any $\epsilon > 0$ has a positive discrete spectrum. We consider weak eigenfunctions.

The function $u^{\epsilon,k}$ is called a weak eigenfunction of problem (1), (2) if $u^{\epsilon,k} \in H_0^1(\Omega)$ and for some λ_k^ϵ it satisfies the integral identity

$$\int_\Omega \left(-\sum_{j=1}^{n} \frac{\partial u^{\epsilon,k}}{\partial x_j} \frac{\partial v}{\partial x_j} + \lambda_k^\epsilon \rho_\epsilon u^{\epsilon,k} v \right) dx = 0$$

for any function v from $H_0^1(\Omega)$.

As before, the space $H_0^1(\Omega)$ is the completion of $C_0^\infty(\Omega)$ (the set of infinitely differentiable functions with compact support in Ω) in the norm

$$\| u \|_1 = \left[\int_\Omega \sum_{j=1}^{n} \left(\frac{\partial u}{\partial x_j} \right)^2 dx \right]^{½}.$$

We assume that $\| u^{\epsilon,k} \|_1 = 1$ and that

$$0 < \lambda_1^\epsilon \leqslant \lambda_2^\epsilon \leqslant \dots \leqslant \lambda_k^\epsilon \leqslant \dots,$$

where each eigenvalue is counted as many times as its multiplicity.

We study the behaviour of λ_k^ϵ as $\epsilon \to 0$. The case $m = 3$, $n = 3$ has been considered in [41]. A similar problem for the system of elasticity was studied in [42].

Let $n \geqslant 3$. Consider the eigenvalue problem

$$\Delta u^k + \lambda_k u^k = 0 \quad \text{in} \quad \Omega, \tag{3}$$

$$u^k = 0 \quad \text{on} \quad \Omega,$$

$$0 < \lambda_1 \leqslant \lambda_2 \leqslant \dots \leqslant \lambda_k \leqslant \dots,$$

where each eigenvalue is also counted as many times as its multiplicity. Set

$$(u,v)_\omega = \int_\omega uv dx, \qquad [u,v]_\omega = \int_\omega \frac{\partial u}{\partial x_j} \frac{\partial v}{\partial x_j} dx.$$

Theorem 1. Let $n \geqslant 3$, $m < 2$. Then for eigenvalues λ_k^ϵ of problem (1), (2) and eigenvalues λ_k of problem (3) the following estimate is valid

$$|(\lambda_k^\epsilon)^{-1} - (\lambda_k)^{-1}| \leqslant C\epsilon^{2-m}, \quad k = 1,2,\dots,$$

where the constant C does not depend on ϵ and k.

Proof. According to the variational principle (the minimax principle), we have

$$|(\lambda_k^\epsilon)^{-1} - (\lambda_k)^{-1}| \leq \sup_{u \in H_0^1(\Omega)} \frac{((\rho_\epsilon - 1)u,u)_\Omega}{[u,u]_\Omega}. \tag{4}$$

It follows from the imbedding theorem that

$$\|u\|_{L^p(\Omega)} \leq C_1 \|u\|_1, \quad C_1 = \text{const}, \quad u \in H_0^1(\Omega), \tag{5}$$

if $1 - \frac{n}{2} + \frac{n}{p} \geq 0$. From (4) and (5) we obtain

$$|(\lambda^\epsilon)^{-1} - (\lambda_k)^{-1}| \leq \sup_{u \in H_0^1(\Omega)} \epsilon^{-m}[u,u]_\Omega^{-1}\Big[\int_{|x|<\epsilon} u^2 p\,dx\Big]^{\frac{1}{p}}\Big[\int_{|x|<\epsilon} \hbar_\epsilon^q dx\Big]^{\frac{1}{q}} \leq C_2\epsilon^{2-m},$$

where the constant C_2 does not depend on ϵ and $1 - \frac{n}{2} + \frac{n}{2p} = 0$, $\frac{1}{p} = 1 - \frac{2}{n}$, $\frac{1}{q} = \frac{2}{n}$.

Lemma 1. Let $n \geq 3$, $m \geq 2$. Then for the eigenvalues of problem (1), (2) we have

$$\lambda_k^\epsilon \geq C_0\epsilon^{m-2}, \tag{6}$$

where the constant C_0 does not depend on ϵ and k.

Proof. From (1), (2) we get

$$-[u^{\epsilon,k},u^{\epsilon,k}]_\Omega + \lambda_k^\epsilon(u^{\epsilon,k},u^{\epsilon,k})_\Omega + \epsilon^{-m}\lambda_k^\epsilon(\hbar_\epsilon u^{\epsilon,k},u^{\epsilon,k}) = 0. \tag{7}$$

According to the Friedrichs inequality we have

$$(u^{\epsilon,k},u^{\epsilon,k})_\Omega \leq C_1[u^{\epsilon,k},u^{\epsilon,k}]_\Omega, \quad C_1 = \text{const}.$$

Let $\lambda_k^\epsilon C_1 < \frac{1}{2}$. Then for $q = \frac{n}{2}$, $\frac{1}{p} + \frac{1}{q} = 1$ we obtain from (7) that

$$\frac{1}{2} \leq \lambda_k^\epsilon \epsilon^{-m}(\hbar_\epsilon u^{\epsilon,k}, u^{\epsilon,k})_\Omega$$

$$\leq \lambda_k^\epsilon \epsilon^{-m}\Big[\int_\Omega (u^{\epsilon,k})^{2p}dx\Big]^{\frac{1}{p}} \Big[\int_{|x|<\epsilon} \hbar_\epsilon^q dx\Big]^{\frac{1}{q}} \leq \lambda_k^\epsilon C_2 \epsilon^{2-m},$$

where C_2 = const, since (7) and (5) are valid and $[u^{\epsilon,k}, u^{\epsilon,k}]_\Omega = 1$, $\lambda_k^\epsilon C_3 < \frac{1}{2}$. Therefore $\lambda_k^\epsilon \geq C_3\epsilon^{m-2}$. The lemma is proved.

Suppose that $n \geq 3$, $m \geq 2$ and consider the eigenvalue problem in $\mathbb{R}^n$

$$\Lambda_k \Delta u^k + \hbar(x)u^k = 0 \tag{8}$$

in the space H with the norm $\|u\|_H = ([u,u]_{\mathbb{R}^n})^{1/2}$. We obtain the space H as the completion of $C_0^\infty(\mathbb{R}^n)$ in the norm $\|u\|_H$. The fact that $[u,u]_{\mathbb{R}^n}^{1/2}$ is the norm in H is due to the Hardy inequality [43]. We can rewrite (8) in the form

$$-\Lambda_k[u^k, v]_{\mathbb{R}^n} + (\hbar u^k, v)_{\mathbb{R}^n} = 0, \quad u^k, v \in H. \tag{9}$$

We can also write (9) in the form

$$-\Lambda_k[u^k, v]_{\mathbb{R}^n} + [Au^k, v]_{\mathbb{R}^n} = 0, \quad u^k, v \in H.$$

It is not difficult to prove that A is a positive, self-adjoint, compact operator. In fact for any u, v ∈ H we have

$$|[Au,v]_{\mathbb{R}^n}| \equiv |(\hbar u,v)_{\mathbb{R}^n}| \leq \Big[\int_{|x|<1} u^2|x|^{-2}dx\Big]^{1/2}\Big[\int_{|x|<1} \hbar^2 v^2 |x|^2 dx\Big]^{1/2}.$$

According to the Hardy inequality we see that

$$\int_{\mathbb{R}^n} u^2|x|^{-2}dx \leq C[u,u]_{\mathbb{R}^n}, \quad u \in H, \quad C = \text{const}. \tag{10}$$

Therefore

$$|[Au,v]| \leq C_1[u,u]_{\mathbb{R}^n}^{1/2}[v,v]_{\mathbb{R}^n}^{1/2}, \quad u, v \in H.$$

Using inequality (10) one can easily prove that A is a compact operator. Therefore

$$\Lambda_1 \geq \Lambda_2 \geq \ldots \geq \Lambda_k \geq \ldots > 0. \tag{11}$$

Theorem 2. Let $m \geq 2$, $n \geq 3$. Then for any eigenvalue (11) of problem (9) there exists a sequence of eigenvalues $\lambda^{\epsilon}_{p(\epsilon)}$ of problem (1), (2) such that $\epsilon^{m-2}(\lambda^{\epsilon}_{p(\epsilon)})^{-1} \to \Lambda_k$ as $\epsilon \to 0$ and

$$\left|\frac{\epsilon^{m-2}}{\lambda^{\epsilon}_{p(\epsilon)}} - \Lambda_k\right| \leq C_1\epsilon^{1/2}, \quad C_1 = \text{const}. \tag{12}$$

Proof. We have from (1), (2) that

$$-(\lambda^{\epsilon}_{p(\epsilon)})^{-1}[u^{\epsilon,p(\epsilon)},v]_{\Omega} + (u^{\epsilon,p(\epsilon)},v)_{\Omega} + \epsilon^{-m}(\hbar_{\epsilon}u^{\epsilon,p(\epsilon)},v)_{\Omega} = 0, \tag{13}$$

$$v \in H, \quad u^{\epsilon,p(\epsilon)} \in H.$$

Let $\psi(x)$ be a function from $C^{\infty}_0(\mathbb{R}^n)$, $\psi(x) = 1$ for $|x| < \frac{1}{2}$, $\psi(x) = 0$ for $|x| > 1$. We assume that the ball $|x| < 1$ belongs to Ω. Let us change the variables in (1), (2) setting $\xi = \frac{x}{\epsilon}$. We obtain

$$-[u^{\epsilon,p(\epsilon)},v]_{\Omega_{\epsilon}} + \lambda^{\epsilon}_{p(\epsilon)}\epsilon^{-2}(u^{\epsilon,p(\epsilon)},v)_{\Omega_{\epsilon}} + \epsilon^{2-m}\lambda^{\epsilon}_{p(\epsilon)}(\hbar u^{\epsilon,p(\epsilon)},v)_{\Omega_{\epsilon}} = 0 \tag{14}$$

for $u^{\epsilon,p(\epsilon)}, v \in H^1_0(\Omega_{\epsilon})$ where $\Omega_{\epsilon} = \{x : \epsilon x \in \Omega\}$. We can write the relation (14) in the form

$$-\epsilon^{m-2}(\lambda^{\epsilon}_{p(\epsilon)})^{-1}[u^{\epsilon,p(\epsilon)},v]_{\Omega_{\epsilon}} + \epsilon^{m}(u^{\epsilon,p(\epsilon)},v)_{\Omega_{\epsilon}} + (\hbar u^{\epsilon,p(\epsilon)},v)_{\Omega_{\epsilon}} = 0, \tag{15}$$

$$[u^{\epsilon,p(\epsilon)}, u^{\epsilon,p(\epsilon)}]_{\Omega_{\epsilon}} = 1.$$

Let u^k be an eigenfunction corresponding to the eigenvalue of Λ_k of the problem (9). It is easy to see that

$$- \Lambda_k [u^k\psi_\epsilon, v]_{\Omega_\epsilon} + \epsilon^m (u^k\psi_\epsilon, v)_{\Omega_\epsilon} + (\hbar u^k\psi_\epsilon, v)_{\Omega_\epsilon}$$

$$= \Lambda_k (\frac{\partial u^k}{\partial x_j}, v\frac{\partial \psi_\epsilon}{\partial x_j})_{\Omega_\epsilon} - \Lambda_k (u^k \frac{\partial \psi_\epsilon}{\partial x_j}, \frac{\partial v}{\partial x_j})_{\Omega_\epsilon} + \epsilon^m (u^k\psi_\epsilon, v)_{\Omega_\epsilon} , \tag{16}$$

where $\psi_\epsilon(x) = \psi(\epsilon x)$. We now estimate the norm of the right-hand side of (16) in $H_0^1(\Omega_\epsilon)$. One can prove that this norm does not exceed $C_2\epsilon^{1/2}$. Then applying Lemma 1 of Section 1 to (15) (16) we get (12).

<u>Theorem 3</u>. Let $n \geq 3$, $m > 2$, $\epsilon^{m-2}(\lambda^\epsilon_{p(\epsilon)})^{-1} \to \Lambda$ as $\epsilon \to 0$. Then Λ is an eigenvalue of problem (9).

<u>Proof</u>. Let us change the independent variables in (1), (2) by setting $\xi = \frac{x}{\epsilon}$ and suppose that $[u^{\epsilon,k}, u^{\epsilon,k}]_{\Omega_\epsilon} = 1$. We obtain

$$\Lambda^{\epsilon,p(\epsilon)} \Delta u^{\epsilon,p(\epsilon)} + (\epsilon^m + \hbar) u^{\epsilon,p(\epsilon)} = 0 \quad \text{in } \Omega_\epsilon,$$

$$u^{\epsilon,p(\epsilon)} = 0 \quad \text{on } \partial\Omega_\epsilon,$$

$$\Lambda^{\epsilon,p(\epsilon)} = \epsilon^{m-2} (\Lambda^\epsilon_{p(\epsilon)})^{-1}.$$

Let $v \in C_0^\infty(\mathbb{R}^n)$. Then for sufficiently small ϵ we have

$$- \Lambda^{\epsilon,p(\epsilon)} [u^{\epsilon,p(\epsilon)}, v]_{\Omega_\epsilon} + \epsilon^m (u^{\epsilon,p(\epsilon)}, v)_{\Omega_\epsilon} + (\hbar u^{\epsilon,p(\epsilon)}, v)_{\Omega_\epsilon} = 0, \tag{17}$$

$$[u^{\epsilon,p(\epsilon)}, u^{\epsilon,p(\epsilon)}]_{\Omega_\epsilon} = 1.$$

It is easy to see that $\epsilon^m(u^{\epsilon,p(\epsilon)}, v) \to 0$, since according to the Friedrichs inequality

$$\| u^{\epsilon,p(\epsilon)} \|_{L^2(\Omega_\epsilon)} \leq C_0 \epsilon^{-1}. \tag{18}$$

We set $v = u^{\epsilon,p(\epsilon)}$ in (17). Then we obtain

$$- \Lambda^{\epsilon,p(\epsilon)}[u^{\epsilon,p(\epsilon)},u^{\epsilon,p(\epsilon)}]_{\Omega_\epsilon} + \epsilon^{m}(u^{\epsilon,p(\epsilon)},u^{\epsilon,p(\epsilon)})$$

$$+ (\hbar u^{\epsilon,p(\epsilon)},u^{\epsilon,p(\epsilon)})_{\Omega_\epsilon} = 0.$$

It follows from (18) that $\epsilon^{m}(u^{\epsilon,p(\epsilon)},u^{\epsilon,p(\epsilon)})_{\Omega_\epsilon} \leq C_1\epsilon^{m-2}$ and since $m > 2$, $\epsilon^{m}(u^{\epsilon,p(\epsilon)},u^{\epsilon,p(\epsilon)})_{\Omega_\epsilon} \to 0$ as $\epsilon \to 0$. Thus we have

$$M_1 \leq (\hbar u^{\epsilon,p(\epsilon)},u^{\epsilon,p(\epsilon)})_{\Omega_\epsilon} \leq M_2, \tag{19}$$

since $[u^{\epsilon p(\epsilon)},u^{\epsilon,p(\epsilon)}]_{\Omega_\epsilon} = 1$, where the positive constants M_1, M_2 do not depend on ϵ. According to the Hardy inequality (10), the sequence $u^{\epsilon,p(x)}$ is compact in $L^2(\omega)$ and is weakly compact in $H^1(\omega)$ for any bounded domain ω. Let u be a limit function for $u^{\epsilon,p(x)}$ as $\epsilon \to 0$. Then $u \not\equiv 0$, since due to (19) we have $(u,u)_B > 0$, where $B = \{x: |x| < 1\}$. On passing to the limit in (17) we get that Λ is an eigenvalue of (9). The theorem is proved.

It follows from Lemma 1 and Theorem 1 that for any k,

$$C_1\epsilon^{m-2} \leq \lambda_k^{1} \leq C_2\epsilon^{m-2}, \quad m \geq 2, \quad k = 1,2,\ldots,$$

where the positive constants C_1,C_2 do not depend on ϵ but do depend on k.

<u>Theorem 4</u>. Let $n \geq 3$, $m = 2$ and $\lambda^{\epsilon}_{p(\epsilon)} \to \lambda$ as $\epsilon \to 0$. Then either λ is an eigenvalue of problem (3) or λ^{-1} is an eigenvalue of problem (9). All eigenvalues λ_k of problem (3) and all values Λ_k^{-1}, where Λ_k is an eigenvalue of problem (9), are limit points for some sequences of $\lambda^{\epsilon}_{p(\epsilon)}$ of problem (1), (2).

<u>Proof</u>. Assume that $\lambda^{\epsilon}_{p(\epsilon)} \to \lambda$ as $\epsilon \to 0$ and

$$(u^{\epsilon,p(x)},u^{\epsilon,p(\epsilon)})_{\Omega} \to 0 \quad \text{as} \quad \epsilon \to 0. \tag{20}$$

In this case $\Lambda = \lambda^{-1}$ is an eigenvalue of problem (9). Let us rewrite the problem (1), (2) in the form (17) and take $[u^{\epsilon,p(\epsilon)},u^{\epsilon,p(\epsilon)}]_{\Omega_\epsilon} = 1$ in the new variables $\xi = \frac{x}{\epsilon}$. Because of the last condition and (20) we have $\epsilon^2(u^{\epsilon,p(\epsilon)},u^{\epsilon,p(\epsilon)})_{\Omega_\epsilon} \to 0$ as $\epsilon \to 0$. On passing to the limit in (17) as $\epsilon \to 0$, we get λ^{-1} is an eigenvalue of (9). It is proved in Theorem 2 that for every eigenvalue Λ_k of problem (9), Λ_k^{-1} is a limit point for a sequence $\lambda^\epsilon_{p(\epsilon)}$ of eigenvalues of problem (1),(2) as $\epsilon \to 0$. Suppose that $\lambda^\epsilon_{p(\epsilon)} \to \lambda$ and $(u^{\epsilon,p(\epsilon)},u^{\epsilon,p(\epsilon)})_\Omega \geq a = \text{const} > 0$ as $\epsilon \to 0$. Then λ is an eigenvalue of problem (3). In fact, we have

$$-[(u^{\epsilon,p(\epsilon)},v]_\Omega + \lambda^\epsilon_{p(\epsilon)}(u^{\epsilon,p(\epsilon)},v)_\Omega + \lambda^\epsilon_{p(\epsilon)}\epsilon^{-2}(\hbar_\epsilon u^{\epsilon,p(\epsilon)},v)_\Omega = 0,$$

$$[(u^{\epsilon,p(\epsilon)},u^{\epsilon,p(\epsilon)}]_\Omega = 1,$$

for any function $v \in H^1_0(\Omega)$. According to the Friedrichs inequality, the sequence $u^{\epsilon,p(\epsilon)}$ is compact in $L^2(\Omega)$ and is weakly compact in $H^1_0(\Omega)$. Let us prove that $\epsilon^{-2}(\hbar_\epsilon u^\epsilon p(\epsilon),v)_\Omega \to 0$ as $\epsilon \to 0$. We have

$$|(\epsilon^{-2}\hbar_\epsilon u^{\epsilon,p(\epsilon)},v)_\Omega| \leq \epsilon^{-2}\|u^{\epsilon,p(\epsilon)}\|_{L^p(\Omega)}\left[\int_{|x|<\epsilon} \hbar^q_\epsilon v^q dx\right]^{\frac{1}{q}}$$

$$\leq C_1\epsilon^{-2}\|u^{\epsilon,p(\epsilon)}\|_{L^p(\Omega)} \max_{\overline{\Omega}} |v| \epsilon^{(n+2)/2}$$

$$\leq C_2\epsilon^{(n-2)/2}$$

if $\dfrac{1}{p} = \dfrac{1}{2} - \dfrac{1}{n}$, $\dfrac{1}{q} = \dfrac{n+2}{2n}$, since for such p the inequality (5) is valid.

We prove now that every eigenvalue λ_k of problem (3) is a limit point for some sequence $\lambda^\epsilon_{p(\epsilon)}$ as $\epsilon \to 0$. In order to apply Lemma 1 of Section 1 we consider the equality

$$-\lambda_k^{-1}[u^k,v]_\Omega + (u^k,v)_\Omega + \epsilon^{-2}(\hbar_\epsilon u^k,v)_\Omega = \epsilon^{-2}(\hbar_\epsilon u^k,v)_\Omega.$$

It is easy to see that

$$|\epsilon^{-2}(\hbar_\epsilon u^k,v)_\Omega| \leq \epsilon^{-2}\|v\|_{L^p(\Omega)}\left[\int_{|x|<\epsilon} \hbar^2_\epsilon (u^k)^2 dx\right]^{\frac{1}{q}} \leq C_3\epsilon^{(n-2)/2}\|v\|_1 ,$$

if $\frac{1}{p} = \frac{1}{2} - \frac{1}{n}$, $\frac{1}{q} = \frac{1}{2} + \frac{1}{n}$, since u^k is a smooth function.

Using the methods applied above for the case $n \geq 3$, one can study problem (1), (2) in the case $n = 2$. We get the following theorems.

Theorem 5. Let $n = 2$, $m < 2$. Then for eigenvalues λ_k^ϵ of problem (1), (2), and eigenvalues λ_k of problem (3) we have

$$|(\lambda_k^\epsilon)^{-1} - (\lambda_k)^{-1}| \leq C\epsilon^\gamma, \quad k = 1,2,\ldots,$$

where the constant $C(\gamma)$ does not depend on ϵ and k, and γ is any constant less than $2 - m$.

Let us consider now the case $n = 2$, $m \geq 2$. In the space $\mathcal{H} = \left\{u : [u,u]_{\mathbb{R}^2} < \infty, \int_{|x|<1} \hbar u dx = 0\right\}$ we consider the eigenvalue problem

$$\Lambda_k \Delta u^k + \hbar u^k = 0 \quad \text{in } \mathbb{R}^2. \tag{21}$$

The problem (21) in $\mathbb{R}^2$ can be written in the form

$$- \Lambda_k [u^k, v]_{\mathbb{R}^2} + (\hbar u^k, v)_{\mathbb{R}^2} = 0, \quad u^k, v \in \mathcal{H},$$

or

$$- \Lambda_k [u^k, v]_{\mathbb{R}^2} + [Au^k, v]_{\mathbb{R}^2} = 0, \quad u^k, v \in \mathcal{H}. \tag{22}$$

It is easy to prove that A is a positive, self-adjoint, compact operator in $\mathcal{H}$. Therefore the problem (22) has a positive discrete spectrum.

Theorem 6. Let $n = 2$, $m \geq 2$. Then for any eigenvalue Λ_k of problem (22) there exists a sequence of eigenvalues $\lambda_{p(\epsilon)}^\epsilon$ of problem (1), (2) such that

$$(\lambda_{p(\epsilon)}^\epsilon)^{-1} \epsilon^{m-2} \to \Lambda_k \quad \text{as } \epsilon \to 0.$$

Theorem 7. If $\epsilon^{m-2}(\lambda_{p(\epsilon)}^\epsilon)^{-1} \to \Lambda$ as $\epsilon \to 0$, $m > 2$, $n = 2$ and $\lambda_{p(\epsilon)}^\epsilon$ is an eigenvalue of problem (1), (2), then Λ is an eigenvalue of problem (22).

<u>Theorem 8</u>. Let $n = 2$, $m = 2$. If $\lambda^{\epsilon}_{p(\epsilon)} \to \lambda$ as $\epsilon \to 0$, $\lambda \neq 0$ and $\lambda^{\epsilon}_{h(\epsilon)}$ is an eigenvalue of problem (1), (2), then either λ is an eigenvalue of problem (3) or $\lambda^{-1} = \Lambda$ is an eigenvalue of problem (22) in $\mathcal{H}$.

Let us consider the case $n = 1$ which corresponds to the vibration of a string.

<u>Theorem 9</u>. Let $n = 1$, $0 < m < 1$. Then

$$|(\lambda^{\epsilon}_k)^{-1} - (\lambda_k)^{-1}| \leq C\epsilon^{1-m}, \quad k = 1, 2, \ldots,$$

where the constant C does not depend on k and ϵ; λ^{ϵ}_k is an eigenvalue of problem (1), (2) and λ_k is an eigenvalue of problem (3).

This theorem is proved in the same way as Theorem 1. In the case $n = 1$ we take $\Omega = (a,b)$, $a < 0$, $b > 0$.

Consider now the case $n = 1$, $m = 1$. We study the eigenvalue problem

$$\Lambda_k \frac{d^2u^k}{dx_1^2} + u^k + M\delta u^k = 0 \quad \text{on} \quad (a,b), \quad u(a) = u(b) = 0, \qquad (23)$$

where δ is the Dirac delta-function concentrated in $x = 0$. We consider the problem (23) in the space $H^1_o(\Omega)$. This problem can be written in the form

$$-\Lambda_k[u^k,v]_\Omega + [Au^k,v]_\Omega = 0, \quad u,v \in H^1_o(\Omega),$$

where

$$[Au,v] \equiv (u,v)_\Omega + Mu(0)v(0), \qquad M = \int_{-1}^{+1} \hbar(x)dx.$$

It can be easily proved that A is a positive, self-adjoint, compact operator in $H^1_o(\Omega)$. Let

$$\Lambda_1 \geq \Lambda_2 \geq \ldots \geq \Lambda_k \geq \ldots > 0 \qquad (24)$$

be the spectrum of the operator A. Each eigenvalue in (24) is counted as many times as its multiplicity. One can prove that

$\|A_\epsilon - A\| \leq C_1\alpha(\epsilon)$, where the constant C_1 does not depend on ϵ; $\alpha(\epsilon) \to 0$ as $\epsilon \to 0$ and

$$[A_\epsilon u, v]_\Omega \equiv ((1 + \epsilon^{-1}\hbar_\epsilon)u, v)_\Omega, \quad u, v \in H_0^1(\Omega).$$

Using the H.Weyl theorem [40] one can prove the following result.

Theorem 10. Let $n = 1$, $m = 1$. Then

$$|(\lambda_k^\epsilon)^{-1} - \Lambda_k| \leq C_1\alpha(\epsilon), \quad k = 1, 2, \ldots,$$

where Λ_k is an eigenvalue (24) of problem (23).

Lemma 2. Let $n = 1$, $m > 1$. Suppose that $u^{\epsilon,p(\epsilon)}$ is a sequence of eigenfunctions of problem (1), (2) such that $u^{\epsilon,p(\epsilon)} \to u(x)$ uniformly on (a,b) as $\epsilon \to 0$, and for the related eigenvalues $\lambda_{p(\epsilon)}^\epsilon$ the inequality $\lambda_{p(\epsilon)}^\epsilon \geq C_0$ is valid, where the constant C_0 does not depend on ϵ. Then $u(0) = 0$.

Lemma 3. Let $n = 1$, $m > 1$. Then for eigenvalues of problem (1), (2) the estimate

$$\lambda_k^\epsilon \geq C\epsilon^{m-1}$$

is valid, where the constant C does not depend on ϵ and k.

Consider the eigenvalue problem in $H_0^1(\Omega)$:

$$\Lambda_k \frac{d^2u^k}{dx_1^2} + M\delta u^k = 0, \quad u^k(a) = u^k(b) = 0. \tag{25}$$

It is easy to see that the eigenvalue problem (25) has only one positive eigenvalue Λ_0.

Theorem 11. Suppose that $n = 1$, $1 < m < 2$. Then $\epsilon^{m-1}(\lambda_1^\epsilon)^{-1} \to \Lambda_0$ as $\epsilon \to 0$, $\lambda_k^\epsilon \to \lambda_{k-1}$ as $\epsilon \to 0$ and $k > 1$, where λ_k is an eigenvalue of the problem

$$\frac{d^2u^k}{dx_1^2} + \lambda_k u^k = 0 \quad \text{on} \quad (a,0) \cup (o,b),$$

$$u^k(a) = u^k(b) = u(0) = 0, \tag{26}$$

$$0 < \lambda_1 \leq \lambda_2 \leq \ldots \leq \lambda_k \leq \ldots .$$

In order to study the case $n = 1$, $m \geq 2$, consider the eigenvalue problem

$$\Lambda_k \frac{d^2u^k}{dx_1^2} + \hbar u^k = 0 \quad \text{on} \quad (-1,1), \qquad \frac{du^k(-1)}{dx_1} = \frac{du^k(1)}{dx_1} = 0, \qquad (27)$$

in the space $\mathcal{H} = \{u: [u,u]_\omega < \infty,\ \omega = (-1,1),\ \int_{-1}^{1} \hbar u dx = 0\}$.

The problem (27) in the space $\mathcal{H}$ has a positive discrete spectrum

$$\Lambda_1 \geq \Lambda_2 \geq \ldots \geq \Lambda_k \geq \ldots > 0. \qquad (28)$$

Theorem 12. Let $n = 1$, $m > 2$. Then $\lambda_k^\epsilon \to 0$ as $\epsilon \to 0$. For any eigenvalue Λ_k of problem (27) there exists a sequence $\lambda^\epsilon_{p(\epsilon)}$ such that $\epsilon^{m-2}(\lambda^\epsilon_{p(\epsilon)})^{-1} \to \Lambda_k$ as $\epsilon \to 0$. If $\epsilon^{m-2}(\lambda^\epsilon_{p(\epsilon)})^{-1} \to \Lambda$ as $\epsilon \to 0$, then Λ is an eigenvalue of problem (27). Moreover $\epsilon^{m-1}(\lambda_1^\epsilon)^{-1} \to \Lambda_0$ as $\epsilon \to 0$, where Λ_0 is the eigenvalue of problem (25).

Theorem 13. Let $n = 1$, $m = 2$. Then $\epsilon^{m-1}(\lambda_1^\epsilon)^{-1} \to \Lambda_0$ as $\epsilon \to 0$, where Λ_0 is the eigenvalue of problem (25). If $\lambda^\epsilon_{p(\epsilon)} \to \lambda \neq 0$ as $\epsilon \to 0$, then λ is either an eigenvalue of problem (26) or $\lambda^{-1} = \Lambda$ is an eigenvalue of problem (27). Every eigenvalue Λ_k of problem (27) is a limit point for some sequence $\lambda^\epsilon_{p(\epsilon)}$ as $\epsilon \to 0$.

The above methods can be applied to study similar problems in the case of higher order elliptic equations, some other boundary conditions, several concentrated masses, or a mass concentrated at a boundary point, as well as masses concentrated on a submanifold Γ, $\Gamma \subset \Omega$..

REFERENCES

1. Benoussan A., Lions J.-L., Papanicolaou G. Asymptotic Analysis for Periodic Structures. North Holland Publ. Co. (1978).
2. Sanchez-Palencia E. Non-Homogeneous Media and Vibration Theory. Lect. Notes in Phys. 127. Springer Verlag. (1980).

3. Bakhvalov N.S., Panasenko G.P., Homogenization of Processes in Periodic Media. Moscow: Nauka. (1984).
4. Oleinik O.A. On homogenization problems. In Trends and Applications of Pure Mathematics to Mechanics. Lecture Notes in Phys. 195, pp. 248-272. Springer Verlag. (1984).
5. Oleinik O.A., Shamaev A.S., Yosifian G.A. On the convergence of the energy, stress tensors and eigenvalues in homogenization problems of elasticity. Z. angew. Math. Mech., 65, no. 1, 13-17. (1985).
6. Tartar L. Etude des oscillations dans les equations aux derivees partielles non lineaires. In Trends and Applications of Pure Mathematics to Mechanics. Lecture Notes in Phys., 195, pp. 384-412. Springer Verlag. (1984).
7. Duvaut G. Homogeneisation et Materiaux Composites. In Trends and Applications of Pure Mathematics to Mechanics. Lecture Notes in Phys., 195, pp. 35-62. (1984).
8. Tartar L. Homogeneisation. Cours Peccot au College de France. Paris. (1977).
9. Kozlov S.M., Oleinik O.A. Zhikov V.V., Kha T'en Ngoan. Averaging and G-convergence of differential operators. Russian Math. Surveys, 34, no. 5 (1979).
10. Lions J.-L. Remarques sur l'homogeneisation. In Computing Methods in Applied Sciences and Engineering VI. Ed. by R. Glowinsky and J.-L. Lions, INRIA, North Holland, pp. 299-315, (1984).
11. Pobedria B.E. Mechanics of Composite Materials. Moscow: Moscow University. (1984).
12. Oleinik O.A., Yosifian G.A. On homogenization of the elasticity system with rapidly oscillating coefficients in perforated domains. In N.E. Kochin and Advances in Mechanics, pp. 237-249. Moscow: Nauka. (1984).
13. Cioranescu D., Paulin S.J. Homogenization in open sets with holes. J. Math. Anal. and Appl., 71, no. 2 pp.590-607. (1979).
14. Oleinik O.A., Shamaev A.S., Yosifian G.A. Problemes d'homogeneisation pour le systeme de l'elasticite lineaire a coefficients oscillant non-uniformement. C.R. Acad. Sci. A. 298, pp. 273-276. (1984)
15. Vanninathan M. Homogeneisation des valeurs propres dans les milieux perfores. C.R. Acad. Sci. A, 287, pp. 823-825. (1978).
16. Kesavan S. Homogenization of elliptic eigenvalue problems. Appl. Math. Optim. Part I, 5, pp. 153-167, Part II, 5 pp. 197-216. (1979).
17. Oleinik O.A., Shamaev A.S., Yosifian G.A. Homogenization of eigenvalues and eigenfunctions of the boundary value problems in perforated domains for elliptic equations with non-uniformly oscillating coefficients. In Current Topics in Partial Differential Equations, pp. 187-216. Tokyo: Kinokuniya Co. Ltd. (1986).
18. Oleinik O.A., Shamaev A.S., Yosifian G.A. Homogenization of eigenvalues and eigenfunctions of the boundary value problem of elasticity in a perforated domain. Vestnik Mosc. Univ., ser. I., Math., Mech., no. 4, pp.53-63. (1983).
19. Marchenko V.A., Khruslov E.Y. Boundary Value Problems in Domains with a Finely Granulated Boundary. Kiev: Naukova Dumka. (1974).

20. Marcellini P. Convergence of second order linear elliptic operators Boll. Un. Mat. Ital., B(5), 15. (1979).
21. Spagnolo S. Convergence in energy for elliptic operators. Proc. Third Sympos. Numer. Solut. Partial Differential Equations, College Park, Md., pp. 469-498. (1976).
22. Cioranescu D., Murat F. L.Un terme etranger venu d'ailleurs. In Non-linear Partial Differential Equations and Their Applications. College de France Seminar, vol.II, pp. 98-138, Vol.III, pp. 154-178. Pitman. (1981,83).
23. Oleinik O.A., Shamaev A.S., Yosifian G.A. On the asymptotic expansion of solutions of the Dirichlet problem for elliptic equations and the elasticity system in a perforated domain. Dokl. AN SSSR, 284, no.5, pp. 1062-1066. (1985).
24. Oleinik O.A., Shamaev A.S., Yosifian G.A. Asymptotic expansion of eigenvalues and eigenfunctions of the Sturm-Liouville problem with rapidly oscillating coefficients. Vestnik Moscow Univ., ser.I, Math., Mech., no. 6, pp.37-46. (1985).
25. Oleinik O.A., Shamaev A.S., Yosifian G.A. On eigenvalues of boundary value problems for the elasticity system with rapidly oscillating coefficients in a perforated domain. Matem. Sbornik, (to appear). (1987).
26. Kozlov S.M., Oleinik O.A., Zhikov V.V. On G-convergence of parabolic operators. Russian Math. Surveys, 36, no.1, (1981).
27. De Giorgi E., Spagnolo S. Sulla convergenza degli integrali dell'energia per operatori ellittici del 2 ordine. Boll. Un. Mat. Ital., (4), 8, pp. 391-411. (1973).
28. Oleinik O.A., Panasenko G.P., Yosifian G.A. Homogenization and asymptotic expansions for solutions of the elasticity system with rapidly oscillating periodic coefficients. Applicable Analysis. 15, no. 1-4, pp. 15-32. (1983).
29. Oleinik O.A., Panasenko G.P., Yosifian G.A. An asymptotic expansion for solutions of the elasticity system in perforated domains. Matem. Sbornik. 120, no. 1, pp. 22-41. (1983).
30. Shaposhnikova T.A. On strong G-convergence for a sequence of the elasticity systems. Vestnik Moscow Univ., ser. I, Math., Mech., no. 5, pp. 29-34. (1984).
31. Oleinik O.A., Zhikov V.V. On homogenization of the elasticity system with almost-periodic coefficients. Vestnik Moscow Univ. ser. I, Math., Mech., no. 6, pp.62-70. (1982).
32. Oleinik O.A., Zhikov V.V. On the homogenization of elliptic operators with almost periodic coefficients. Rend. Semin. Mat. e Fis. Milano, 52, pp. 149-166. (1982).
33. Oleinik O.A., Shamoaev A.S., Yosifian G.A. On homogenization problems for the elasticity system with non-uniformly oscillating coefficients. In Mathematical Analysis B.79, Teubner-Texte zur Mathematik, pp.192-202. (1985).
34. Lions J.-L. Some Methods in the Mathematical Analysis of Systems and Their Control. Beijing, China: Science Press. New York: Gordon & Breach. (1981).
35. Lions J.-L. Asymptotic expansions in perforated media with a periodic structure. The Rocky Mountain J. of Math., 10, no. 1, pp. 125-140. (1980).

36. Oleinik O.A., Shamaev A.S., Yosifian G.A. On homogenization of elliptic operators describing processes in stratified media. Uspekhi Mat. Nauk, 41, no. 3, pp. 185-186. (1986).
37. Oleinik O.A., Shamaev A.S., Yosifian G.A. On the homogenization of stratified structures. Volume dedicated to the 60th birthday of J.-L. Lions (to appear). (1987).
38. Oleinik O.A., Shamaev A.S., Yosifian G.A. On asymptotic expansions of solutions of the Dirichlet problem for elliptic equations in perforated domains. In Non-linear partial differential equations and their applications. College de France Seminar, 8 (to appear). (1987).
39. Vishik M.I., Liusternik L.A. Regular degeneration and boundary layer for differential equations with a small parameter. Uspekhi Mat. Nauk, 12, no. 5, pp. 3-122. (1957).
40. Riesz F., Sz.-Nagy B. Lecons d'Analyse Fonctionnelle. Budapest: Academiai Kiado. (1952).
41. Sanchez-Palencia E. Perturbation of eigenvalues in thermoelasticity and vibration of systems with concentrated masses. In Trends and Applications of Pure Mathematics to Mechanics. Lecture Notes in Phys. 195, pp. 346-368. Springer Verlag. (1984).
42. Sanchez-Palencia E., Tchatat H. Vibration de systems elastiques avec des masses concentrees. Rend. Sem. Mat. Univ. Politech. Torino. 42, no. 3, pp. 43-63. (1984).
43. Kondratiev V.A., Oleinik O.A. Sur un probleme de Sanchez-Palencia. C.R. Acad. Sci. A., 299, no. 15, pp. 745-748. (1984).

Conservation Laws in Continuum Mechanics

Peter J. Olver
School of Mathematics, University of Minnesota,
Minneapolis, MN 55455
USA

Preface.

This article reviews some recent work on the conservation laws of the equations of continuum mechanics, with especial emphasis on planar elasticity. The basic material on conservation laws and symmetry groups of systems of partial differential equations is given an extensive treatment in the author's book, [6], so this paper will only give a brief overview of the basic theory. Some of the applications appear in the published papers cited in the references, while others are more recent.

This research was supported in part by NSF Grant DMS 86-02004.

1. Conservation Laws of Partial Differential Equations.

The equations of non-dissipative equilibrium continuum mechanics come from minimizing the energy functional

$$\mathcal{W}[u] = \int_\Omega W(x,u^{(n)})\, dx \qquad (1)$$

Here the independent variables $x=(x^1,\ldots,x^p) \in \Omega$ represent the material coordinates in the body, and the dependent variables $u=(u^1,\ldots,u^p)$ the deformation, where p=2 for planar theories, while p=3 for fully three-dimensional bodies. In the absence of body forces, the stored energy W will usually depend just on x and the deformation gradient ∇u, but may, in a theory of higher grade material, depend on derivatives of u up to order n, denoted $u^{(n)}$. Smooth minimizers will satisfy the Euler-Lagrange equations

$$E_\nu(W) = 0, \qquad \nu = 1, \ldots, p, \tag{2}$$

which, in the case of continuum mechanics, form a strongly elliptic system of partial differential equations of order 2n. Strong ellipticity implies that this system is totally nondegenerate (in the sense of [6; Definition 2.83]).

Given the system of partial differential equations (2), a *conservation law* is a divergence expression

$$\mathrm{Div}\, P = \sum_{i=1}^{p} D_i P_i = 0 \tag{3}$$

which vanishes on all solutions to (1), where the p-tuple $P(x,u^{(m)})$ can depend on x, u and the derivatives of u. For static problems, conservation laws provide path-independent integrals, which are of use in determining the behavior at singularities such as cracks or dislocations. For dynamic problems, conservation laws provide constants of the motion, such as conservation of mass or energy.

Two conservation laws are *equivalent* if they differ by a sum of trivial conservation laws, of which there are two types. In the first type of triviality, the p-tuple P itself vanishes on all solutions to (2), while the second type are the *null divergences*, where the identity (3) holds for all functions u=f(x) (not just solutions to the system). As trivial laws provide no new information about the solutions, we are only interested in equivalence classes of nontrivial conservation laws.

An elementary integration by parts shows that any conservation law for the nondegenerate system (2) is always equivalent to a conservation law in *characteristic form*

$$\mathrm{Div}\, P = Q.E(W) = \sum_{\nu=1}^{p} Q_\nu E_\nu(W) \tag{4}$$

where the p-tuple $Q=(Q_1,\ldots,Q_p)$ is the *characteristic* of the conservation law. A characteristic is called *trivial* if it vanishes on solutions to (2), and two characteristics are equivalent if they differ by a trivial characteristic. For nondegenerate systems of partial differential equations, each conservation law is uniquely determined by its characteristic, up to equivalence.

Theorem. If the system (2) is nondegenerate, then there is a one-to-one correspondence between (equivalence classes of) nontrivial conservation laws and (equivalence classes of) nontrivial characteristics.

2. Symmetries and Noether's Theorem.

A *generalized vector field* is a first order differential operator

$$\mathbf{v} = \sum_{i=1}^{p} \xi^i(x,u^{(m)}) \frac{\partial}{\partial x^i} + \sum_{\alpha=1}^{p} \varphi_\alpha(x,u^{(m)}) \frac{\partial}{\partial u^\alpha}.$$

If the coefficients ξ^i and φ_α depend only on x and u, then **v** generates a one-parameter group of *geometrical* transformations, which solve the system of ordinary differential equations

$$\frac{dx^i}{d\varepsilon} = \xi^i(x,u), \qquad \frac{du^\alpha}{d\varepsilon} = \varphi_\alpha(x,u).$$

For general **v**, the group transformations are nonlocal, and determined as solutions of a corresponding system of evolution equations.

The vector field **v** is a *symmetry* of the system (2) if and only if the infinitesimal invariance condition

$$\mathrm{pr}\,\mathbf{v}\,[E_\nu(W)] = 0, \qquad \nu = 1,\ldots,p,$$

holds on all solutions to (2). Here pr **v** denotes the *prolongation* of **v**, which determines how **v** acts on the derivatives of u. An elementary lemma says that we can always replace **v** by the simpler *evolutionary vector field*

$$\mathbf{v}_Q = \sum_{\alpha=1}^{p} Q_\alpha(x,u^{(m)}) \frac{\partial}{\partial u^\alpha}$$

where the *characteristic* $Q = (Q_1, \ldots, Q_p)$ of $\mathbf{v}$ is defined by

$$Q_\alpha = \varphi^\alpha - \sum_{i=1}^{p} \xi^i \cdot \frac{\partial u^\alpha}{\partial x^i}$$

(See [6; Chapter 5] for the explicit formulas.) The infinitesimal invariance condition

$$\mathrm{pr}\,\mathbf{v}_Q\,[E_\nu(W)] = 0, \text{ whenever } E(W) = 0, \quad \nu = 1, \ldots, p, \tag{5}$$

constitutes a large system of elementary partial differential equations for the components of the characteristic Q. Fixing the order of Q, the defining equations (5) can be systematically solved so as to determine the most general symmetry of the given order of the system.

An evolutionary vector field $\mathbf{v}_Q$ is a *trivial symmetry* of (2) if the characteristic $Q(x,u^{(m)})$ vanishes on all solutions to (2). Two symmetries are *equivalent* if they differ by a trivial symmetry. Clearly we are only interested in determining classes of inequivalent symmetries of a given system of partial differential equations.

More restrictively, the evolutionary vector field $\mathbf{v}_Q$ is called a *variational symmetry* of the variational problem (1) if the infinitesimal invariance condition

$$\mathrm{pr}\,\mathbf{v}_Q\,(W) = \mathrm{Div}\,B \tag{6}$$

holds for some p-tuple $B(x,u^{(k)})$. Every variational symmetry of a variational integral (1) is a symmetry of the associated Euler-Lagrange equations (2), but the converse is *not* always true. (The most common counterexamples are scaling symmetry groups.) It is easy to check which of the symmetries of the Euler-Lagrange equations satisfy the additional variational criterion (6); see also [6; Proposition 5.39].

Noether's Theorem provides the connection between variational symmetries of a variational integral and conservation laws of the associated Euler-Lagrange equations $E(W) = 0$.

Theorem. Suppose we have a variational integral (1) with non-degenerate Euler-Lagrange equations (2). Then a p-tuple $Q(x,u^{(m)})$ is the characteristic of a conservation law for the Euler-Lagrange equations (2) if and only if it is the characteristic of a variational symmetry of (1). Moreover, equivalent conservation laws correspond to equivalent variational symmetries and vice versa.

Thus there is a one-to-one correspondence between equivalence classes of nontrivial variational symmetries and equivalence classes of nontrivial conservation laws. The proof rests on the elementary integration by parts formula

$$\mathrm{pr}\, v_Q(W) = Q \,.\, E(W) + \mathrm{Div}\, A, \tag{7}$$

for some p-tuple $A = (A_1, \ldots, A_p)$ depending on Q and W. (There is an explicit formula for A, but it is a bit complicated; see [6; Proposition 5.74].) Comparing (7) and the symmetry condition (6), we see that

$$\mathrm{Div}(B - A) = Q \,.\, E(W),$$

and hence $P = B - A$ constitutes a conservation law of (2) with characteristic Q. The nontriviality follows from the theorem of section 1.

3. Finite Elasticity.

As an application of the general theory, we consider the case of an elastic material, so the stored energy function $W(x,\nabla u)$ depends only on the deformation gradient. We show how simple symmetries lead to well-known conservation laws. Material frame indifference implies that W is invariant under the Euclidean group

$$u \to Ru + a$$

of rotations R and translations a. The translational invariance is already implied by the fact that W does not depend explicitly on u, while rotational invariance requires that $W(x,R.\nabla u) = W(x,\nabla u)$ for all rotations R. The conservation laws coming from translational invariance are just the Euler-Lagrange equations themselves

$$\sum_{i=1}^{p} D_i\{ \frac{\partial W}{\partial u_i^\alpha} \} = 0,$$

written in divergence form. The rotational invariance provides p(p-1)/2 further conservation laws

$$\sum_{i=1}^{p} D_i\{ u^\alpha \frac{\partial W}{\partial u_i^\beta} - u^\beta \frac{\partial W}{\partial u_i^\alpha} \} = 0.$$

If the material is homogeneous, then W does not depend on x, and we have the additional symmetry group of translations

$$x \rightarrow x + b$$

in the material coordinates. There are thus p additional conservation laws

$$\sum_{i=1}^{p} D_i\{ \sum_{\alpha=1}^{p} u_j^\alpha \frac{\partial W}{\partial u_i^\alpha} - \delta_j^i W \} = 0,$$

whose entries form the components of Eshelby's celebrated energy-momentum tensor. If the material is isotropic, then W is invariant under the group of rotations in the material coordinates, so $W(\nabla u.R) = W(\nabla u)$ for all rotations R. There are an additional p(p-1)/2 conservation laws

$$\sum_{i=1}^{p} D_i\{ \sum_{\alpha=1}^{p} [x^j u_k^\alpha - x^k u_j^\alpha] \frac{\partial W}{\partial u_i^\alpha} - [\delta_j^i x^k - \delta_k^i x^j] W \} = 0.$$

Scaling symmetries can produce conservation laws under the assumption that W is a homogeneous function of the deformation gradient

$$W(\lambda . \nabla u) = \lambda^n . W(\nabla u), \quad \lambda > 0.$$

The scaling group $(x,u) \to (\lambda x, \lambda^{(n-p)/n}u)$ is a variational symmetry group, leading to the conservation law

$$\sum_{i=1}^{p} D_i\Big\{ \sum_{\alpha=1}^{p} \Big[\frac{n-p}{n}u^\alpha - \sum_{j=1}^{p} x^j u_j^\alpha\Big] \frac{\partial W}{\partial u_i^\alpha} + x^i W \Big\} = 0.$$

Of course, stored energy functions which are invariant under the scaling symmetry group are rather special. If one writes out the above divergence in the more general case, then we obtain the divergence identity

$$\sum_{i=1}^{p} D_i\Big\{ \sum_{\alpha=1}^{p} \Big[u^\alpha - \sum_{j=1}^{p} x^j u_j^\alpha\Big] \frac{\partial W}{\partial u_i^\alpha} + x^i W \Big\} = pW.$$

This was used by Knops and Stuart, [3], to prove the uniqueness of the solutions corresponding to homogeneous deformations. This latter identity is closely related to the general dentities determinedby Pucci and Serrin, [10]. Indeed the general formula used by Pucci and Serrin to determine their identities is a special case of the integration by parts formula (7) in the case that the characteristic Q comes from a geometrical vector field. Particular choices of the coefficient functions ξ^i and φ_α lead to the particular identities that are used to study eigenvalue problems and uniqueness of solutions, generalizing earlier ideas of Rellich and Pohozaev.

4. Linear Planar Elasticity.

Although the general structure of symmetries and conservation laws for many of the variational problems of continuum mechanics remains an open problem, the case of linear planar elasticity, both isotropic and anisotropic, is now well understood. In this case, the stored energy function $W(\nabla u)$ is a quadratic function of the deformation gradient, which is usually written in terms of the strain tensor $e=(\nabla u+\nabla u^T)/2$. We have

$$W(\nabla u) = \sum c_{ijkl}\, e_{ij}\, e_{kl}, \tag{8}$$

where the constants c_{ijkl} are the elastic moduli which describe the physical properties of the elastic material of which the body is composed. The elastic moduli must satisfy certain inequalities stemming from the Legendre-Hadamard strong ellipticity condtion. This states that the quadratic stored energy function $W(\nabla u)$ must be positive definite whenever the deformation gradient is a rank one tensor, i.e. $\nabla u = a \otimes b$ for vectors a, b. Following [7], we define the symbol of the quadratic variational problem with stored energy (8) to be the biquadratic polynomial $Q(x,u) = W(x \otimes u)$ obtained by replacing ∇u by the rank one tensor $x \otimes u$. In this case, the Legendre-Hadamard strong ellipticity condtion requires that

$$Q(x,u) > 0 \quad \text{whenever} \quad x \neq 0, \text{ and } u \neq 0. \tag{9}$$

The symmetry of the stress tensor and the variational structure of the equations impose the symmetry conditions

$$c_{ijkl} = c_{jikl} = c_{ijlk}, \qquad c_{ijkl} = c_{klij}.$$

on the elastic moduli, which are equivalent to the symmetry condition

$$Q(x,u) = Q(u,x)$$

on the symbol.

For each fixed u, $Q(x,u)$ is a homogeneous quadratic polynomial in x, and so we can form its discriminant $\Delta_x(u)$ (i.e. b^2-4ac), which will be a homogeneous quartic polynomial in u. The nature of the roots of $\Delta_x(u)$ provides the key to the structure of the problem. First, the Legendre-Hadamard condition (9) requires that $\Delta_x(u)$ has all complex roots. There are then only two distinct cases.

Theorem. Let $W(\nabla u)$ be a strongly elliptic quadratic planar variational problem, and let $\Delta_x(u)$ be the discriminant of its symbol. Then exactly one of the following possibilities holds.

1. <u>The Isotropic Case</u>. If $\Delta_x(u)$ has a complex conjugate pair of double roots, then there exists a linear change of variables

$$x \to Ax, \quad u \to Bu, \qquad A, B \text{ invertible } 2\text{x}2 \text{ matrices}$$

which changes W into an isotropic stored energy function.

2. The Orthotropic Case. If $\Delta_x(u)$ has two complex conjugate pairs of simple roots, then there exists a linear change of variables

$$x \to Ax, \quad u \to Bu, \qquad A, B \text{ invertible } 2\text{x}2 \text{ matrices}$$

which changes W into an orthotropic (but not isotropic) stored energy function.

(Recall, [2], that an orthotropic elastic material is one which has three orthogonal planes of symmetry. Thus, this theorem states that any planar elastic material is equivalent to an orthotropic (possibly isotropic) material, and so has three (not necessarily orthogonal) planes of symmetry. The analogous result is not true in three dimensions, cf. [1].)

This theorem is a special case of a general classification of quadratic variational problems in the plane, [7], and results in the construction of "canonical elastic moduli" for two-dimensional elastic media, [8]. One consequence is that in planar linear elasticity, there are, in reality, only two independent elastic moduli, since one can rescale any orthotropic stored energy to one whose elastic moduli have the "canonical form"

$$c_{1111} = c_{2222} = 1, \quad c_{1122} = c_{2211} = \alpha, \quad c_{1212} = \beta, \quad c_{1112} = c_{1222} = 0.$$

Thus the constants α and β play the role of canonical elastic moduli, with the special case $2\alpha + \beta = 1$ corresponding to an isotropic material. This confirms a conjecture made in [5]. Extensions to three-dimensional materials are currently under investigation.

Although isotropic and more general orthotropic materials have similar looking Lagrangians, the structure of their associated conservation laws is quite dissimilar. (For simplicity of notation, we write (x,y) for the independent variables and (u,v) for the dependent variables from now on.)

Theorem. Let $\mathcal{W}[u]$ be a strongly elliptic quadratic planar variational problem, with corresponding Euler-Lagrange equations $E(W) = 0$.

1. The Isotropic Case. If W is equivalent to an isotropic material, then there exists a complex linear combination z of the variables (x, y), a complex linear combination w of the variables (u, v), and two complex linear combinations ξ, η of the components of the deformation gradient (u_x, u_y, v_x, v_y) with the properties:

a) The two Euler-Lagrange equations can be written as a single complex differential equation in form

$$D_z \eta = 0.$$

(Recall that if $z = x + iy$, then the complex derivative D_z is defined as $(D_x - iD_y)$.)

b) Any conservation law is a real linear combination of the Betti reciprocity relations, the complex conservation laws

$$\mathrm{Re}[\, D_z F\,] = 0,$$

and

$$\mathrm{Re}[D_z\{(\xi + z)G_\eta + \overline{G}\}] = 0,$$

where $F(z, \eta)$ and $G(z, \eta)$ are arbitrary complex analytic functions of their two arguments, and the extra conservation law

$$\mathrm{Re}[\, D_z\{w\eta - iz\eta^2\}\,] = 0.$$

2. The Orthotropic Case. If W is equivalent to an orthotropic, non-isotropic material, then there exist two complex linear combinations z, ζ of the variables (x, y), and two corresponding complex linear combinations ξ, η of the components of the deformation gradient (u_x, u_y, v_x, v_y) with the properties:

a) The two Euler-Lagrange equations can be written as a single complex differential equation in either of the two forms

$$D_z \xi = 0,$$

or

$$D_\zeta \eta = 0.$$

b) Any conservation law is a real linear combination of the Betti reciprocity relations, and the complex conservation laws

$$\mathrm{Re}[\, D_z F \,] = 0, \quad \text{and} \quad \mathrm{Re}[\, D_\zeta G \,] = 0,$$

where $F(z,\xi)$ and $G(\zeta,\eta)$ are arbitrary complex analytic functions of their two arguments.

Thus one has the striking result that in *both* isotropic and anisotropic planar elasticity, there are three infinite families of conservation laws. One family is the well-known Betti reciprocity relations. The other two are determined by two arbitrary analytic functions of two complex variables. However, the detailed structure of these latter two families is markedly different depending upon whether one is in the isotropic or truly anisotropic (orthotropic) case. The two orthotropic families degenerate to a single isotropic family, but a second family makes its appearance in the isotropic case. In addition, the isotropic case is distinguished by the existence of one extra anomalous conservation law, the significance of which is not fully understood. The details of the proof of this theorem in the isotropic case have appeared in [4; Theorem 4.2] (although there is a misprint, corrected in an Errata to [4] appearing recently in the same journal); the anisotropic extension will appear in [9].

I suspect that a similar result even holds in the case of nonlinear planar elasticity, but have not managed to handle the associated "vector conformal equations", cf. [5]. Extensions to three-dimensional elasticity have only been done in the isotropic case; see [4] for a complete classification of the conservation laws there. In this case, beyond Betti reciprocity, there are just a finite number of conservation laws, some of which were new.

References

[1] Cowin, S.C. and Mehrabadi, M.M., On the identification of material symmetry for anisotropic elastic materials, IMA preprint series #204, Institute for Mathematics and Its Applications, University of Minnesota, 1985.

[2] Green, A.E. and Zerna, W., *Theoretical Elasticity*, The Clarendon Press, Oxford, 1954.

[3] Knops, R.J. and Stuart, C.A. Quasiconvexity and uniqueness of equilibrium solutions in nonlinear elasticity, *Arch. Rat. Mech. Anal.* **86** (1984), 234-249.

[4] Olver, P.J., Conservation laws in elasticity. II. Linear homogeneous isotropic elastostatics, *Arch Rat. Mech. Anal.* **85** (1984), 131-160.

[5] Olver, P.J., Symmetry groups and path-independent integrals, in: *Fundamentals of Deformation and Fracture*, B.A. Bilby, K.J. Miller and J.R. Willis, eds., Cambridge Univ. Press, New York, 1985, pp.57-71.

[6] Olver, P.J., *Applications of Lie Groups to Differential Equations*, Graduate Texts in Mathematics, vol. 107, Springer-Verlag, New York, 1986.

[7] Olver, P.J., The equivalence problem and canonical forms for quadratic Lagrangians, *Adv. in Math.*, to appear

[8] Olver, P.J., Canonical elastic moduli, preprint.

[9] Olver, P.J., Conservation laws in elasticity. III. Linear planar anisotropic elastostatics, preprint.

[10] Pucci, P. and Serrin, J., A general variational identity, *Indiana Univ. Math. J.* **35** (1986), 681-703.

ON GEOMETRIC AND MODELING PERTURBATIONS IN PARTIAL DIFFERENTIAL EQUATIONS

L E Payne
Department of Mathematics
Cornell University
Ithaca
New York 14853, U S A

I. INTRODUCTION

This paper discusses certain types of stability questions that have been largely ignored in the literature, i.e. continuous dependence on geometry and continuous dependence on modeling. Although we shall consider these questions primarily in the context of ill-posed problems we shall briefly indicate some difficulties that might arise under geometric and/or modeling perturbations in well posed problems.

In setting up and analyzing a mathematical model of any physical process it is inevitable that a number of different types of errors will be introduced e.g. errors in measuring data, errors in determining coefficients, etc. There will also be errors made in characterizing the geometry and in formulating the mathematical model. In most standard problems the errors made will induce little error in the solution itself, but for ill-posed problems in partial differential equations this is no longer true.

Throughout this paper we shall assume that a "solution" to the problem under consideration exists in some accepted sense, but in the case of ill-posed problems such a "solution" will invariably fail to depend continuously on the data and geometry. We must appropriately constrain the solution in order to recover the continuous dependence (see [8]); however, appropriate restrictions are often difficult to determine. In the first place any such constraint must be both mathematically and physically realizable. At the same time a given constraint must simultaneously stabilize against all possible errors that may be made in setting up the mathematical model of the physical problem. Since a constraint restriction has the effect of making an otherwise linear problem nonlinear, one must use care in treating the various errors separately and superposing the effects. In any case a constraint restriction which stabilizes the problem against errors in one type of data might not at the same time stabilize the problem against errors in other types of data, errors in geometry or errors made in setting up the model equation.

Although we are concerned here primarily with the question of errors in geometry we must point out that modeling errors are likely to be more serious, simply because it is impossible to characterize modeling errors precisely. Hence it is never certain that a constraint requirement which stabilizes the problem against other sources of error will also stabilize it against modeling errors. This means that predictions based on results for the model problem will have to be verified in the physical context which is being modeled before these predictions can be considered reliable.

In order to keep the arguments from becoming excessively involved we shall not in our study of the question of continuous dependence on geometry simultaneously try to deal with all other types of errors that may have been introduced in setting up our mathematical models. Except for modeling errors we could in fact deal with these other error sources, but in the interests of simplicity we consider only special types of errors. As reference to previous work on continuous dependence on geometry we mention the paper of Crooke and Payne [6] and the recent results of Persens [12]. Papers that might be thought of as investigations of continuous dependence on modeling include those of Payne and Sather [11], Adelson [1,2], Ames [3,4] and Bennett [5]. These latter papers are merely illustrative examples which indicate that in some cases the types of constraint restriction used to stabilize problems against errors in Cauchy data may not in fact stabilize against errors in modeling.

Since there is no general theory for handling ill-posed problems for partial differential equations we consider here a number of relatively simple special examples. By way of comparison we also give a few specific related examples of well posed problems subject to errors in modeling and geometry. Also since we are interested primarily in techniques, we shall assume throughout that classical "solutions" exist. Extensions of these results to weak solutions will be obvious.

In the subsequent text we shall adopt the convention of summing over repeated indices, and a comma will denote differentiation.

II GEOMETRIC AND MODELING ERRORS IN STANDARD PROBLEMS

In this section a few illustrative examples are given which demonstrate the effects of geometric and modeling errors in a number of well-posed problems.

A. Modeling errors for the Navier-Stokes equations

Let us suppose that the problem we wish to solve is the following: We seek a vector u_i, $i = 1,2,3$, which satisfies

$$\left.\begin{aligned} \frac{\partial u_i}{\partial t} + u_j\, u_{i,j} &= \nu\Delta u_i + p_{,i} \\ u_{j,j} &= 0 \end{aligned}\right\} \quad \text{in } D\times(0,T), \tag{2.1}$$

$$u_i = 0 \quad \text{on } \partial D\times[0,T),$$

$$u_i(x,0) = \epsilon g_i(x), \qquad x \in D,$$

where ν is a physical constant, p is an unknown scalar (the pressure), g_i is prescribed, and D is a bounded region in $\mathbb{R}^3$ with smooth boundary ∂D. If $\epsilon <<< 1$ we would like to know whether the solution of (2.1) is approximated well enough by ϵv_i, where v_i is a solution of

$$\left.\begin{aligned} \frac{\partial v_i}{\partial t} &= \nu\Delta v_i + q_{,i} \\ v_{j,j} &= 0 \end{aligned}\right\} \quad \text{in } D\times(0,T), \tag{2.2}$$

$$v_i = 0 \quad \text{on } \partial D\times[0,T),$$

$$v_i(x,0) = g_i(x), \quad x \in D.$$

As mentioned earlier it is assumed that g, ν, and D are such that classical solutions of both problems exist. We make the assumption that

$$\int_D \left\{ |g|^2 + |\nabla g|^2 \right\} dx \leq M^2. \tag{2.3}$$

To compare u_i and ϵv_i set

$$w_i = u_i - \epsilon v_i, \tag{2.4}$$

and let

$$\phi(t) = \int_D |w(x,t)|^2 dx \equiv \|w(t)\|^2. \tag{2.5}$$

Clearly

$$\frac{d\phi}{dt} = 2\left(w, \frac{\partial w}{\partial t}\right)$$

$$= 2\int w_i\left\{\nu\Delta w_i + (p - \epsilon q)_{,i} - u_j u_{i,j}\right\}dx$$

$$= -2\nu\int_D w_{i,j}w_{i,j}dx - 2\epsilon\int_D w_i u_j v_{i,j}dx$$

$$= -2\nu\int_D w_{i,j}w_{i,j}dx - 2\epsilon\int_D w_i w_j v_{i,j}dx - 2\epsilon^2\int_D w_i v_j v_{i,j}dx. \quad (2.6)$$

Here we have carried out the obvious integration by parts and application of differential equations and boundary conditions.

We now introduce the Sobolev constant γ defined as

$$\gamma = \inf_{\psi_i \in H_1^0(D)} \frac{\|\psi\|\ \|\nabla\psi\|^3}{\int_D[\psi_i\psi_i]^2dx} \quad (2.7)$$

which, upon use of Schwarz's inequality in (2.6), yields

$$\frac{d\phi}{dt} \leq -2\nu\|\nabla w\|^2 + 2\epsilon\gamma^{-1/2}\|\nabla w\|^{3/2}\ \|w\|^{1/2}\|\nabla v\|$$

$$+ 2\epsilon^2\gamma^{-1/2}\|\nabla w\|^{3/4}\ \|w\|^{1/4}\ \|\nabla v\|^{7/4}\|v\|^{1/4}. \quad (2.8)$$

To bound $\|\nabla v\|$ note that

$$\frac{d}{dt}\int_D v_{i,j}v_{i,j}dx = 2\int_D v_{i,j}v_{i,jt}dx$$

$$= -\frac{2}{\nu}\int_D v_{i,t}v_{i,t}dx. \quad (2.9)$$

An application of Schwarz's inequality on the right then leads to

$$\frac{d}{dt}\int_D v_{i,j}v_{i,j}dx \leq -\frac{2}{\nu}\frac{\left[\int_D v_i v_{i,t}dx\right]^2}{\|v(t)\|^2}$$

$$\leq -\frac{2\lambda_1}{\nu}\int_D v_{i,j}v_{i,j}dx \quad (2.10)$$

where λ_1 is the first eigenvalue of

$$\begin{aligned} \Delta u + \lambda u &= 0 \quad \text{in } D \\ u &= 0 \quad \text{on } \partial D. \end{aligned} \tag{2.11}$$

An integration of (2.10) yields

$$\|\nabla v\|^2 \leq \|\nabla g\|^2 e^{-2\lambda_1 t}. \tag{2.12}$$

In a similar way it can be shown that

$$\|v\|^2 \leq \|g\|^2 e^{-2\lambda_1 t}. \tag{2.13}$$

Returning to (2.8), we now have

$$\begin{aligned} \frac{d\phi}{dt} \leq &- 2\nu \|\nabla w\|^2 + 2\epsilon\gamma^{-1/2}\lambda_1^{-1/4} \|\nabla g\| \|\nabla w\|^2 e^{-\lambda_1 t} \\ &+ 2\epsilon^2\gamma^{-1/2}\lambda_1^{1/8} \|g\|^{1/4} \|\nabla g\|^{7/4} \|\nabla w\|^2 e^{-2\lambda_1 t}, \end{aligned} \tag{2.14}$$

or

$$\begin{aligned} \frac{d\phi}{dt} \leq &- 2(\nu - \epsilon\gamma^{-1/2}\lambda_1^{-1/4} \|\nabla g\| e^{-\lambda_1 t} - \beta) \|\nabla w\|^2 \\ &+ \frac{\gamma^{-1/2}\epsilon^4\lambda_1^{1/2}}{2\beta} \|g\|^{1/2} \|\nabla g\|^{7/2} e^{-4\lambda_1 t} \end{aligned} \tag{2.15}$$

for any positive constant β. Now assuming that

$$\epsilon \leq \lambda_1^{1/4}\nu/(\gamma^{-1/2}M) \tag{2.16}$$

we choose β as

$$\beta = \frac{1}{2}(\nu - \epsilon\gamma^{-1/2}M\lambda_1^{-1/4}) \tag{2.17}$$

and obtain

$$\frac{d\phi}{dt} \leq - 2\beta\lambda_1\phi + \left[\frac{\epsilon^4}{\nu\lambda_1^{1/4} - \epsilon\gamma^{-1/2}M}\right] M^4 e^{-4\lambda_1 t}. \tag{2.18}$$

An integration yields

$$\phi(t) \leq \left[\frac{\epsilon^4 M^4}{\nu\lambda_1^{1/4} - \epsilon\gamma^{-1/2}M}\right] \cdot \begin{cases} \dfrac{|e^{-4\lambda_1 t} - e^{-2\beta\lambda_1 t}|}{2\lambda_1|\beta - 2|}, & \beta \neq 2 \\ te^{-4\lambda_1 t}, & \beta = 2 \end{cases} \tag{2.19}$$

which shows that $\phi(t)$ is the product of a term $O(\epsilon^4)$ and a decaying exponential in time. Inequality (2.16) is of course a Reynold's Number hypothesis. Using the triangle inequality it follows that

$$\|v(t)\| - \|\phi(t)\| \leq \|u(t)\| \leq \|v(t)\| + \|\phi(t)\|, \tag{2.20}$$

and an application of (2.19) on the left and right thus leads to upper and lower bounds for $\|u(t)\|$. One could have obtained a slightly different result for (2.19) by first integrating by parts the last term of (2.6). However, the order of ϵ would have been the same.

B. Errors in initial geometry for the Heat Equation

In this problem we assume that not all Cauchy data were measured precisely at time $t = 0$, but that in fact the data were measured along a surface $t = \epsilon f(x)$ in (x,t)-space. If, for instance,

$$|f(x)| < 1, \tag{2.21}$$

is it possible to solve the problem with the measured data prescribed on $t = 0$ and thus obtain a close approximation to the actual physical problem being modeled?

To make this precise let u be a solution of

$$\begin{aligned} \frac{\partial u}{\partial t} - \Delta u &= 0 \qquad \text{in } D\times(0,T) \\ u &= 0 \qquad \text{on } \partial D\times[0,T) \\ u(x,0) &= u_0(x) \qquad \text{in } D, \end{aligned} \tag{2.22}$$

where D is a bounded region in $\mathbb{R}^N$ with smooth boundary ∂D. The problem we should be solving asks for the solution v of

$$\begin{aligned} \frac{\partial v}{\partial t} - \Delta v &= 0 \qquad f(x) < t < T, \qquad x \in D \\ v &= 0 \qquad f(x) \leq t < T, \qquad x \in \partial D \\ v(x,\epsilon f(x)) &= u_0(x), \qquad x \in D. \end{aligned} \tag{2.23}$$

The first potential difficulty arises from the fact that $t = \epsilon f(x)$ is not a characteristic surface and thus the problem (2.23) for v is not a standard well-posed problem. Nevertheless we wish to compare the solution v of (2.23) (assumed to exist) with the solution u of (2.22).

Again set

$$w = (u - v) \tag{2.24}$$

and observe that from standard properties of solutions of initial-boundary value problems for the heat equation it follows that for $\epsilon \leq t < T$,

$$\|w(t)\|^2 \leq \|w(\epsilon)\|^2 e^{-2\lambda_1(t-\epsilon)}. \tag{2.25}$$

We, therefore, need to bound $\|w(\epsilon)\|$ in terms of the data.

Now define

$$\tilde{u}(x,t) = \begin{cases} u(x,t), & t > 0 \\ u_0(x), & t \leq 0, \end{cases} \tag{2.26}$$

$$\tilde{v}(x,t) = \begin{cases} v(x,t), & t > \epsilon f(x) \\ u_0(x), & t \leq \epsilon f(x). \end{cases} \tag{2.27}$$

Then

$$\|w(\epsilon)\|^2 = 2\int_{-\epsilon}^{\epsilon}\int_D \tilde{w}\tilde{w}_{,\eta}dxd\eta \leq \frac{8\epsilon}{\pi}\int_{-\epsilon}^{\epsilon}\|w_{,\eta}\|^2 d\eta. \tag{2.28}$$

Using the arithmetic-geometric mean inequality we have

$$\|w(\epsilon)\|^2 \leq \frac{16\epsilon}{\pi}\left\{\int_0^{\epsilon}\|u_{,\eta}\|^2 d\eta + \int_D\int_{\epsilon f(x)}^{\epsilon}[v_{,\eta}]^2 d\eta dx\right\}. \tag{2.29}$$

Now

$$\int_0^{\epsilon}\|u_{,\eta}\|^2 d\eta = \int_0^{\epsilon}(u_{,\eta}\Delta u)d\eta$$

$$\leq \frac{1}{2}\int_D |\nabla u_0(x)|^2 dx, \tag{2.30}$$

and

$$\int_D \int_{\epsilon f(x)}^{\epsilon} (v_{,\eta})^2 d\eta dx = \int_D \int_{\epsilon f(x)}^{\epsilon} v_{,\eta} \Delta v \, d\eta dx \tag{2.31}$$

$$\leq \int_\Sigma \left\{ v_{,t} v_{,i} n_i - \frac{1}{2} v_{,i} v_{,i} n_t \right\} ds,$$

where Σ is the surface $t = \epsilon f(x)$, $x \in D$. We may rewrite (2.31) as

$$\int_D \int_{\epsilon f(x)}^{\epsilon} (v_{,\eta})^2 d\eta dx \leq \int_\Sigma v_{,i} \left[v_{,t} n_i - v_{,i} n_t \right] ds + \frac{1}{2} \int_\Sigma v_{,i} v_{,i} n_t ds. \tag{2.32}$$

Now assume that

$$n_t < - \delta \quad \text{on } \Sigma \tag{2.33}$$

for some positive δ. Then (2.32) may be rewritten as

$$\int_D \int_{\epsilon f(x)}^{\epsilon} (v_{,\eta})^2 d\eta dx \leq \int_\Sigma v_{,i} \left[\left[n_i \frac{\partial}{\partial t} - n_t \frac{\partial}{\partial x_i} \right] u_o(x) \right] ds - \frac{\delta}{2} \int_\Sigma v_{,i} v_{,i} ds$$

$$\leq \frac{1}{2\delta} \int_\Sigma \left[n_i u_{o,t} - n_t u_{o,i} \right] \left[n_i u_{o,t} - n_t u_{o,i} \right] ds. \tag{2.34}$$

Inserting (2.30) and (2.34) into (2.29) we find that

$$\| w(\epsilon) \|^2 \leq \tilde{M}^2 \epsilon, \tag{2.35}$$

where $\tilde{M}^2$ is the indicated bound on the data. Thus for $t \geq \epsilon$ it follows that

$$\| w(t) \| \leq \tilde{M}^2 \epsilon e^{-2\lambda_1 (t - \epsilon)} \tag{2.36}$$

a result which clearly implies continuous dependence on the initial geometry.

No analogous result for the Navier-Stokes equations has been derived.

C. Errors in spatial geometry for the Heat Equation

In this example it is assumed that some error was made in characterizing the spatial geometry and as a result one looks for the solution u of

$$\begin{aligned} \frac{\partial u}{\partial t} &= \Delta u \quad \text{in } D_1 x(0,T) \\ u &= 0 \quad \text{on } \partial D_1 x[0,T) \\ u(x,0) &= g(x) \quad \text{in } D_1, \end{aligned} \tag{2.37}$$

instead of the solution v of the actual model problem

$$\begin{aligned} \frac{\partial v}{\partial t} &= \Delta v \quad \text{in } D_2 x(0,T) \\ v &= 0 \quad \text{on } \partial D_2 x[0,T) \\ v_2(x,0) &= \tilde{g}(x) \text{ in } D_2. \end{aligned} \tag{2.38}$$

For simplicity of presentation we assume that $g(x)$ and $\tilde{g}(x)$ are identical on $D = D_1 \cap D_2$.

Setting

$$w = u_1 - u_2 \tag{2.39}$$

and restricting the problem to D we have from standard a priori inequalities that

$$\|w(t)\|^2 \leq k\int_0^t \oint_{\partial D} \left[w(x,\eta)\right]^2 ds d\eta, \tag{2.40}$$

for some computable constant k depending only on the geometry of D. Here the norm is the ordinary L_2 norm over the domain D. Now Crooke and Payne [6] showed that

$$\int_0^t \oint_{\partial D} \left[w(x,\eta)\right]^2 dx d\eta \leq k_1 \delta \left[\int_0^t \int_{D_1} |\nabla u|^2 dx d\eta + \int_0^t \int_{D_2} |\nabla u|^2 dx d\eta\right] \tag{2.41}$$

for a computable constant k_1. Here δ is essentially the maximum distance to the boundary in $(D_1 \cup D_2) \cap D^C$. But from our previous computations (2.12) we now know that if $\tilde{\lambda}_1$ is the first eigenvalue for $D_1 \cup D_2$ then

$$\|w(t)\|^2 \leq k k_1 \delta \hat{M}^2 e^{-2\tilde{\lambda}_1 t} \tag{2.42}$$

where $\hat{M}$ represents the indicated bound on the initial data.

We have presented three illustrative examples of the effects of modeling errors and geometric errors on the solutions of well-posed problems. In the next section we shall see that the behavior is radically different for analagous ill-posed Cauchy problems.

III STABILIZING AGAINST GEOMETRICAL ERRORS IN ILL POSED PROBLEMS

In the previous section we gave examples of two different types of errors in geometry, i.e. errors in the time geometry and errors in the spatial geometry. For ill-posed problems very little work has appeared in the literature. The first paper to study the question of stabilizing an ill-posed problem against errors in the initial time geometry was that of Knops and Payne [7] who investigated the question in the context of classical elastodynamics. Analogous results for the backward heat equation were announced in [10].

The first paper dealing with the stabilization of ill-posed problems against errors in the spatial geometry was that of Crooke and Payne [6] who derived the appropriate stabilization inequalities for the initial-boundary value problem for the backward heat equation with Dirichlet boundary conditions. Other problems have subsequently been investigated by Persens [12].

We first present a sharpened version of the result of Knops and Payne in a special case.

A. Continuous Dependence on Initial Geometry

As in Problem B of Section 2 we consider a problem in which initial data were actually taken on the surface $t = \epsilon f(x)$ where $|f(x)| < 1$. The problems whose solutions we now wish to compare are

$$\rho(x)\frac{\partial^2 u_i}{\partial t^2} = (c_{ijkl}\, u_{k,l})_{,j} \qquad \text{in } D\times(0,T)$$

$$u_i = 0 \qquad \text{on } \partial D\times(0,T) \tag{3.1}$$

$$u_i(x,0) = g_i(x), \quad \frac{\partial u_i}{\partial t}(x) = h_i(x), \quad x \in D$$

and

$$\rho(x)\,\frac{\partial^2 v_i}{\partial t^2} = \left[c_{ijkl}(x)v_{k,l}\right] \qquad \text{for } \epsilon f(x) < t < T, \quad x \in D$$

$$v_i = 0 \qquad \text{for } \epsilon f(x) \leq t < T,\ x \in D \qquad (3.2)$$

$$v_i(x,\epsilon f(x)) = g_i(x),\ \frac{\partial v_i}{\partial t}(x,\epsilon f(x)) = h_i(x), \qquad x \in D,$$

under the assumptions that $\rho(x) > 0$ and

$$\text{i)}\quad c_{ijkl} = c_{klij}, \qquad (3.3)$$
$$\text{ii)}\quad c_{ijkl}\,\psi_{ij}\psi_{kl} \leq 0,$$

for all tensors ψ_{ij} i.e. symmetry and negative semidefiniteness of the strain energy. As the constraint assumption we prescribe that

$$\int_0^T\!\!\int\left[\rho u_{i,t}u_{i,t}\right]dxdt + \int_D\int_{\epsilon f}^T\left[\rho v_{i,t}v_{i,t}\right]dxdt \leq M^2. \qquad (3.4)$$

Using (3.4) we now derive a continuous dependence inequality for the L_2 integral of $w_i \equiv u_i - v_i$.

A straight forward application of the Lagrange identity method (see [8]) leads, for $t < \frac{T}{2}$, to

$$\int_D \rho w_i(x,t)w_i(x,t)dx \leq \frac{1}{2}\int_D \rho\left[w_i(x,\epsilon)w_i(x,\epsilon) + w_i(x,\epsilon)w_i(x,2t-\epsilon)\right]dx$$
$$+\frac{1}{2}\int_\epsilon^{2t-\epsilon}\int_D \rho w_{i,\eta}(x,\eta)w_i(x,\eta)dxd\eta. \qquad (3.5)$$

Now observe that

$$\int_D \rho\left[w_i(x,2t-\epsilon)w_i(x,2t-\epsilon)\right]dx = \int_D \rho w_i(x,\epsilon)w_i(x,\epsilon)dx$$
$$+ 2\int_\epsilon^{2t-\epsilon}\int_D \rho w_i(x,\eta)\,\frac{\partial}{\partial t}\,w_i(x,\eta)dxd\eta. \qquad (3.6)$$

Using the fact that w_i vanishes on ∂D, the Poincare inequality for such functions and (3.4), we easily conclude that the last integral in

(3.6) is bounded by a multiple of M. Similarly

$$\int_{\epsilon}^{2t-\epsilon}\int_D \rho w_i(w,\eta)w_i(x,\eta)dxd\eta \leq CM^2 \tag{3.7}$$

Thus from (3.5) it follows after a use of Schwarz's inequality that

$$\int_D \rho w_i(x,t)w_i(x,t)dx \leq \int_D \rho w_i(x,\epsilon)w_i(x,\epsilon)dx + C_1M\left\{\int_D \rho w_i(x,\epsilon)w_i(x,\epsilon)dx\right\}^{1/2}$$
$$+ C_2Mt\left\{\int_D \rho w_{i,t}(x,\epsilon)w_{i,t}(x,\epsilon)dx\right\}^{1/2}. \tag{3.8}$$

Let us now continue u_i and v_i as follows:

$$\tilde{u}_i = \begin{cases} u_i(x,t), & 0 < t < T, \quad x \in D, \\ g_i(x), & t \leq 0, \quad x \in D, \end{cases} \tag{3.9}$$

$$\tilde{v}_i = \begin{cases} v_i(x,t), & \epsilon f(x) < t < T, \quad x \in D, \\ g_i(x), & t \leq \epsilon f(x). \quad x \in D. \end{cases} \tag{3.10}$$

Then for x fixed, $\tilde{w}_i = \tilde{u}_i - \tilde{v}_i$ is a continuous function of t and at $t = \epsilon$, $w_i \equiv \tilde{w}_i$. Thus

$$\int_D \rho(x)w_i(x,\epsilon)w_i(x,\epsilon)dx = 2\int_{-\epsilon}^{\epsilon}\int_D \rho\tilde{w}_i(x,\eta)\tilde{w}_{i,\eta}(x,\eta)dxd\eta$$

$$\leq \frac{8\epsilon}{\pi}\int_{-\epsilon}^{\epsilon}\int_D \rho\tilde{w}_{i,\eta}(x,\eta)\tilde{w}_{i,\eta}(x,\eta)dxd\eta \tag{3.11}$$

$$\leq \frac{16\epsilon}{\pi}\left\{\int_0^{\epsilon}\int_D \rho u_{i,\eta}u_{i,\eta}dxd\eta + \int_D\int_{\epsilon f(x)}^{\epsilon} \rho v_{i,\eta}v_{i,\eta}dxd\eta\right\}.$$

Here the arguments are the same as in (2.28).

Clearly, the integrals in braces may be bounded in terms of M^2 so

$$\int_D \rho(x)w_i(x,\epsilon)w_i(x,\epsilon)dx \leq \frac{16\epsilon M^2}{\pi}. \tag{3.12}$$

We must now bound the last term in (3.8). Proceeding as before, defining continuations as in (3.9) and (3.10) for $u_{i,t}$ and $v_{i,t}$ we obtain

$$\int_D \rho w_{i,t}(x,\epsilon)w_{i,t}(x,\epsilon)dx \leq \frac{16\epsilon}{\pi}\left\{\int_0^\epsilon\int_D \rho u_{i,\eta\eta}u_{i,\eta\eta}dxd\eta + \iint_{D\,\epsilon f}^{\epsilon} \rho v_{i,\eta\eta}v_{i,\eta\eta}dxd\eta\right\}. \tag{3.13}$$

To bound the second integral on the right observe that

$$\begin{aligned}\iint_{D\,\epsilon f}^{\epsilon} \rho v_{i,\eta\eta}v_{i,\eta\eta}dxd\eta \leq & \iint_{D\,\epsilon f}^{T}\left(\frac{T-\eta}{T-\epsilon}\right)^2\left[\rho v_{i,\eta\eta}v_{i,\eta\eta} - c_{ijkl}v_{i,j\eta}v_{k,l\eta}\right]dxd\eta \\ & - \left(\frac{T+\epsilon}{T-\epsilon}\right)^2\iint_{\Sigma}\left[\rho v_{i,\eta}v_{i,\eta\eta}n_t - c_{ijkl}v_{i,\eta}v_{k,l\eta}n_j\right]ds \\ & + \frac{2}{(T-\epsilon)^2}\iint_{D\,\epsilon f}^{T}(T-\eta)\rho v_{i,\eta}v_{i,\eta\eta}dxd\eta \end{aligned} \tag{3.14}$$

where Σ designates the surface $t = \epsilon f(x)$.

Using the arithmetic-geometric mean inequality we have

$$\iint_{D\epsilon f}^{T}\left(\frac{T-\eta}{T-\epsilon}\right)^2\left[\rho v_{i,\eta}v_{i,\eta} - c_{ijkl}v_{i,j\eta}v_{k,l\eta}\right]dxd\eta \tag{3.15}$$

$$\begin{aligned}&\leq \frac{4}{(T-\epsilon)^2}\iint_{D\,\epsilon f}^{T}\rho v_{i,\eta}v_{i,\eta}\,dxd\eta + 2\frac{(T+\epsilon)^2}{(T-\epsilon)^2}\int_\Sigma\Big[\rho v_{i,\eta}v_{i.\eta\eta}n_t \\ &\qquad\qquad - c_{ijkl}v_{i,\eta}v_{k,l\eta}n_j\Big]dx \\ &\leq \frac{4}{(T-\epsilon)^2}M^2 + 2\frac{(T+\epsilon)^2}{(T-\epsilon)^2}\int_\Sigma^{2} c_{ijkl}v_{i,\eta}\left\{v_{k,kl}n_t - v_{k,l\eta}n_j\right\}ds \\ &\qquad + 2\frac{(T+\epsilon)^2}{(T-\epsilon)^2}\int_\Sigma c_{ijkl,j}v_{k,l}v_{i,\eta}ds.\end{aligned}$$

But clearly the last two terms are data terms. Thus we conclude, after a similar argument involving the first integral on the right of (3.13), that

$$\int_D \rho w_{i,t}(x,\epsilon)w_{i,t}(x,\epsilon)dx \leq C_2M^2\epsilon. \tag{3.16}$$

Combining this with (3.12) and inserting into (3.8) we obtain finally, for $t \leq T/2$,

$$\int_D \rho w_i(x,t) w_i(x,t) dx \leq CM^2 \epsilon^{1/2}, \tag{3.17}$$

which clearly displays the Hölder continuous dependence on the data for solutions constrained by (3.4).

With a more careful analysis we could have established a result of type (3.17) requiring only L_2 bounds of u_i and v_i over space time. Note that we have tacitly assumed enough smoothness of the surface $t = \epsilon f(x)$ and compatibility of the data on ∂D at $t = 0$ to ensure that the differential equation is satisfied on the initial data surfaces.

B. Continuous Dependence on Spatial Geometry

In this section are listed some recent results of Crooke and Payne [6] and Persens [12].

i) Backward heat equation with Dirichlet conditions

The problem considered by Crooke and Payne asked for the comparison of solutions of initial-boundary value problems (with Dirichlet boundary conditions) for two neighboring domains D_1 and D_2 that did not vary with time. Specifically they sought to compare solutions of the two problems defined by

$$\begin{aligned} \frac{\partial u_\alpha}{\partial t} + \Delta u_\alpha &= 0 \qquad \text{in } D_\alpha \times (0,T), \\ u_\alpha &= f_\alpha \qquad \text{on } \partial D_\alpha \times [0,T), \\ u_\alpha(x,0) &= g_\alpha \qquad x \in D_\alpha, \end{aligned} \tag{3.18}$$

for $\alpha = 1,2$. The L_2 constraint imposed by the authors was that

$$\max_{0 \leq t \leq T} \left\{ \int_{D_1} u_1^2 dx + \int_{D_2} u_2^2 dx \right\} \leq M^2. \tag{3.19}$$

Under this constraint they showed that for

$$D = D_1 \cap D_2 \tag{3.20}$$

$$\|u_1 - u_2\|_D^4 \leq k_1 M^2 \|g_1 - g_2\|_D^2 + k_2 M^2 \int_0^T \|f_1 - f_2\|_{\partial D_1}^2 d\eta + k_3 M^4 \delta, \tag{3.21}$$

where $\| \ \|_D$ and $\| \ \|_{\partial D_1}$ denote L_2 norms over D and ∂D_1 respectively, the norm on the left being taken at time t for $t \leq T/4$. Here δ is a measure of the maximum distance to the boundary in $(D_1 \cup D_2) \cap D^C$ and $\hat{f}_2$ indicates a well defined extension of the data f_2 (defined on ∂D_2) to the boundary ∂D_1. The details of the establishment of (3.21) are quite complicated and therefore not reproduced here. This result was extended by Persens [12] to the case of domains D_1 and D_2 that vary with t. The results were essentially the same as in the previous case with a slightly modified definition of some of the terms.

ii) Backward heat equation with Neumann conditions

In this problem analyzed by Persens [12] he considered instead of (3.18)

$$\frac{\partial u_\alpha}{\partial t} + \Delta u_\alpha = 0 \quad \text{in } D_\alpha x(0,t),$$
$$\frac{\partial u_\alpha}{\partial \nu_\alpha} = f_\alpha \quad \text{on } \partial D_\alpha x[0,T), \tag{3.22}$$
$$u_\alpha(x,0) = g_\alpha(x) \quad \text{in } D_\alpha,$$

for $\alpha = 1,2$, and D_α not varying with time. In this problem, by employing the constraint

$$\int_0^T \left\{ \int_{D_1} u_1^2 dx + \int_{D_2} u_2^2 dx \right\} d\eta \leq M^2, \tag{3.23}$$

he obtained the stability result

$$\int_0^T \|u_1 - u_2\|_D^2 d\eta \leq k_1 M \|g_1 - g_2\|_D + k_2 M \left[\int_0^{4t} \|f_1 - \hat{f}_2\|_{\partial D_1}^2 d\eta \right]^{½} + k_3 M^2 \delta + k_4 M^2 \sigma, \tag{3.24}$$

for $0 \leq t \leq T/4$. Here the notation is as before with the addition of the term σ which is an explicit measure of the maximum deviation between the outward normal directions at points on ∂D_1 and associated points on ∂D_2. The quantity σ will be small if this maximum deviation is small.

iii) The Cauchy problem for the Poisson equation

Here the comparison was between solutions of

$$\Delta u_\alpha = F_\alpha \text{ in } D_\alpha,$$
$$u_\alpha = f_\alpha \text{ on } \Sigma_\alpha, \tag{3.25}$$
$$\frac{\partial u_\alpha}{\partial \nu_\alpha} = g_\alpha \text{ on } \Sigma_\alpha, \quad \alpha = 1,2,$$

where Σ_α is a smooth portion of ∂D_α and Σ_1 is assumed to be a small perturbation of Σ_2. No data are given on the remainder of ∂D_2. As is usual in problems of this type one defines

$$D = D_1 \cap D_2 \tag{3.26}$$

and then defines subdomains D^β of D by a family of appropriately chosen surfaces $p(x) = \beta$. Using the constraint

$$\int_{D_1} u_1^2 dx + \int_{D_2} u_2^2 dx \leq M^2. \tag{3.27}$$

Persens obtained the continuous dependence inequality

$$\|u_1 - u_2\|^2_{D^\beta} \leq k_1 M^{2\nu(\beta)} \Big[a_1 \|f_1 - \hat{f}_2\|^2_{\Sigma_1} + a_2 \|\text{grad}_S (f_1 - \hat{f}_2)\|^2_{\Sigma_1}$$
$$+ a_3 \|g_1 - \hat{g}_2\|^2_{\Sigma_1} + a_4 M^2 \delta^{1/2}\Big]^{1-\nu(\beta)}, \tag{3.28}$$

where the various terms are defined as before and $\nu(\beta)$ is an explicit function of β which satisfies for $0 \leq \beta < 1$,

$$0 \leq \nu(\beta) < 1. \tag{3.29}$$

Persens [12] also considered the problem of continuous dependence on geometry for the Dirichlet initial boundary value problem of linear elastodynamics without a definiteness assumption on the elasticities. In this problem the constraint restriction was more severe, but results were quite similar to those obtained by Crooke and Payne [6] for the backward heat equation.

IV CONTINUOUS DEPENDENCE ON MODELING: SOME EXAMPLES

As mentioned earlier it is impossible to characterize the error made in setting up a mathematical model of a physical problem. For ill-posed problems we know that any error in modeling may lead to instabilities unless "solutions" are adequately constrained. Since the modeling errors are unknown we can only give examples and determine constraints sufficient to ensure stability against the modeling errors indicated in the examples.

The first study of this type of modeling error in an ill-posed problem was perhaps that of Payne and Sather [11] who compared the solution of the backward heat equation with that of a singularly perturbed well-posed hyperbolic problem. A few years later Adelson [1,2] considered a number of quasilinear Cauchy problems for elliptic systems of which the following is typical. One wishes to compare an appropriately constrained solution v of

$$\left.\begin{aligned} b\epsilon\Delta v + v &= u \\ \Delta u &= 0 \end{aligned}\right\} \text{ in } D \subset \mathbb{R}^n$$

$$\left.\begin{aligned} u = f,\ \operatorname{grad} u &= \underset{\sim}{g} \\ v = \hat{f},\ \operatorname{grad} v &= \underset{\sim}{\hat{g}} \end{aligned}\right\} \text{ on } \Sigma \tag{4.1}$$

where the Cauchy surface Σ is as in (3.25), ϵ is a small positive number and b is a constant (positive or negative), with an appropriately constrained solution w of

$$\begin{aligned} \Delta w &= 0 \text{ in } D \\ \left.\begin{aligned} w &= \underset{\sim}{f} \\ \operatorname{grad} w &= \underset{\sim}{g} \end{aligned}\right\} &\text{ on } \Sigma \end{aligned} \tag{4.2}$$

assuming of course that the data of w are close to that of v. Using L_2 constraints, Adelson obtained comparison results similar to (3.28) in the case $b < 0$. (Here of course $\delta \equiv 0$). For $b > 0$ more severe constraints on v were required and the resulting continuous dependence inequality was less sharp.

The results of Ames [3,4] might also be interpreted as continuous dependence on modeling results. Ames developed comparison inequalities relating, for instance, solutions of the Dirichlet initial-boundary value problem for the backward heat equation with solutions of various related well posed singular perturbation problems introduced when employing the quasireversibility method for stabilizing ill-posed problems.

Finally, we mention some recent results of Bennett [5].

i) Cauchy problem for the minimal surface equation

Here one is concerned with the solution u of the problem

$$\left[\left[1+|\nabla u|^2\right]^{-\frac{1}{2}}u_{,j}\right]_{,j}=0 \quad \text{in } D\subset \mathbb{R}^n,$$

$$\left.\begin{aligned} u &= \epsilon f(x) \\ \operatorname{grad}u &= \epsilon \underset{\sim}{g}(x)\end{aligned}\right\} \text{ on } \Sigma, \tag{4.3}$$

where Σ is defined as before to be a smooth portion of ∂D. Bennett compared the solution of this problem with that of

$$\Delta h = 0 \text{ in } D$$

$$\left.\begin{aligned} h &= \hat{f}(x) \\ \operatorname{grad}h &= \hat{\underset{\sim}{g}}(x)\end{aligned}\right\} \text{ on } \Sigma, \tag{4.4}$$

assuming

$$\int_\Sigma \left\{|f-\hat{f}|^2+|g-\hat{g}|^2\right\}ds \leq K\epsilon^\gamma \tag{4.5}$$

for some positive γ. Setting

$$w = u - \epsilon h \tag{4.6}$$

and imposing constraints of the form

$$\int_D |\operatorname{grad}u|^6\left[1+\epsilon^2|\operatorname{grad}u|^2\right]dx \leq M_1^2\epsilon^4 \tag{4.7}$$

and

$$\int_D \left[u^2+\epsilon^2h^2\right]dx \leq M_2^2, \tag{4.8}$$

he derived the following continuous dependence result for $\gamma = 2$

$$\int_{D^\beta} w^2dx \leq \hat{K}\epsilon^{4\nu(\beta)}, \quad 0<\beta<\beta_1<1.$$

where $\nu(\beta)$ is as in (3.28). He actually obtained results for a range of values of γ. The constant $\hat{K}$ of course depends on the M.

We note that in order to stabilize the Dirichlet initial-boundary value problem for the backward heat equation against most other sources of error it was sufficient to impose an L_2

constraint. Regarding the minimal surface equation problem as a modeling perturbation of the backward heat equation we note that here a much stronger constraint was imposed. Admittedly (4.6) is only a sufficient constraint, but it seems highly unlikely that continuous dependence could be established under a significantly weaker constraint on u.

A second problem considered by Bennett is of some interest,

ii) <u>The end problem for the one dimensional nonlinear heat equation</u>

In this example one wishes to compare the solution of

$$\frac{\partial u}{\partial t} = \frac{\partial}{\partial x}\left[\rho\left(\left[\frac{\partial u}{\partial x}\right]^2\right)\frac{\partial u}{\partial x}\right], \quad 0 < x < a, \quad t > t,$$

$$\left.\begin{aligned} u &= \epsilon f(t) \\ \frac{\partial u}{\partial x} &= \epsilon g(t) \end{aligned}\right\} \text{ on } x = 0,\ t_1 < t < t_2, \tag{4.9}$$

with that of the ill posed problem

$$\frac{\partial h}{\partial t} - \Delta h = 0, \quad 0 < x < a, \quad t > t_1$$

$$\left.\begin{aligned} h &= \hat{f}(t) \\ \frac{\partial h}{\partial x} &= \hat{g}(t) \end{aligned}\right\} \text{ on } x = 0,\ t_1 < t < t_2. \tag{4.10}$$

Here he postulated that $\rho(q^2)$ satisfy

$$1 + \left[\rho + 2q^2\rho'\right]^{-2} = O(q^2) \tag{4.11}$$

and assumed that the data satisfied for some δ $(0 \leqslant \delta < 6)$

$$\int_{t_1}^{t_2}\left\{\left[f - \hat{f}\right]^2 + \left[g - \hat{g}\right]^2 + \left[\frac{\partial f}{\partial t} - \frac{\partial \hat{f}}{\partial t}\right]^2\right\}d\tau = O(\epsilon^{4-d}). \tag{4.12}$$

The imposed constraints were

$$\int_{t_1}^{t_2}\int_0^a \left(\frac{\partial u}{\partial t}\right)^2\left(\frac{\partial u}{\partial x}\right)^4 dxdt = O(\epsilon^{6-d}) \tag{4.13}$$

and

$$\int_{t_1}^{t_2}\int_0^a\left\{u^2 + \left(\frac{\partial u}{\partial x}\right)^2 + \epsilon^2\left[h^2 + \left(\frac{\partial h}{\partial x}\right)^2\right]\right\}dxd\eta = O(1). \tag{4.14}$$

It was then possible to derive a stability inequality of the form

$$\int_{D^\beta} [u - \epsilon h]^2 dxdt = O\left[\epsilon^{(6-d)(1-\beta)}\right], \tag{4.15}$$

where D^β is the region defined by the surface $p(x) = \beta$.

We again point out the relative severity of the constraint (4.13). We remark also that with the constraint (4.14) it is possible to derive a Hölder continuous dependence result relating the L_2 integral of h over D^β to L_2 integrals of its data (see Payne [9]).

Bennett [5] considered other examples, but since they are somewhat more difficult to describe we refer the interested reader to his paper. These examples seem to indicate that modeling errors may be much more critical than other possible types of errors one might make in setting up a mathematical problem to analyze a physical situation. Consequently we must be cautious in presenting the results of an analysis of a given mathematical ill-posed problem to those who would use these results in a physical context. Because we cannot be certain that we have adequately constrained the solution of our mathematical problem against modeling errors our results for the model problem may give incorrect predictions.

BIBLIOGRAPHY

1. Adelson, L., Singular perturbations of improperly posed problems, SIAM J. Math. Anal. 4 (1973) pp. 344-366.
2. Adelson, L., Singular perturbation of an improperly posed Cauchy problem, SIAM J. Math. Anal. 5 (1974) pp. 417-424.
3. Ames, K., On the comparison of solutions of properly and improperly posed Cauchy problems for first order systems, SIAM J. Math. Anal. 13 (1982) pp. 594-606.
4. Ames, K., Comparison results for related properly and improperly posed problems for second order operator equations, J. Diff. Eqtns 44 (1982) pp. 383-399.
5. Bennett, A., Continuous dependence on modeling in the Cauchy problem for second order partial differential equations Ph.D. Dissertation, Cornell University (1986).
6. Crooke, P.S. and Payne, L.E. Continuous dependence on geometry for the backward heat equation, Math. Meth. in the Appl. Sci. 6 (1984) pp. 433-448.
7. Knops, R.J. and Payne, L.E., Continuous data dependence for the equations of classical elastodynamics, Proc Camb. Phil. Soc. 66 (1969) pp. 481-491.
8. Payne, L.E., Improperly posed problems in partial differential equations, Reg. Conf. in Appl. Math. #22 SIAM (1975).

9. Payne, L.E., Improved stability estimates for classes of improperly posed Cauchy problems, Appl. Analysis 19 (1985) pp. 63-74.

10. Payne, L.E., On stabilizing ill-posed problems against errors in modeling, Proceedings of Conference on Ill Posed and Inverse Problems, Strobl Austria (1986) (to appear)

11. Payne, L.E. and Sather, D., On singular perturbations of non-well-posed problems, Annali di Mat Pura ed Appl. 75 (1967) pp. 219-230.

12. Persens, J., On stabilizing ill-posed problems in partial differential equations under perturbations of the geometry of the domain, Ph.D. dissertation Cornell (1986).

THE APPEARANCE OF OSCILLATIONS IN OPTIMIZATION PROBLEMS

L. TARTAR
Centre d'Etudes de Limeil-Valenton
B.P. 27
94190 Villeneuve Saint Georges
FRANCE

It has been observed by Lurie (1970) that some optimization problems do not have a solution in a classical sense (for a review on this question in engineering problems the reader is referred to Armand _et al._ 1983). Approximate solutions of these problems often tend to develop progressively faster oscillations. This leads to serious difficulties in the numerical as well as in the analytic treatment of such problems. In recent years new methods, motivated by the study of composite materials, have been developed to overcome these difficulties. The aim of these lectures is to present in an informal way the main ideas and indicate their relevance to applications. For the more technical details and some of the proofs the reader is referred to Murat & Tartar (1985), Tartar (1985) and to the recent work of Kohn & Strang (1986) which contains an extensive bibliography on related questions.

1 AN OPTIMAL DESIGN PROBLEM

We start with the following problem in optimal design. Let Ω be a bounded region in the plane which is filled with two (isotropic) heat conducting materials. Given the total proportion of each material and a heat source term we want to arrange the two materials in such a way that the minimum amount of heat is kept in the region Ω.

The governing equation for this problem is the stationary heat equation.

$$- \operatorname{div}(a(x)\operatorname{grad}u) = f \text{ in } \Omega \qquad \text{(H)}$$

where $a(x)$ denotes the conductivity of the material at point x, u the temperature and f the heat source term.

Equation (H) has to be complemented by appropriate boundary conditions, like Dirichlet conditions

$$u(x) = u_0(x) \text{ on } \partial\Omega \text{ (prescribed temperature)} \tag{D}$$

or Neumann conditions

$$\frac{\partial u}{\partial n}(x) = 0 \text{ on } \partial\Omega \text{ (no heat flux through the boundary).} \tag{N}$$

In addition we suppose that a(x) takes only two values α and β corresponding to the two given materials. Furthermore the proportion of area where a(x) takes the value α is prescribed. Under these constraints we want to solve

<u>Problem 1</u> : Minimize $\int_\Omega u(x)dx$

We can also consider more general problems. If we want to maximize the rate at which heat flows out of Ω (say for $u = 0$ on $\partial\Omega$) we are led to an eigenvalue problem instead of (H). Let λ_1 be the first eigenvalue for the problem

$$-\operatorname{div}(a(x)\operatorname{grad} v) = \lambda_1 v \text{ in } \Omega \text{ ; } v = 0 \text{ on } \partial\Omega. \tag{E}$$

The optimization problem becomes

<u>Problem 2</u> : Maximize λ_1.

With the opposite goal of having the maximum amount of heat kept in the region Ω we would consider

<u>Problem 3</u> : Maximize $\int_\Omega u(x)dx$.

How can we tackle these problems numerically? To fix ideas let us consider Problem 1. One idea is to improve a given design by an iterative procedure. Given a certain arrangement of the two materials we consider a small perturbation of this configuration and calculate the derivative of $\int_\Omega u(x)dx$ with respect to the perturbation parameter. If this derivative is negative we can find a configuration with a lower value of

$\int_\Omega u(x)dx$ and repeat the procedure for the new configuration. If for all possible perturbations the derivative is nonnegative we have found a (local) minimum of $\int_\Omega u(x)dx$. This method is known as a gradient method.

The constraint requiring the proportion of the two materials to remain fixed enters by restricting the class of perturbations that we have to consider.

Applied to our problem the gradient method has two important shortcomings which severely restrict its applicability.

The first difficulty arises if we start with a design where one material occupies a connected component while the optimal design, however, requires the division of this material between several subregions. These two configurations cannot be connected by successive (small) perturbations and hence we cannot calculate the derivative required for the gradient method.

Since our problem involves solving the partial differential equation (H) we can numerically approximate the optimal design only on a mesh of finite size. On the other hand good approximations of the optimal design may develop a progressively finer structure. This is the second difficulty that will be discussed in more detail later.

Intuitively this phenomenon can be understood by considering the case where Ω is a circle, the heat supply is constant and the temperature is zero at the boundary. To minimize $\int_\Omega u(x)dx$ we have to make sure that the heat generated in Ω can flow easily to the boundary. One suggestion is to concentrate the good conductor in the middle with some "fingers" of material stretching towards the boundary. The important point to notice is that the design could be improved by choosing more but smaller "fingers". To understand this intuitively imagine the "good" material to be an ideal conductor. Then the temperature in the good material will be zero. The temperature at a point of the "bad" material will be the smaller the closer this point is to the "good" material. Hence choosing many "fingers" will force the temperature to be small. Therefore we expect the optimal design to have infinitely many "fingers" which is not, of course, an admissible shape, and so the design problem may have no classical solution.

Applying traditional numerical methods with mesh size 1/n leads to a solution with more fingers as n goes to infinity. Hence it is not possible to find the shape of the "fingers" numerically because they have a finer structure than the mesh. Furthermore the solutions for a mesh size tending to zero do not converge in the usual sense.

Therefore better analytical methods are needed to deal with problems of this type. It turns out that we can find an explicit analytical solution for the above problem by introducing the local proportion of the "good" conductor as a new variable. Before discussing this result in more detail we will develop the main ideas of a new approach by considering an easier one dimensional model problem which nevertheless inherits the main difficulties of the original problem.

2 AN EASY MODEL PROBLEM

We consider the following model problem

$$\text{Minimize } J(u) = \int_0^T (|y|^2 - |u|^2)dt \tag{M}$$

where y is related to the control variable u by the state equation

$$\frac{dy}{dt} = u(t) \; ; \; y(0) = 0 \tag{S}$$

and u $\quad U_{ad}$. The set U_{ad} of admissible functions reflects the constraints imposed on u. We assume

$$U_{ad} = \{u \mid u(t) \quad K, \quad t \quad (0,T)\} \tag{C}$$

and we discuss two examples.

Example 1 : $K = K_o \equiv [-1,+1]$

Example 2 : $K = K_1 \equiv \{-1,0,+1\}$

The important difference between these two examples is that in Example 1, K (and hence U_{ad}) is convex, whereas this is not the case in Example 2. Even worse, U_{ad} in Example 2 contains functions (e.g. $u_1 \equiv -1$ and $u_2 \equiv +1$) which cannot be joined by any (continuous) path which lies in U_{ad}. This is not merely an academic difference but severely restricts the applicability of the gradient method for solving the problem numerically as indicated in Section 1.

We first consider Example 1. To obtain necessary conditions for optimality we have to calculate the derivative $\frac{\partial J}{\partial u}$. Here u is not an element of R^n but of a function space, say $L^\infty(0,T)$, the space of bounded measurable functions. Therefore functional analysis is needed to give a precise meaning to $\frac{\partial J}{\partial u}$. Roughly speaking, the derivatives as usual give a relation between a small change in u and the corresponding change in J. In our example J is just a quadratic functional of u and y, where y depends linearly on u. An explicit calculation therefore gives

$$J(u + \varepsilon\delta u) = J(u) + \varepsilon\delta J(u) + O(\varepsilon^2), \tag{1}$$

where

$$\delta J(u) = 2 \int_0^T (y.\delta y - u.\delta u)dt \tag{2}$$

and

$$\frac{d(\delta y)}{dt} = \delta u \ ; \ \delta y(0) = 0. \tag{3}$$

We want to write δJ as a linear function of δu. To eliminate δy we introduce the <u>adjoint state equation</u>

$$- \frac{dp}{dt} = y \ ; \ p(T) = 0. \tag{S*}$$

A simple integration by parts shows $\int_0^T (y.\delta y)dt = \int_0^T (p.\delta u)dt$ and we obtain

$$\delta J(u) = 2 \int_0^T ((p - u).\delta u)dt. \tag{4}$$

A necessary condition for u to be a minimum is that $\delta J(u)$ is nonnegative, otherwise we can lower J as seen from (1). This conclusion, however, only holds if $u + \varepsilon\delta u$ is admissible, at least for small positive ε. Therefore the constraint (C) introduced by U_{ad} restricts the possible choices for δu in (4). More precisely we have

$$\left|\begin{array}{l} u(x) = +1 \ \text{ implies } \delta u(x) < 0, \\ u(x) = -1 \ \text{ implies } \delta u(x) > 0, \\ \delta u(x) \text{ has arbitrary sign if } -1 \quad u(x) \quad +1. \end{array}\right. \tag{5}$$

Using these admissible δu in (4) we obtain from the requirement $\delta J > 0$

$$\left|\begin{array}{l} u(x) = +1 \text{ implies } p(x) - u(x) < 0 \ \text{ i.e } p(x) < +1, \\ u(x) = -1 \text{ implies } p(x) - u(x) > 0 \ \text{ i.e } p(x) > -1, \\ -1 \quad u(x) \quad +1 \text{ implies } p(x) - u(x) = 0 \text{ i.e } p(x) = u(x). \end{array}\right. \tag{6}$$

To sum up we have shown that <u>if</u> (M) has a solution $u \quad U_{ad}$ then there exists a function p such that

$$\frac{dy}{dt} = u \ ; \ y(0) = 0, \tag{S}$$

$$-\frac{dp}{dt} = y \ ; \ p(T) = 0, \tag{S*}$$

$$u(x) \quad F(p(x)), \tag{N.C}$$

where $F(p(x))$ is the multivalued function given by (6).

We point out that there exist functions u and p such that the necessary condition (N.C) holds. We can, for example, choose $p \equiv u \equiv 0$.

To solve (M) numerically we can again use the gradient method. If an iteration step leads to u outside U_{ad} then we simply project this point onto the convex set U_{ad} and continue the iteration with the projected function. The fact that we can apply a numerical scheme and we know functions u which satisfy the necessary conditions for optimality may lead us to believe that (M) has a solution which will be approached by the numerical scheme. This is, however, false.

<u>Problem (M) has no solution</u>. To see this we show that the following assertions hold.

$$\text{(i) } J(u) \quad -T, \quad u \quad U_{ad}. \tag{7}$$

$$\text{(ii)} \left|\begin{array}{l} \text{There exists a sequence } \{u_n\}_{n \in N}, \ u_n \quad U_{ad} \\ \\ \text{such that } \lim_{n \to \infty} J(u_n) = -T. \end{array}\right. \tag{8}$$

Therefore the optimal value for $J(u)$ is $-T$ but it cannot be realized by any admissible function.

To prove (i) we first note that $J(u) = \int_o^T(|y^2|-|u|^2)dt \geq -T$ since $|u(t)| \leq 1$. If equality holds then $y(t) = 0$ and $|u(t)| = 1$ for (almost) every t $(0,T)$. This contradicts (S).

To prove (ii) we choose functions u_n with progressively faster oscillations. Let

$$\left| \begin{array}{l} u_n(t) = +1 \text{ for } t \quad (\frac{2k}{2n}T, \frac{2k+1}{2n}T), \\ \\ u_n(t) = -1 \text{ for } t \quad (\frac{2k+1}{2n}T, \frac{2k+2}{2n}T), \end{array} \right. \tag{9}$$

where k $\{0,1,\ldots,n-1\}$. The state equation (S) yields $|y_n(t)| \leq \frac{T}{2n}$ and hence $J(u_n) \leq -T + T(\frac{T}{2n})^2$ so that (ii) follows.

The reason that (M) has no solution is that the "optimal" solution would have to satisfy two incompatible conditions, namely $u = 0$ (to ensure $y = 0$) and $u^2 \equiv 1$. The approximate solutions develop fast oscillations to avoid this incompatibility. In our case we have $u_n^2 \equiv 1$ and we claim that u_n is close to zero. Obviously u_n is not close to zero in the sense that $|u_n-0|$ is small. Nevertheless the primitive y_n of u_n is small and it is y_n which enters the optimization problem. Hence we are led to measure the distance between two functions by means of their primitives rather than by the difference of their values. We define a new distance between two (bounded measurable) functions f and g by

$$d(f,g) = \sup_{0 \leq t \leq T} \left| \int_o^T (f-g)dt \right|. \tag{10}$$

By the claim that u_n is close to zero we simply mean that $d(u_n,0) \to 0$ as $n \to \infty$. The idea of using a new measure for the proximity of two functions is fundamental for many applications and the terms "weak topology" or "weak convergence" are often used in this context. We will come back to this point later on. We first want to address the existence question again. For the reader familiar with functional analysis we remark that for an equibounded sequence u_n it is equivalent to say that $d(u_n,0) \to 0$ or that u_n converges to zero in the $L^\infty(0,T)$ weak * topology.

We have seen that an optimal solution of (M) would have to satisfy the incompatible conditions $u^2 \equiv 1$ and $u \equiv 0$ and therefore does not exist. We will now show that (M) has a solution if we enlarge the space of admissible functions. The idea is to relax the relation between u and u^2. Now u^2 will be replaced by an independent variable v and instead of the equality $v = u^2$ we will only require that there exists a sequence u_n of admissible functions such that $d(u_n,u) \to 0$ and $d(u_n^2,v) \to 0$ as $n \to \infty$. It is natural to ask which pairs (u,v) are now admissible. Obviously (u,u^2) is admissible if u is. The explicit sequence u_n given by (9) leads to the (nontrivial) admissible pair (0,1). In fact there is an easy geometric description of all admissible (u,v). Define $\underline{K}_o$, the set of "classical values", by

$$\underline{K}_o = \{(x,y) \mid y = x^2,\ x \quad K_o=[-1,+1]\}$$

and let $\tilde{K}_o$ be the convex hull of $\underline{K}_o$. Then (u,v) is admissible if and only if $(u(t),v(t)) \quad \tilde{K}_o$ for all $t \quad (0,T)$. The same holds true for K_1 instead of K_o. Now we can state the

Relaxed problem

$$\text{Minimize } \tilde{J}(u,v) = \int_o^T (y^2 - v)dt \tag{$\tilde{M}$}$$

where

$$\frac{dy}{dt} = u \ ;\ y(0) = 0 \tag{S}$$

and $(u,v) \quad \tilde{U}_{ad} = \{(u,v) \mid (u(t),v(t)) \quad \tilde{K}\}$ and $\tilde{K}$ equals $\tilde{K}_o$ or $\tilde{K}_1$ in Example 1 or 2 respectively.

The relaxed problem has a solution. To prove this we note that $\tilde{J}(u,v) > -T$ (since $v < 1$) and use the fact that now $(u,v) \equiv (0,1)$ is an admissible choice for which $\tilde{J}(u,v) = -T$. We can obtain also an abstract proof by remarking that $\tilde{J}$ is a convex functional on a convex set and then using the result that (under appropriate technical hypotheses on the underlying spaces) continuous convex functionals on convex bounded closed sets attain their minimum. The (unique) solution (0,1) of ($\tilde{M}$) does not lie on the curve of classical values (u,u^2) which tells us again that the original problem (M) has no solution. Furthermore (0,1) is a convex combination of the two classical values (-1,1) and (1,1). Therefore the solution of the relaxed problem ($\tilde{M}$) is somehow a mixture of two classical values, a fact exactly reflected in the sequence u_n defined in (9) which we chose as a good approximate solution for the original problem.

Let us emphasize that enlargement of the class of admissible controls does not only enables us to prove existence but perhaps more importantly in the present context the *generalized solution* also *contains information about good approximate solutions for the original problem.*

Furthermore we are also able to *obtain better necessary conditions for optimality* for the original problem since we have enlarged the class of admissible perturbations. Without going into the details, we remark that we find exactly Pontryagin's (maximum/minimum) principle. Expressed in terms of p given by the adjoint state equation (S^*) the necessary conditions for optimality now read

$$\left| \begin{array}{l} u(x) = +1 \quad \text{implies } p(x) \quad 0 \\ u(x) = -1 \quad \text{implies } p(x) \quad 0 \\ \text{if } -1 \quad u(x) \quad +1 \text{ then no } p(x) \text{ is admissible.} \end{array} \right. \tag{11}$$

In contrast to (6) there exists no pair (u,p) satisfying (S,S^*) such that (11) can be satisfied. This is good since we know that the original problem has no solution.

Let us *summarize the main ideas* used to tackle the one dimensional model problem (M).

We enlarge the class of admissible controls by using a different notion of proximity of two functions (a different topology).

The enlarged problem turns out to be convex, and existence of a solution is easily shown.

Convexity allows us to obtain stronger necessary conditions, which in particular exclude the existence of a classical solution.

The optimal solution of the generalized problem contains information on how to find a sequence of approximate solutions to the original problem. The approximate solutions have to develop progressively faster oscillations.

For the sake of definiteness we conclude this section by stating some lemmas.

Definition A : Let Ω be a bounded set in R^N and let a_n be a sequence of measurable functions on Ω, satisfying $|a_n(x)| \leq C$. We say that a_n converges to a function a in L^∞ weak * if and only if

$$\Big| \int_\omega a_n dx \to \int_\omega a \, dx \quad \text{as } n \to \infty \tag{12}$$

for all bounded open sets ω of Ω.

Remark B : if $\Omega = (0,T)$ this is equivalent to $d(a_n,a) \to 0$.

Lemma C : Given any function $\theta : \Omega \to R$, $0 \leq \theta(x) \leq 1$, there exists a sequence χ_n of characteristic functions (i.e. measurable functions which only take the values 0 and 1) such that $\chi_n \to \theta$ in L^∞ weak *

Lemma D : Let v_n be a sequence of functions, $v_n : \Omega \to R^p$, such that $v_n(x) \quad K$ where K is a closed bounded subset of R^p. If $v_n \to v_\infty$ in L^∞ weak * then $v_\infty(x) \quad$ convK (the convex hull of K). Conversely for a given (measurable) function $v : \Omega \to R^p$, satisfying $v(x) \quad$ convK there exists a sequence of functions v_n such that $v_n(x) \quad K$ and $v_n \to v$ in L^∞ weak *.

Remark E : Lemma D implies the characterization of admissible pairs given above : set $v = (v_1,v_2)$ with $v_1 = u$ and $v_2 = u^2$.

3 A FEW GENERAL REMARKS ON MINIMIZATION PROBLEMS

The problems stated in Sections 1 and 2 can be phrased in the following general framework. Given a set Z of admissible controls z and a functional $F : Z \to R$, we want to minimize F, i.e. we want to find a $z_o \quad Z$ such that

$$F(z_o) = \inf_{z \quad Z} F(z). \tag{13}$$

In the last section we have seen that such a problem may not have a solution even when $\inf_{z \quad Z} F(z)$ is finite. There are, however, easy abstract conditions which ensure that F attains its minimum. To formulate the result we assume that Z is equipped with a notion of convergence.

Theorem F : Assume that $\inf_{z \quad Z} F(z)$ is finite and that the following conditions hold :

(i) (Compactness of Z) From any sequence z_n we can extract a subsequence $z_{n'}$, which converges to a $z \quad Z$.

(ii) (Lower-semicontinuity of F) For any sequence z_n which converges to z we have

$$F(z) \leq \liminf_{n\to\infty} F(z_n). \tag{14}$$

Then there exists $z_o \quad Z$ such that

$$F(z_o) = \inf_{z \quad Z} F(z). \tag{15}$$

The proof of this theorem is obvious. Choose a sequence z_n such that $F(z_n)$ approaches $\inf_{z \quad Z} F(z)$. By (i) we can extract a convergent subsequence which converges to a z_o. By (ii) $F(z_o) \leq \inf_{z \quad Z} F(z)$, hence equality must hold.

The condition (ii) is weaker than continuity. We only require that F does not jump upwards if we pass to the limit, whereas continuity rules out all jumps. To illustrate this, consider the example $Z = [-1,+1]$, $F_1(z) = F_2(z) = z^2$ if $z \neq 0$, $F_1(0) = -1$, $F_2(0) = 1$. Neither function is continuous; F_1 is lower-semicontinuous and attains its minimum, whereas F_2 is not lower-semicontinuous and does not attain its minimum.

To apply this theorem we have to find a notion of convergence which is well adapted to the minimization problem. To satisfy (i) it is favourable to have "many" convergent sequences whereas (ii) forces us to have not "too many" convergent sequences.

The importance of the L^∞ weak * convergence introduced earlier stems from the following compactness result.

<u>Lemma G</u> : Let $\Omega \quad R^N$ be a bounded open set. Let f_n be a sequence of functions, $f_n : \Omega \to R^p$ such that $|f_n(x)| \leq C$ for all $x \quad \Omega$ and all $n \quad N$. Then there exists a subsequence (denoted by the same symbols) such that $f_n \to f$ in L^∞ weak *.

There is no such lemma for the usual uniform convergence.

4 OPTIMAL DESIGN REVISITED

We consider again the problem

$$\text{Minimize} \int_\Omega u\,dx \tag{M'}$$

where

$$- \operatorname{div}(a(x)\operatorname{grad}u) = f \text{ in } \Omega \ ; \ u = 0 \text{ on } \partial\Omega \tag{S'}$$

and a A_{ad}. The set A_{ad} of all admissible controls consists of all measurable functions which take only the values α and β, and satisfy the additional constraint

$$\int_\Omega a dx = \gamma \tag{C'}$$

where $\alpha \text{meas}\Omega < \gamma < \beta \text{meas}\Omega$.

To choose an optimal design means to choose an optimal control a. The constraint (C') means that the total proportion of the materials α and β is given.

As indicated in the first section, this problem will, in general, have no solution. Therefore we will enlarge the class of admissible controls, in analogy to the treatment of the one dimensional problem in Section 2. Physically, this can be motivated by using composites made of the two materials to achieve the optimal design. Mathematically, the main difficulty is to define a suitable notion of convergence for conductivity tensors in order to characterize these composites.

We begin by remarking that the L^∞ weak * convergence introduced earlier is not appropriate for the controls a. Roughly speaking, this is due to the fact that we are dealing with the conductivity tensor aI (where I denotes the identity on R^2) rather than with a scalar quantity. To illustrate this point let us consider a material which consists of alternating layers of the two given materials α and β with relative proportions θ and $1-\theta$. If the width $1/n$ of the layers goes to zero we expect to obtain some averaged material.

Parallel to the layers, the conductivity will be $\theta\alpha + (1-\theta)\beta$ (addition of conductivity) whereas perpendicular to the layers it will be $(\theta\alpha^{-1} + (1-\theta)\beta^{-1})^{-1}$ (addition of resistivity). This suggests that a sequence of laminates made of two isotropic materials can approach (in a sense to be made precise later) an anisotropic material. Passing to the L^∞ weak * limit would just lead to a scalar function (in fact the constant function $\theta\alpha + (1-\theta)\beta$) falsely suggesting that the limit material is isotropic.

To study this phenomenon more precisely let us look at the temperature u_n in a cross section C of the laminate. We expect u_n to be continuous but its derivative will have jumps at the interface of two layers.

At this stage we should mention that we understand equation (S') cannot hold in the usual sense since, due to the jumps of a(x), a(x)gradu(x) will, in general, not be differentiable. We will require that (S') holds in a weaker sense, namely in the sense of distributions. Without going into the details we just remark that this implies that (S') holds in the usual sense in the parts where a(x) does not jump. If a(x) has a jump across a smooth surface then $a(x)\frac{\partial u}{\partial n}(x)$ has still to be continuous, where n denotes the normal vector to the surface. Furthermore the tangential derivative of u will be continuous if we choose a suitable function space for u (namely the Sobolev space H^1).

Returning to the laminate, we remark that the precise form of u_n will not be identified by macroscopic measurements. We will see a smooth function u_o instead. The derivative of u_n has jumps whereas the derivative of u_o has none, so that they cannot be close to each other in the usual sense. But they are close to each other in a weak sense; for example, the sequence u_n given by (9) is close to zero. To be definite, we will use weak convergence in $L^2(\Omega)$ (the space of square integrable functions) to describe this behaviour.

We can now make precise what we mean by saying that two materials are close to each other, thereby identifying a suitable convergence for the controls a_n.

<u>Definition H</u> : Let $A_n(x)$ be a sequence of symmetric conductivity tensors which satisfy

$$0 \quad \alpha \leq \text{eigenvalues of } A_n(x) \leq \beta \quad \infty. \tag{16}$$

We say that A_n is convergent to A in the sense of homogenization (H-convergent) if for every f $L^2(\Omega)$ the solution u_n (belonging to $H^1_o(\Omega)$) of

$$- \operatorname{div}(A_n(x)\operatorname{grad}u_n) = f \text{ in } \Omega \text{ ; } u_n = 0 \text{ on } \partial\Omega \qquad (S'_n)$$

converges weakly in $H^1(\Omega)$ to the solution u $H^1(\Omega)$ of

$$- \operatorname{div}(A(x)\operatorname{grad}u) = f \text{ in } \Omega \text{ ; } u = 0 \text{ on } \partial\Omega. \qquad (S'_\infty)$$

By weak convergence in $H^1(\Omega)$ is meant

$$\operatorname{grad}u_n \to \operatorname{grad}u \text{ (weakly in } L^2(\Omega)^2) \qquad (17)$$

Using Definition H we regard two materials (conductivity tensors) as being close to each other if this notion of convergence is true for the solutions of the corresponding equations. Materials which are close in this sense can hardly be distinguished by macroscopic measurements of the temperature u. In our case we have simply $A_n(x) = a_n(x)I$. Important is the following :

<u>Compactness theorem</u>

Let 0 $\alpha < a_n(x) < \beta$ ∞. Then the sequence A_n contains a H-convergent subsequence converging to a <u>symmetric</u> conductivity tensor A.

Furthermore we have a

<u>Continuity theorem</u>

Let u_n be the solution of

$$- \operatorname{div}(A_n(x)\operatorname{grad}u_n) = f \text{ in } \Omega \text{ ; } u_n = 0 \text{ on } \partial\Omega. \qquad (S'_n)$$

Assume that $A_n(x)$ is symmetric and H-convergent to A(x). Then $F_1(u_n) = \int_\Omega u_n dx$ converges to $F_1(u)$ where u is the solution of

$$- \operatorname{div}(A(x)\operatorname{grad}u) = f \text{ in } \Omega \text{ ; } u = 0 \text{ on } \partial\Omega. \qquad (S'_\infty)$$

Having identified a suitable notion of convergence for the controls we proceed as in the one dimensional model problem. We <u>enlarge the class of admissible controls by the limits of originally admissible controls</u>. Precisely we set

$$\tilde{A}_{ad} = \{A \mid A(x) \text{ is an H-limit of } a_n(x)I, \text{ where } a_n \quad A_{ad}\}. \qquad (18)$$

Then the abstract theorem of Section 3 ensures that the <u>(relaxed) problem</u>

$$\text{Minimize} \int_\Omega u dx \qquad (M')$$

where

$$- \operatorname{div}(A(x)\operatorname{grad}u) = f \text{ in } \Omega \ ; \ u = 0 \text{ on } \partial\Omega \qquad (S')$$

and A $\tilde{A}_{ad}$ always has a solution.

Physically, this means that we can find an optimal design, provided we are allowed to use not only the pure materials α and β but also all composite (i.e. H-limits) which can be made from them. An example where these composites are needed is given below.

Knowing existence, how can we calculate the solution? We will derive necessary conditions in the same way as in the one dimensional problem. Given a "candidate" A we consider a small perturbation $A_\varepsilon = A + \varepsilon\delta A$. Solving (S') gives $u_\varepsilon = u + \varepsilon\delta u + O(\varepsilon^2)$, and for the functional F_1 which it is required to minimize we obtain $F_1(u_\varepsilon) = F_1(u) + \varepsilon(\delta F_1)(u) + O(\varepsilon^2)$. Substituting this into (M') and (S') and by neglecting quadratic terms in ε we obtain

$$\delta F_1 = \int_\Omega (\delta u)dx \qquad (19)$$

where δu is the solution of

$$- \operatorname{div}\{A(x)\operatorname{grad}(\delta u) + \delta A(x)\operatorname{grad}u\} = 0 \text{ in } \Omega \ ; \ \delta u = 0 \text{ on } \partial\Omega. \qquad (20)$$

To make this formal calculations rigorous a proper functional analysis setting is needed. To express δF_1 in term of δA we introduce the adjoint state equation

$$- \operatorname{div}(A(x)\operatorname{grad}p) = 1 \text{ in } \Omega \ ; \ p = 0 \text{ on } \partial\Omega. \qquad (S'^*)$$

Multiplying (S') by p, integrating over Ω and applying integration by parts we obtain $\int_\Omega \{(A(x)\operatorname{grad}\delta u.\operatorname{grad}p) + (\delta A(x)\operatorname{grad}u.\operatorname{grad}p)\}dx = 0$, i.e

$$- \int_\Omega \delta u.\operatorname{div}(A(x)\operatorname{grad}p)dx + \int_\Omega (\delta A(x)\operatorname{grad}u.\operatorname{grad}p)dx = 0. \qquad (21)$$

Here (,) denotes the scalar product in R^2 and we used the symmetry of A, i.e. $(Aw.z) = (Az.w)$. Now (19) and (S'^*) imply

$$\delta F_1 = - \int_\Omega (\delta A(x)\text{gradu.gradp})dx \qquad (22)$$

The <u>necessary condition</u> for A to be optimal is

$$\delta F_1 \geq 0 \text{ whenever } A + \varepsilon\delta A \quad \tilde{A}_{ad} \qquad (\text{N.C}')$$

for all sufficient small positive ε.

5 CHARACTERIZATION OF ALL COMPOSITE MATERIALS

In order to exploit the necessary condition (N.C') we need to have a sufficiently large supply of admissible variations δA, hence we need to know the structure of $\tilde{A}_{ad}$, the set of controls corresponding to composite materials made from α and β. Recall that $A \quad \tilde{A}_{ad}$ if there exists a sequence $a_n \quad A_{ad}$ such that $a_n I$ H-converges to A. By Lemmas C and G we know that (after possibly passing to a subsequence) $a_n \rightarrow a$ (in L^∞ weak *) and $a(x) = \theta(x)\alpha + (1-\theta(x))\beta$. It turns out that the local conductivity A(x) can be characterized in terms of $\theta(x)$, the local proportion of material α.

<u>Theorem I</u> : For every θ, $0 \leq \theta \leq 1$, there exists a convex set M_θ of symmetric 2x2 tensors such that $A \quad \tilde{A}_{ad}$ if and only if for (almost) every $x \quad \Omega$, $A(x) \quad M_{\theta(x)}$. Moreover all matrices in M_θ have eigenvalues between $\lambda_-(\theta)$ and $\lambda_+(\theta)$ where

$$\left| \begin{array}{l} \lambda_-(\theta) = (\theta\alpha^{-1} + (1-\theta)\beta^{-1})^{-1} \\ \\ \lambda_+(\theta) = \theta\alpha + (1-\theta)\beta \end{array} \right. \qquad (23)$$

and these bounds can be attained. More precisely M_θ is given by

$$M_\theta = \{\text{symmetric 2x2 tensors } M \mid \text{eigenvalues of } M \quad K_\theta\} \qquad (24)$$

where $K_\theta \quad R^2$ can be characterized as follows. We have $(\lambda_1, \lambda_2) \quad K_\theta$ if and only if

$$\left|\begin{array}{l} \frac{1}{\lambda_1-\alpha}+\frac{1}{\lambda_2-\alpha} \leq \frac{1}{\lambda_-(\theta)-\alpha}+\frac{1}{\lambda_+(\theta)-\alpha} \\ \frac{1}{\beta-\lambda_1}+\frac{1}{\beta-\lambda_2} \leq \frac{1}{\beta-\lambda_-(\theta)}+\frac{1}{\beta-\lambda_+(\theta)} \end{array}\right. \qquad (25)$$

From this description of K_θ we deduce the convexity of K_θ which implies the convexity of M_θ. We also deduce the bounds on the eigenvalues for $M \quad M_\theta$. These bounds are attained by a composite made of layers as we saw earlier.

6 SHARP NECESSARY CONDITIONS AND EXPLICIT SOLUTIONS

We will now exploit (N.C') using the characterization theorem of the previous section. Let $A \quad \tilde{A}_{ad}$ be our "candidate" for optimal solution. Then there exists $\theta(x)$ such that $A(x) \quad M_{\theta(x)}$ for (almost) every $x \quad \Omega$. Let B be a tensor-valued function such that $B(x) \quad M_{\theta(x)}$. By the convexity of M_θ we know that $(1-\eta)A(x) + \eta B(x) \quad M_{\theta(x)}$ for all $\eta \quad (0,1)$. Therefore we can use (22) with $\delta A = B-A$ obtaining

$$\delta F_1 = -\int_\Omega ((B(x)-A(x))\text{grad}u.\text{grad}p)dx \geq 0. \qquad (26)$$

We conclude that

$$-((B(x)-A(x))\text{grad}u.\text{grad}p) \geq 0 \text{ for (almost) every } x \quad \Omega. \qquad (27)$$

This conclusion is standard in the calculus of variations. If (27) was false, there would exist a set ω (of positive measure) and a B such that $((B(x)-A(x)\text{grad}u.\text{grad}p) \quad 0$ for all $x \quad \omega$. We set $\tilde{B}(x) = B(x)$ for $x \quad \omega$, $\tilde{B}(x) = A(x)$ for $x \quad \Omega \quad \omega$. Substituting $\tilde{B}$ into (26) leads to a contradiction. We rewrite (27) as

$$(B\text{grad}u.\text{grad}p) \leq (A\text{grad}u.\text{grad}p) \text{ for all } B \quad M_\theta \qquad (27')$$

provided $A \quad M_\theta$.

A must have eigenvalues $\lambda_-(\theta)$ and $\lambda_+(\theta)$ as can be shown by the following simple geometric argument. First define two vectors e and f by $\text{grad}u = |\text{grad}u|(e+f)$ and $\text{grad}p = |\text{grad}p|(e-f)$; then $(A\text{grad}u.\text{grad}p)$ is equal to $|\text{grad}u|.|\text{grad}p|((Ae.e)-(Af.f))$ which, by Theorem I is less or equal than $|\text{gradu}|.|\text{gradp}|(\lambda_+(\theta)|e|^2 - \lambda_-(\theta)|f|^2)$. Choose B to have eigenvalues $\lambda_+(\theta)$ and $\lambda_-(\theta)$ with corresponding eigenvectors e and f respectively. Then (27') can only hold if the same property is true for A.

This means that the optimal material can (locally) be approximated by simply layering the materials α and β and the layers have to lie in a certain direction.

To avoid technicalities we make the following assumption on the heat source term :

$$f \equiv 1 \tag{28}$$

In this case the equations (S') and (S'*) are identical and hence

$$u \equiv p. \tag{29}$$

Then gradu is an eigenvector of A with eigenvalue $\lambda_+(\theta)$. Physically, this means that the layers lie parallel to gradu so that the <u>conductivity is maximal in the direction of the heat flux</u>. Our problem reduces to the following

$$\text{Minimize } \tilde{F}_1(\theta) = \int_\Omega u dx, \tag{$\tilde{M}'$}$$

where u is given by

$$- \text{div}(\lambda_+(\theta)\text{gradu}) = 1 \text{ in } \Omega \text{ ; } u = 0 \text{ on } \partial\Omega \tag{S'}$$

and θ satisfy the constraints

$$0 \leq \theta(x) \leq 1 \text{ and } \alpha \int_\Omega \theta dx + \beta \int_\Omega (1-\theta)dx = \gamma. \tag{C'}$$

This <u>new problem is formulated in terms of the local proportions $\theta(x)$ for material α</u>, as we promised would do in the first section. Equation ($\tilde{S}'$) is equivalent to the fact that u maximizes

$$\int_\Omega (- \lambda_+(\theta)|\text{gradv}|^2 + 2v)dx \tag{30}$$

under the constraints $v = 0$ on $\partial\Omega$. Moreover, using $(\tilde{S}')$ we obtain

$$\max_{v} \int_{\Omega} (-\lambda_{+}(\theta)|\mathrm{grad}v|^{2} + 2v)dx = \int_{\Omega} u dx = \tilde{F}_{1}(\theta). \tag{31}$$

Hence $(\tilde{M}')$-$(\tilde{S}')$-$(\tilde{C}')$ are equivalent to the saddle point problem

$$\min_{\theta} \max_{v} \int_{\Omega} (-\lambda_{+}(\theta)|\mathrm{grad}v|^{2} + 2v)dx \tag{32}$$

where $v = 0$ on $\partial\Omega$ and

$$0 \leq \theta(x) \leq 1 \text{ and } \int_{\Omega} \theta dx = \frac{\beta \mathrm{meas}\Omega - \gamma}{\beta - \alpha}. \tag{33}$$

The integral in (31) is strictly concave in v and affine, thus convex, in θ. Therefore (31) has a solution which can be numerically approximated by standard methods. Due to the concavity/convexity no progressively faster oscillations appear. If $\theta(x)$ $\{0,1\}$ for x Ω for all solutions of (31) then our original problem (M')-(S')-(C'), where we were only allowed to use pure materials α and β, has no solution. If we then look for approximate solutions of the original problem we see that they try to approximate a composite material at the point x by forming progressively finer layers.

This is the reason for the appearance of oscillations in the approximate solutions for the original problem. Allowing composite materials eliminates the reason for such oscillations. Therefore extending the set of admissible controls does not only resolve the academic question of existence but also leads to a more stable problem. We may, however, ask whether the generalized solution has any practical relevance. To give some ideas in that direction we solve (31) explicitly when Ω is a disk of radius R. Starting from (31) the usual calculation gives

$$\delta F_{1} = -\int_{\Omega} \delta\lambda_{+}(\theta)|\mathrm{grad}u|^{2}dx = (\beta-\alpha)\int_{\Omega} \delta\theta(x)|\mathrm{grad}u|^{2}dx \geq 0 \tag{34}$$

for all admissible $\delta\theta$. Equation (34) gives a necessary condition for an optimal solution. But the relaxed problem has a solution, hence at least one solution of (34) solves our relaxed optimization problem. By concavity/convexity all the solutions of (34) lead to the same value of F_1, hence all solve the problem. Moreover u is uniquely determined by (34) because of strict concavity.

The integral constraint (33) on θ can be taken into account by introducing a Lagrange multiplier C_1. It follows that

$$\int_\Omega \delta\theta \left|\text{gradu}\right|^2 dx - C_1 \int_\Omega \delta\theta dx \geqslant 0 \tag{35}$$

where θ is only restricted by $0 \leqslant \theta(x) \leqslant 1$. Hence

$$\left|\begin{array}{l} \theta(x) = 1 \text{ implies } \left|\text{gradu}\right|^2 \leqslant C_1, \\ \theta(x) = 0 \text{ implies } \left|\text{gradu}\right|^2 \geqslant C_1, \\ 0 \quad \theta(x) \quad 1 \text{ implies } \left|\text{gradu}\right|^2 = C_1. \end{array}\right. \tag{36}$$

We now recall that Ω is a disk and show that $(\tilde{S}')$ and (36) can in fact be satisfied by radially symmetric functions. Let $u \equiv u(r)$ and $\theta \equiv \theta(r)$; setting $w(r) \equiv \lambda_+(\theta(r))u'(r)$ we obtain from $(\tilde{S}')$

$$-\left(w'(r) + \frac{1}{r}w(r)\right) = 1, \tag{37}$$

i.e. $(rw)' = -r$. Hence

$$w(r) = -\frac{r}{2} + \frac{C}{r}, \tag{38}$$

where C is an integration constant. Now $C = 0$, because otherwise $w(r)$ is too singular (more precisely $\tilde{w}(x)$, defined by $\tilde{w}(x) = w(|x|)$, is not in $L^2(\Omega)$). It follows that

$$\lambda_+(\theta(r))u'(r) = -\frac{r}{2}. \tag{39}$$

From (36) it follows that, for the optimal solution, Ω can be divided into 3 zones : inner zone

$$0 \leqslant r \leqslant 2\alpha\, C_1^{\frac{1}{2}} \text{ with } u'(r) = -\frac{r}{2\alpha} \text{ and } \lambda_+(\theta(r)) = \alpha, \tag{40}$$

intermediate zone

$$2\alpha\, C_1^{\frac{1}{2}} < r < 2\beta\, C_1^{\frac{1}{2}} \text{ with } u'(r) = -\, C_1^{\frac{1}{2}} \text{ and } \lambda_+(\theta(r)) = \frac{r}{2\, C_1^{\frac{1}{2}}}, \tag{41}$$

and outer zone

$$2\beta\, C_1^{\frac{1}{2}} < r < R \text{ with } u'(r) = -\frac{r}{2\beta} \text{ and } \lambda_+(\theta(r)) = \beta. \tag{42}$$

Since $u'(0) = 0$ (by radial symmetry) the inner zone consists entirely of the bad conductor whereas the outer zone consists entirely of the good conductor. They are joined by an intermediate zone where we have a mixture of both materials, and the proportion of the good conductor is an increasing affine function of the radius. We remark that the outer zone and part of the intermediate zone will disappear if the total proportion of good material is too small.

Comparing with our wrong guess of placing the good conductor in the middle we see that our abstract approach has given us nontrivial and very precise information concerning the optimal design problem. As a by-product we have shown that the problem admits no classical solution since the generalized solution is unique and would only be classical if θ took only the values 0 and 1.

In real life we can only use the two materials and no mixtures of them. Nevertheless the generalized solution provides a good idea how to arrange the materials in the intermediate zone. We should use inward pointing triangular spikes of the good material such that the proportion of material α in the circle $|x| = r$ approximates $\theta(r)$.

Furthermore we see that the generalized solution preserves the radial symmetry of the problem, whereas none of the approximations suggested above does.

7 CONCLUDING REMARKS

We have seen that the optimization design problem posed in Section 1 has in general no solution and is difficult to tackle numerically. Our approach was to consider a generalized problem by admitting the use of composite materials. We are led to determine the local proportion of each material. This turns out to be an analytically and numerically well-posed problem. From the solution of the generalized problem we can deduce whether our original problem has an exact solution and obtain approximate solutions if it has not. The main mathematical tools introduced a suitable notion of convergence for conductivity tensors, called "convergence in the sense of homogenization or H-convergence".

Acknowledgements. I want to thank here Stefan Müller for having written the first version of this article from his notes taken during the three lectures that I gave at the symposium.

Reference list.

Armand J.L. & Lurie K.A. & Cherkaev A.V. (1983). Optimal control theory and structural design, In Optimum Structure Design, Vol. 2, (R.H. Gallagher, E. Atrek, K. Ragsdell and O.C. Zienkiewicz eds.), J. Wiley and Sons, New-York.

Kohn R.V. & Strang G. (1986). Optimal design and relaxation of variational problems, I, II, III, Comm. Pure Appl. Math. 39, 1986, pp. 113-137, 139-182 and 353-377.

Lurie K.A. (1970). On the optimal distribution of the resistivity tensor of the working substance in a magnetohydrodynamic channel, J. Appl. Mech. (PMM) 34, 255-274.

Murat F. & Tartar L.C. (1985). Calcul des variations et homogénéisation," In Les Méthodes de l'Homogéneisation : Théorie et Applications en Physique, Coll. de la Dir. des Etudes et Recherches de Elec.de France 57, Eyrolles (Paris) 1985. 319-369.

Tartar L.C. (1985). Estimations fines de coefficients homogénéisés, In Ennio de Giorgi Colloquium, (P. Krée ed.) Research Notes in Mathematics 125, Pitman (Londres) 1985. 168-187.

PART II

SINGLE INVITED LECTURES

SOME MATHEMATICAL PROBLEMS ARISING FROM THE OIL SERVICE INDUSTRY

C. Atkinson
Department of Mathematics, Imperial College, London, England

P.S. Hammond
Schlumberger-Doll Research, Connecticut, USA

M. Sheppard
Schlumberger-Cambridge Research, Cambridge, England

I.J. Sobey
Schlumberger-Cambridge Research, Cambridge, England

1 INTRODUCTION

An outline is given of the mathematical modelling of some problems which occur in various situations in the oil service industry. In each case the objective is to use a mathematical model as a means of interpreting properties of the formation either during drilling or by well tests after drilling has been completed.

We begin with a brief outline of oil well testing. The main aim of such tests is the recovery of a fluid sample and an estimation of the flow capacity of the formation. We start with a description of a standard procedure for single phase flow (Horner's method (1)) and then discuss the complications introduced when the flow equations are coupled with the temperature. A description is also given of some two phase flow problems.

A second example is that of a mathematical model to be used for interpreting measurements while drilling. This consists of studying the axial vibrations generated by a roller cone bit during the drilling process. Through the model the axial force and displacement histories are calculated at the bit and related to the force and acceleration measured at an MWD (measurements while drilling) site just above the drill bit (possibly as far away as 60 ft.) or at the surface. It is, of course, important to identify the relation between the measured quantities and the values of these quantities at the bit. The objective of this interpretation is to determine formation and bit properties as drilling proceeds so as to facilitate the drilling process.

In section 2 we concentrate on the well test problems and in section 3 on the roller cone bit problem. It is worth noting that although each of these problems leads to interesting systems of differential equations the mathematical modeller may not have time to dwell on their niceties, his job being to rationalise the physical problem to such an extent that actual field data can be interpreted. Moreover, this interpretation may be required in real time which places an additional constraint on the modeller.

2 OIL WELL TESTING

As stated in the introduction, the main aims of oil well testing are the recovery of a fluid sample and an estimation of the flow capacity of the formation. A test is conducted by allowing the well to flow for a few hours and then shutting it in, either at the surface or by closing a special downhole valve system. The bottom hole pressure is measured and recorded throughout. Typical data are shown in Figure 1, the bottom hole pressure falls during flow periods and recovers during shut in. In contrast the bottom hole fluid temperature rises during flow.

In order to interpret the bottom hole pressure/time record one requires a mathematical description of fluid flow in a porous medium. This can be obtained by combining mass conservation, a flow law such as Darcy's Law and an equation of state for the fluid. The continuity equation for flow of a fluid in a porous medium can be expressed as

$$\operatorname{div}(\rho \underline{q}) + \frac{\partial}{\partial t}(\varphi\rho) = 0 \tag{2.1}$$

where ρ is the fluid density, φ the porosity of the rock and $\underline{q}$ is the volumetric rate of flow per unit cross sectional area. Darcy's law, valid for laminar flow at low Reynolds numbers, can be expressed as

$$\underline{q} = -\frac{k}{\mu}\nabla p \tag{2.2}$$

where μ is the fluid viscosity, p the fluid pressure and k the permeability of the medium (for simplicity we have assumed isotropic permeability and neglected the influence of gravity). Combining equations (2.1) and (2.2) gives

$$\frac{\partial}{\partial t}(\varphi\rho) = \operatorname{div}\left(\frac{\rho k}{\mu}\operatorname{grad} p\right) \tag{2.3}$$

To complete the description an equation of state for the pore fluid is required. For single phase flow a suitable relation can be expressed as

$$\rho = \rho_0 \exp(c(p-p_0)) \tag{2.4}$$

where ρ_0 is the value of the fluid density at some reference pressure p_0 and c is the fluid compressibility. Equation (2.4) is a good description for isothermal flow of most fluids of small and constant compressibility.

If one assumes constant permeability and porosity, constant and small compressibility and that $|\nabla p|^2$ is negligibly small then (2.3) reduces to

$$\nabla^2 p = \frac{\varphi\mu c}{k}\frac{\partial p}{\partial t} \tag{2.5}$$

This equation is often used in petroleum engineering, though it is worth noting that retaining $|\nabla p|^2$ still allows (2.3) with (2.4) to be written as a linear equation with ρ as the dependent variable. The constant $k/(\varphi\mu c)$ is called the hydraulic diffusivity by analogy with the diffusion equation.

For the case of single-phase gas flow the equation of state (2.4) can be replaced by the ideal gas law, for example, which can be written

$$\rho = \frac{M}{RT}\,p \tag{2.6}$$

where M is the molecular weight of the gas, R the gas law constant and T the absolute temperature. For non-ideal gas flows this equation must be modified. With the equation of state (2.6) the corresponding equation to (2.3) reduces for constant gas viscosity (μ_g) and constant rock properties to

$$\nabla.(p\nabla p) = \frac{\varphi\mu_g}{k}\frac{\partial p}{\partial t} \tag{2.7}$$

where we have again neglected gravity body forces.

In the next subsection 2.1 we will discuss a classical interpretation procedure and then in 2.2 and 2.3 formulate the corresponding problems for the temperature and for two phase flow.

2.1 Interpretation (single phase fluid flow)

As discussed above a test is conducted by allowing the well to flow for a few hours and then shutting it in, Figure 1 shows a typical downhole pressure-time record. The rate of pressure recovery is conventionally interpreted to give the formation permeability-thickness in the following way:

The pressure is assumed to satisfy the diffusion equation (2.5) subject to the boundary condition at the well bore

$$q(t) = -\frac{2\pi k}{\mu} r_w \int_{z=0}^{h} \left(\frac{\partial p}{\partial r}\right)_{(r=r_w,t)} dz \qquad (2.8)$$

q is the volumetric flow rate out of the well and $\frac{\partial p}{\partial r}$ is the pore pressure gradient at $r=r_w$ the well bore radius. For a well assumed open over the entire thickness h of the oil bearing formation in an infinite reservoir of constant porosity and permeability a radially symmetric solution of (2.5) subject to (2.8) can be used to show that

$$P(t)-P(0) \approx \frac{\mu q_0}{4\pi kh} \log\left(\frac{t-t_p}{t}\right) \quad \text{for} \quad t \gg t_p \qquad (2.9)$$

where q(t) is piecewise constant as shown in Figure 1 and q_0 is the flow rate at t=0 as shown and $t=t_p$ is the time at which the well is shut in. From the slope of a graph of P(t) against $\log(\frac{t-t_p}{t})$, formation permeability-thickness product can be estimated. This interpretation procedure is known as Horner's method and is commonly used.

It is worth noting that although a detailed bottom-hole pore pressure profile could be deduced from solving equation (2.5) subject to (2.8) given the time dependent volumetric flow rate q(t) this would depend upon additional parameters such as the well bore radius r_w, porosity φ and compressibility c. Actual pressure records show deviations from such ideal curves especially during the producing period of the well. This deviation is usually attributed to damage to the near well bore region and is modelled as a skin region in which the permeability is different to that in the bulk rock. Nevertheless, the longer time interpretation such as given by equation (2.9) is quite successful. However, given profiles such as shown in Figure 1 it is natural to ask if one can interpret the temperature data also.

2.2 The interpretation of temperature records during well tests (a formulation)

From Figure 1 certain general trends can be noticed: (i) as the well flows and pressure falls, the well bore temperature rises, (ii) during periods when the well is shut in and pressure recovery occurs, the temperature falls, (iii) over a complete test downhole temperature rises by several degrees. The question arises as to what information can be

Fig 1

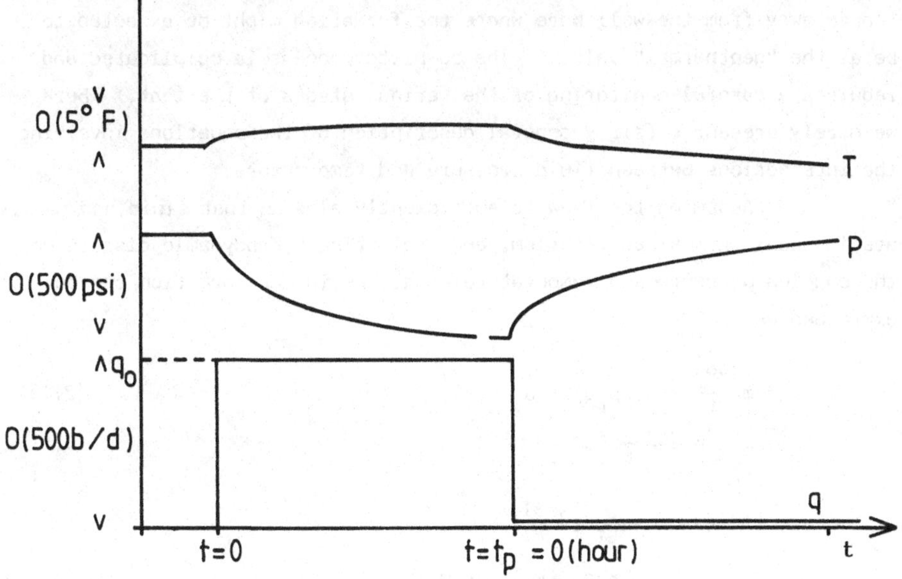

q is the volumetric flow rate out of the well (7 barrels ≈ $1m^3$)

P is the bottomhole pressure, T the bottomhole temperature.

gleaned about the formation properties of the rock from such a test. A number of factors are clearly involved including flow of fluid from distances away from the well bore where the formation might be expected to be at the "geothermal" value. The complete process is complicated and requires a careful monitoring of the various stages of the test. Here we merely present a fairly general description of the equations involving the interactions between fluid pressure and temperature.

Assuming the flow is sufficiently slow so that fluid and matrix are in local thermal equilibrium, and neglecting hydrodynamic dispersion, the coupled pressure and temperature equations in the formation can be expressed as

$$\varphi \frac{\partial \rho_f}{\partial t} + \nabla.(\rho_f \underline{q}) = 0 \tag{2.10}$$

$$q = -\frac{k}{\mu} \nabla p \tag{2.11}$$

$$[(1-\varphi)\rho_r c_r + \varphi\rho_f c_f] \frac{\partial T}{\partial t} + \rho_f c_f \underline{q}.\nabla T$$
$$= \beta T[\varphi \frac{\partial p}{\partial t} + \underline{q}.\nabla p] - \underline{q}.\nabla p + \nabla.(k_H \nabla T) \tag{2.12}$$

together with an equation of state which could be taken as

$$\rho_f = \rho_0 \exp(c(p-p_0) - \beta T) \quad . \tag{2.13}$$

In the above equations φ is the porosity of the rock, ρ_f and ρ_r the density of fluid and rock respectively; c_f and c_r the coefficients of specific heat at constant pressure of fluid and rock respectively; k_H is the thermal conductivity and β a "thermal expansion coefficient". Equations (2.10) and (2.11) are of course the same as (2.1) and (2.2).

To complete the system a boundary condition at the well bore is required. In the neighbourhood of the well bore the formation is damaged by a variety of physical and chemical interactions between constituents of the drilling mud and the formation so there may be permeability impairment in a region near the well bore. This region called the "skin" (we alluded to it at the end of the last subsection) is often replaced by an infinitesimal boundary condition relating p at $r=r_w$ (r_w is the well bore radius) and p_{wb} (the well bore pressure).

An example of a boundary condition at the well bore is the

specification of the volumetric flow rate out of the well $\underline{q}$ such as was considered in the last section. In addition models are required for the heat exchanges between well bore and formation and within the well bore itself. Both of these require careful fluid mechanical analysis, examining for example the relative importance of convection and conduction and the mixing properties of natural convection during shut in and of multiphase flows during drawdown.

2.3 Two phase flow

Below the bubble point of oil in the reservoir, gas flow will begin, one then has to consider the flow of two phases i.e. oil and gas. A more fundamental formulation of the equations for multiphase flow should consider the spatial distribution of each component in the hydrocarbon - water system as a function of time. In the simplified treatment given here all hydrocarbon liquid which is present at atmospheric conditions, as obtained by differential vaporisation, is referred to as oil and the gas phase treated as an entity without regard to its composition. Even with this simplification there are usually three phases to be treated oil, gas and water. Here we describe the simplest model of the simultaneous flow of liquid and gas during a DST (drill stem test). This type of test usually follows the pattern: flow/shut-in, flow/shut-in with the second shut-in period being interpreted. Moreover, the test lasts only a few hours so transient pressure gradients produced in the reservoir are important.

The model to be described derives from a black oil description of flow in the formation and assumes that an ideal gas is dissolved in a non-volatile oil. In the original state the formation contains only the oil phase; initially there is no free gas present and the system is in equilibrium (for a general background of these multiphase problems see e.g. Muskat (1949) or Peaceman (1977)).

At any instant an element of the reservoir will contain certain volumes of oil and gas which when reduced to standard conditions will be modified as a result of gas solubility in the oil and the compressibility of each phase. The ratio of the volume of gas liberated (V_{gs}) from a volume of oil to the oil volume (V_{0s}) (all referred to standard conditions) is the gas solubility factor R_{so}. The use of a formation volume factor to allow for the changes in volume which occur in each phase upon transition from reservoir to standard surface conditions of temperature and pressure is a common procedure. Thus we have

$$B_0 = V_0/V_{0s} \ , \quad B_g = V_g/V_{GS} \tag{2.14}$$

where V_0 is the volume of the oil *phase* at a particular pressure (and temperature) V_{0s} is the volume of oil at standard conditions, V_g the gas *phase* volume and V_{GS} the volume of gas at standard conditions and

$$R_{S0} = \frac{V_{gS}}{V_{0S}} \tag{2.15}$$

is the gas solubility factor.

Under *equilibrium* conditions these functions are given uniquely by the thermodynamic conditions.

To describe the relative size of the oil and gas phases the gas saturation is defined as

$$s = \frac{V_g}{V_g+V_0} \quad . \tag{2.16}$$

Hence when $s = 0$ there is no free gas and when $s = 1$ there is only gas present.

Whereas in single phase flow there is a unique pressure in the formation, in multiphase flow the pressure in different phases may be different because of surface tension effects in the porous medium. Thus defining a gas phase pressure P_g and an oil phase pressure P_0 these are related by

$$P_g - P_0 = P_c \tag{2.17}$$

where P_c is a capillary pressure. Empirically it is known that under equilibrium conditions the capillary pressure is a function of the saturation

$$P_c = P_c(s) \tag{2.18}$$

Little appears to be known about this dependence under non-equilibrium conditions. We assume in a DST that variations in pressure will be large compared to the variations in P_c which we thus take as effectively constant in our discussion below.

If the phase velocities are denoted by $\underline{v}_0$ and $\underline{v}_g$ and a Darcy like flow is assumed in the formation, then

$$\underline{v}_0 = -\frac{k_0}{\mu_0} \nabla P_0$$
$$\underline{v}_g = -\frac{k_g}{\mu_g} \nabla P_g \qquad (2.19)$$

where μ and k represent viscosity and permeability of the two phases.

At present the only practical method for estimating the permeabilities of the two phases is to assume a relative permeability model where the permeabilities are independent functions of the gas saturation. The functional dependence is estimated from observation of steady flow through core samples. Typical models for the relative permeability are

$$k_0 = k\,k_{r0} \quad , \quad k_{r0} = (1-s)^4$$
$$k_g = k\,k_{rg} \quad , \quad k_{rg} = s^3(2-s) \qquad (2.20)$$

where k is some overall permeability. From (2.20) we can see that as the saturation, s, increases the relative mobility of the oil phase decreases and that of the gas phase increases from zero as we might expect.

If we now consider mass conservation of gas, by analogy with (2.1), we must take into account both free gas and gas dissolved in the oil phase hence using Darcy's law (2.19),

$$\frac{\partial}{\partial t}\left\{ \varphi[\frac{s}{B_g} + \frac{(1-s)R_{s0}}{B_0}]\right\} = \nabla.\left[\frac{k_g}{B_g\mu_g}\nabla P_g + \frac{k_0 R_{s0}}{B_0\mu_0}\nabla P_0\right] . \qquad (2.21)$$

Similarly conservation of oil in the oil phase is described by

$$\frac{\partial}{\partial t}[\varphi\,\frac{(1-s)}{B_0}] = \nabla.[\frac{k_0}{B_0\mu_0}\nabla P_0] \qquad (2.22)$$

Note that whilst the general functional dependence of quantities such as k, μ, B, R_{s0} and φ on each other or on the pressure, saturation, temperature, etc. is unknown, adequate approximations can usually be made so that a closed set of equations is obtained.

The relative permeability approximation has already been discussed above. For the others it is important to have accurate PVT (pressure-volume-temperature) data since we are dealing with multiphase flow. A PVT laboratory will normally calculate the *equilibrium* values of the functions B_0, B_g and R_{s0} for a particular sample as the pressure varies. This process may be complicated because a sample at formation conditions may not be available and recomibnation of oil and gas samples at the surface may produce errors. The reservoir fluid viscosity is also measured in such a laboratory.

Consider first the oil phase volume factor B_0. As the quantity of dissolved gas increases the oil is observed to swell. Generally it is observed that as the pressure increases slightly above atmospheric there is a relatively rapid swelling to a value of B_0 around 1.1-1.2 and thereafter the increase is more or less linear until the bubble pressure is reached. At pressures above the bubble point the oil phase is compressed and the decrease in B_0 is close to exponential so that the compressibility $c_0 = -\frac{1}{B_0}\frac{dB_0}{dP}$ is close to being constant. Thus a reasonable approximation is to model B_0 as a function of P which increases linearly up to the bubble point and then decreases exponentially at higher pressures.

If the pressure is increased by two orders of magnitude, the gas volume will decrease by roughly two orders of magnitude. For example a gas volume factor is given by Earlougher (1977) to be approximately

$$B_g(P,T) = .005039\,\frac{zT}{P} \tag{2.23}$$

where T is the temperature in °R and z the real gas deviation factor. Most correlations are based on Standing (1952). Of course whilst the gas volume factor can be approximated by an equation of a given form, the values for any constants must be obtained from PVT analysis of the particular gas under consideration.

From the PVT analysis, the variation in the dissolved gas-oil ratio (R_{s0}) can be obtained as the pressure is lowered below the bubble point and equilibrium conditions are applied. What is not clear is to what extent transient conditions will affect this. In the absence of a better approximation $R_{s0}(P)$ obtained from the equilibrium PVT must be used.

In principle we can now proceed with equations (2.21) and (2.22) assuming $P_c(s)$ is effectively constant so the variations in P_0 and P_g are the same as the variations in a common pressure P. Thus equations (2.21) and (2.22) hold with P_0 and P_g replaced by P and we note that $k_{rg}(s)$, $k_{r0}(s)$, $B_g(P)$, $B_0(P)$, $R_{s0}(P)$ and $\varphi(P)$ are known functions obtained from laboratory experiments. Thus we have two coupled nonlinear equations in the variables s (the gas saturation) and the pressure P.

Numerical simulations of well-test conditions have been made using these coupled equations and investigations made of some of their properties. There is not room here to discuss this matter further and we restrict discussion to some brief comments regarding a simplified case.

Clearly from (2.21) and (2.22) if both oil and gas velocities are negligible then the saturation s can be found as a unique function of the pressure. In this case the right hand sides of (2.21) and (2.22) are both zero so the equations can be integrated directly. If we consider a volume V(P) of the two phases then when the pressure equals the bubble pressure, $P = P_b$, the gas saturation is zero. Thus we obtain

$$s(P) = \frac{B_g(P)[R_{s0}(P_b)-R_{s0}(P)]}{B_0(P)+B_g(P)[R_{s0}(P_b)-R_{s0}(P)]} \tag{2.24}$$

This allows the pressure to be used to determine the local saturation assuming equilibrium conditions. For the full equations with general boundary conditions recourse to large numerical simulation is necessary.

3 THE GENERATION OF AXIAL VIBRATIONS BY A ROLLER CONE BIT

As a conventional tricone bit drills through a particular formation, driven by the torque applied to the bit via the drillstring, the motion of the bit over the rock gives rise to a vibrational signal which is transmitted along the drillstring. In principle, this vibrational signal contains information about three important aspects of the drilling system:- (i) the input impedance of the drillstring, (ii) the state of the bit, (iii) the nature of the formation being drilled. We discuss here a fairly simple model for the generation of axial vibrations at the bit.

Viewed from the bit the drillcollars and heavyweight drillpipe dominate the dynamic response of the drillstring thus the model adopted is that of a simple, viscously damped, uniform cross section prismatic column with an appropriate stress boundary condition at one end to give rise to the required weight on bit (WOB). At the other end the interaction between the bit and the rock needs to be modelled. This is complicated since each of the three cones has three rows of teeth with different numbers of teeth per row, depending on the type of bit; a typical bit might have on the order of ten teeth per row. The interaction of each tooth with the rock is itself a very complicated process leading to a nonlinear response between force and penetration. A simple model is a linear force-displacement curve for each tooth as it penetrates, together with a relation of different slope (much steeper) as the tooth egresses from the rock.

With the above assumptions the equation of motion for the displacement $u(x,t)$ in the drillcollar is given by

$$m \frac{\partial^2 u}{\partial t^2}(x,t) + c \frac{\partial u}{\partial t}(x,t) - EA \frac{\partial u^2}{\partial x^2}(x,t) = F(x,t) \tag{3.1}$$

where $F(x,t)$ is a force distribution, m is the mass of the column per unit length, c is the magnitude of the damping per unit length, E is Young's modulus for steel and A is the cross-sectional area of the column. In this formulation the forces at the ends $x = 0$ and $x = \ell$ are accounted for by including delta function components in $F(x,t)$ of the form $\delta(x)$ and $\delta(x-\ell)$ where ℓ is the column length. The important one of these forces is the bit forcing function $F_b(t)\delta(x)$, this is itself a function of $u(0,t)$ as well as depending on the details of the motion of the bit indentors. The total force acting on the bit will have the form

$$F_b(t) = \sum_i f(u_i, \mathrm{sgn}\, \dot{u}_i) \tag{3.2}$$

where i denotes the ith tooth, the function f reflects the tooth geometry and the failure properties of the rock the function being in general different depending on whether the tooth is indenting or being withdrawn from the rock. To relate u_i to the displacement $u(0,t)$ of equation (3.1) the motion of the bit indentors must be taken into account. This is simply a matter of bit geometry folded in with $u(0,t)$. Coupling (3.2) with (3.1) and imposing free-free boundary conditions on equation (3.1) together with the initial condition u(x,0) enables the displacement history u(x,t) and the force history $F_b(t)$ to be calculated. (Analytical solution of (3.1) proves evasive, however a numerical solution involving a normal mode decomposition generates the displacement and force histories u(x,t), $F_b(0,t)$).

Taking suitable account of bit geometry and rock/bit interaction, solutions of (3.1) prove to be strongly dependent on the bit state (particularly tooth wear) and the rock/indentor interaction which in turn depends on lithology. The solutions thus enable interpretation of bit generated signals in terms of lithology and wear and are consequently of considerable value in an industry in which little or no information concerning the rock/bit interaction is usually available while drilling.

REFERENCES

Earlougher, R.C. (1977). Advances in Well Test Analysis (AIME).

Horner, D.R. (1951). "Pressure Build-up in Wells". Proc. Third World Pet. Cong., E.J. Brill, Leiden (1951), II, 503.

Muskat, M. (1949). Physical Principles of Oil Production. McGraw-Hill, New York.

Peaceman, D.W. (1977). Fundamentals of Numerical Reservoir Simulation. Elsevier.

Standing, M.B. (1952). Volumetric and Phase Behaviour of Oilfield Hydrocarbon Systems (SPE).

RANDOMLY DILUTED INHOMOGENEOUS ELASTIC NETWORKS NEAR THE PERCOLATION THRESHOLD

D. J. Bergman
School of Physics and Astronomy
The Raymond and Beverly Sackler Faculty of Exact Sciences
Tel Aviv University, Tel Aviv 69978, Israel

Abstract. A brief discussion is given of the use of discrete network models for representing a randomly perforated continuum elastic solid near its rigidity or percolation threshold p_c. As the threshold is approached the elastic stiffness moduli tend to zero as $(p-p_c)^T$ where p is the volume fraction of solid material, or fraction of bonds present in the diluted network. Exact and extremely close upper and lower bounds are obtained for the critical exponent T, based on the links-nodes-blobs picture for the percolating backbone and on a variational principle for the network.

Although this conference is devoted to (nonclassical aspects of) continuum mechanics, there are some topics in this field where discrete models are a convenient and even a necessary tool. A case in point is the critical behavior of elastic composites near a percolation threshold. (Excellent reviews of the current status of research on physical properties near a percolation threshold can be found in the volume edited by Deutscher et al. 1983).

Consider a homogeneous solid in which holes are punched at random up to a critical concentration at which the solid falls apart. As this percolation or rigidity threshold of the solid component p_c is approached, the macroscopic stiffness moduli tend to zero with a characteristic exponent

$$C_e \sim (p-p_c)^T \quad , \qquad p \rightarrow p_c^{+} \, , \tag{1}$$

where p is the volume fraction of the solid component. The fracture stress σ_F, i.e., the minimum macroscopic stress required to break the solid apart, will then also tend to zero with a characteristic exponent

$$\sigma_F \sim (p-p_c)^{T_F} \quad . \tag{2}$$

Both of these properties are much easier to investigate on a discrete network than on a continuum solid. (For theoretical and numerical studies of these properties, see Bergman 1985, 1986a,b, Bergman and Kantor 1984, Bergman and Duering 1986, Kantor and Webman 1984). Moreover, this approach is justified by the principle of universality which is found to occur at all critical points of large systems (i.e., systems in which the number of degrees of freedom is large and increases with the volume. See, e.g., Amit 1978.) According to this principle, certain quantities such as the critical exponents T, T_F have values that do not vary within a rather broad universality class of systems: They are independent of details such as the stiffness moduli of the components or whether the system is a discrete or a continuum one. Therefore, the model we will use to discuss this problem is a network of elastic bonds in which each bond is either present with probability p or absent with probability 1-p. Each bond is assumed to have an elastic stretching force constant k_o. But if that were all, we would have a system which might not exhibit solid behavior even somewhat above the usual geometric percolation threshold p_c, since one-dimensional chains of bonds would not have any rigidity. To avoid this problem, we must introduce angular forces as well: In 2D (two-dimensional) systems it is enough to introduce a bending force constant m_o which describes the force required to change the angle between two adjacent bonds. In 3D systems we must add to that a torsion constant which describes the force required to change the azimuthal angle of a bond in the reference frame determined by its nearest neighbor and its next-nearest-neighbor bonds. In higher dimensionalities $d>3$ additional angular forces must be allowed, their total number being $d-1$, in order for the rigidity threshold to coincide with p_c.

The elastic potential energy of the 2D network described above can be written in terms of changes in the bond lengths δb_i and changes of the angular orientations of the bonds $\delta\phi_i$. For small changes the energy is a quadratic function of these changes

$$H = \sum_i \frac{1}{2} k_i \, \delta b_i^{\,2} + \sum_{(ij)=nn} \frac{1}{2} m_{ij} \, (\delta\phi_i - \delta\phi_j)^2 , \tag{3}$$

where the second sum is over pairs (ij) of nearest neighbor bonds. In this formulation the variables δb_i, $\delta\phi_i$ are not independent: Whenever there is a closed loop of bonds, this results in constraint relations among some of these variables. However, when the network is near p_c the percolating cluster becomes very ramified - the loops become rare - and consequently this formulation becomes a very useful one, as we shall show below.

Near the percolation threshold but slightly above it, a useful approach is to consider only the percolating backbone, i.e., those sites (and the bonds connecting them) which are connected to the (opposite) edges of the network by at least two mutually exclusive paths (i.e., paths that have no other sites in common). Only the backbone bonds would carry current or stress, so that the rest of the percolating cluster can be ignored. The backbone is made of nodes, i.e., sites which are connected to the edges by at least three mutually exclusive paths, and the links, each of which connects a pair of neighboring nodes. The average distance between the ends of a link is the percolation correlation length $\xi \sim (p-p_c)^{-\nu}$ (see Skal and Shklovskii 1975) and each link is made of singly connected portions, of total length $L_1 \sim \xi^{1/\nu} \sim (p-p_c)^{-1}$ (see Coniglio 1982), and multiply connected portions, called blobs. From this links-nodes-blobs (LNB) picture, which was developed in a number of stages (see Skal and Shklovskii 1975, Coniglio 1981, Kantor and Webman 1984), many important qualitative and quantitative conclusions can be derived.

Consider a long, randomly twisting chain of bonds in d-dimensions. It is easy to estimate how the elastic coefficients of the chain depend on the microscopic force constants and on the length scales: A change δR in the end-to-end distance L of the chain can be achieved by stretching each of the L bonds in the chain by an amount of order $\delta b \sim \delta R/L$. The elastic energy associated with such a distortion would be

$$E \sim L k_o \left(\frac{\delta R}{L}\right)^2 = \frac{k_o}{L} \delta R^2 \quad . \tag{4}$$

The same change δR could alternatively be achieved by changing each interbond angle by an amount of order $\delta\phi_{ij} \sim \delta R/LL$. The additional factor 1/L arises from the fact that most of these angles are at a distance of order L from the edges, so that a change of angle $\delta\phi$ brings about a change of end-to-end distance by $\delta R \sim L\delta\phi$. The elastic energy associated with these angular distortions is

$$E \sim \mathit{L} m_o \left(\frac{R^2}{L\mathit{L}}\right) = \frac{m_o}{L^2 \mathit{L}} \delta R^2 \tag{5}$$

For long chains, this is clearly much less than the bond stretching energy, so that only the angles will be distorted and the elastic coefficient of the chain k_L will be

$$k_L \sim \frac{m_o}{L^2 \mathit{L}} . \tag{6}$$

When the dimensionality is greater than 2, m_o represents some mean of the various angular force constants. Applying this result to the singly connected portions of a link, we immediately get an upper bound for the elastic coefficient of a link k_ξ (it is an upper bound because it assumes that the blobs do not get distorted, i.e., that they are entirely rigid)

$$k_\xi \lesssim \frac{m_o}{\xi^2 \mathit{L}_1} \sim m_o (p-p_c)^{2\nu+1} . \tag{7}$$

This inequality was first obtained by Kantor and Webman (1984).

In order to get a lower bound for k_ξ we must consider more carefully the properties of the blobs. The linear dimensions of the blobs are $\lesssim \xi$. Each blob is made of simple chains connected together in various ways, where for each of these chains the end-to-end length L satisfies $L \lesssim \xi$, while the total length of a chain L satisfies $\mathit{L} \lesssim L^{1/\nu} \lesssim \xi^{1/\nu}$. The elastic constant of each of these chains k_L satisfies

$$k_L \gtrsim \frac{m_o}{L^2 \mathit{L}} , \tag{8}$$

where the inequality appears rather than equality in order to take into account the fact that for sufficiently short chains the bond stretching force constants may become important. The electrical conductance of such a chain g_L is of course inversely proportional to L

$$g_L \sim \frac{g_o}{\mathit{L}} , \tag{9}$$

and therefore

$$k_L \geqslant \frac{g_L}{L^2} \tag{10}$$

for all the chains making up the blob. When different chains are connected in parallel their elastic coefficients add up, just like the conductances. For most other connections the elastic coefficients also combine similarly to electrical conductances. The only exception to this rule is when two chains are connected in series, but such connections do not occur in a blob. Nor do they occur in a link if we construct it by starting from a singly connected chain stretching from node to node, and then add onto it the thicker portions at the appropriate places. These considerations lead to the result

$$k_\xi \geqslant \frac{g_\xi}{\xi^2} \sim (p-p_c)^{\zeta+2\nu} \tag{11}$$

where ζ is the critical exponent that characterizes the conductance of a link $g_\xi \sim (p-p_c)^\zeta$. This inequality was first obtained by Roux (1986).

The macroscopic elastic moduli C_e are related to k_ξ in the same way as the macroscopic conductivity σ_e is related to g_ξ, namely

$$\begin{aligned} C_e &\sim k_\xi/\xi^{d-2} \sim (p-p_c)^T \\ \sigma_e &\sim g_\xi/\xi^{d-2} \sim (p-p_c)^{\zeta+(d-2)\nu} \equiv (p-p_c)^t \ . \end{aligned} \tag{12}$$

From the inequalities Eqs. (7) and (11) we therefore deduce

$$d\nu+1 \leqslant T \leqslant d\nu+\zeta = t + 2\nu \ , \tag{13}$$

where $t = \zeta+(d-2)\nu$ is the conductivity critical index (see Eq. (12)). The lower bound was discovered by Kantor and Webman (1984) who derived it in the way described here, and the upper bound was discovered by Roux (1986) who used a somewhat different argument than ours.

These bounds are very useful because the indices ν and t are known rather well for d=2-6, and also because they turn out to be quite close, as can be seen from Table 1. The fact that for d=6 they coincide at T=4 reflects the well known fact that for d$\geqslant$6 the blobs are unimportant and all the important properties of the backbone are determined by

Table 1. Upper and lower bounds (columns 5 and 4, respectively) for the elastic modulus critical index T calculated from known values of ν (the correlation length index) and t (the conductivity index) for dimensionalities d from 2 to 6. Error bars are shown where known. Appearance of an integer without a decimal point, or a rational fraction, indicates that the value is known exactly.

d	ν	t	dν+1	t+2ν
2	4/3[a]	1.297 ± 0.007[b]	11/3	3.967 ± 0.007
3	0.89 ± 0.01[c]	1.96 ± 0.1[d]	3.67 ± 0.03	3.74 ± 0.1
4	0.66[e]	2.37[e]	3.64	3.69
5	0.57[e]	2.73[e]	3.85	3.87
6	1/2	3	4	4

[a] Nienhuis 1982

[b] Herrmann et al. 1984, Zabolitzky 1984, Lobb and Frank 1984

[c] Heermann and Stauffer 1981

[d] Derrida et al. 1983

[e] Fisch and Harris 1978

the singly connected portions of the links (see, e.g., Deutscher 1983). It is also noteworthy that good numerical simulations of 2D random networks lead to the result T = 3.96 ± 0.04 which is consistent with the upper bound for T (see Zabolitzky et al. 1986).

We now present an alternative derivation of the upper bound on T for the case d=2. In this case we are able to give a rigorous mathematical proof of the inequality based on a variational principle. We exploit the fact that the angular part of the elastic energy, i.e., the second term in Eq. (3), is similar in form to the expression for the rate of production of Joule heat W of a conducting network

$$W = \sum_{(ij)=nn} \frac{1}{2} g_{ij} (V_i - V_j)^2 , \tag{14}$$

where V_i is the potential at the site i. Moreover, if the potentials at two sites of a connected network are fixed, the potentials at the other sites will be such as to minimize W. Since any other assignment of the internal V_i's will give a larger value for W, we can obtain an upper bound for W by setting $g_{ij}=m_{ij}$ and substituting the angular increments $\delta\phi_i$ for V_i. Some care is required when doing this: The angles $\delta\phi_i$ refer to bonds whereas the potentials V_i refer to sites. Therefore the conducting network that must be considered is the "covering network" to the original elastic network. The covering network is constructed by replacing each bond of the original network by a new site, and connecting a pair of new sites by a new bond if and only if the original pair of bonds had a site in common. The covering network is known to have the same percolation threshold as the original network in addition to having the same critical indices (see, e.g., Shante and Kirkpatrick, 1971).

In order to get a bound for the macroscopic moduli, we consider a large elastic network with an internal boundary in the form of a circle of radius L and an external boundary at a very large distance away (i.e., much greater than either L or ξ). The bonds at the external boundary are fixed, i.e., $\delta\phi_i=\delta b_i=0$ there, while those at the internal boundary are given a rigid rotation, i.e., $\delta b_i=0$ and $\delta\phi_i=\delta\phi_o\neq 0$ there. If $L\gg\xi$, then the elastic energy of the network will be of order

$$E_L \sim C_e L^d \delta\phi_o^2 \quad . \tag{15}$$

We now consider the covering network as a network of conductors $g_{ij}=m_{ij}$, with boundary conditions $V_i=0$ on the external boundary and $V_i=\delta\phi_o$ on the internal boundary. The production rate of Joule heat will then be of order

$$W_L \sim \sigma_e L^{d-2} \delta\phi_o^2 \quad . \tag{16}$$

The different power of L appearing in Eqs. (15) and (16) is a result of the different physical dimensions of the elastic modulus C and the conductivity σ. From the earlier discussion, we deduce that $W_L \lesssim E_L$.

We now allow L to decrease until it becomes less than ξ. When $L\ll\xi$, we expect both E_L and W_L to be independent of L since the internal boundary will now intersect the infinite cluster only at a small number of points, all of them belonging at most to one link. To preserve the correct physical dimensions of Eqs. (15) and (16) and ensure continuity with L, we must replace L by ξ in those equations when $L \lesssim \xi$. Thus we obtain $W_\xi \lesssim E_\xi$, and hence

$$\sigma_e \, \xi^{-2} \sim (p-p_c)^{t+2\nu} \lesssim C_e \sim (p-p_c)^T \; ; \qquad T < t+2\nu \quad . \tag{17}$$

There remains a challenge to extend this type of proof to higher dimensions $d>2$. There also remains a challenge to prove that Eq. (17) is actually an equality for all $2<d<6$, as would appear reasonable from the numerical evidence of Table 1. Such a proof might teach us something new about the structure of the percolating cluster.

This research was supported in part by a grant from the Israel Academy of Sciences and Humanities.

REFERENCES

Amit, D.J. (1978). Field Theory, the Renormalization Group, and Critical Phenomena. McGraw-Hill, London.
Bergman, D.J. (1985). Phys. Rev. B31, 1696.
Bergman, D.J. (1986a). Phys. Rev. B33, 2013.
Bergman, D.J. (1986b). In Fragmentation, Form and Flow in Fractured Media, Vol. 8 of Annals of Israel Physical Society, eds. R. Englman and Z. Jaeger, pp. 266-272. Adam Hilger, Bristol.
Bergman, D.J. and Duering, E. (1986). To appear in Phys. Rev. B - Rapid Communication.
Bergman, D.J. and Kantor, Y. (1984). Phys. Rev. Lett. 53, 511.
Coniglio, A. (1981). In Disordered Systems and Localization, Vol. 149 of Lecture Notes in Physics, eds. C. Costellani, C. DiCastro, and L. Peliti. Springer, Berlin.
Coniglio, A. (1982). J. Phys. A15, 3829.
Derrida, B., Stauffer, D., Herrmann, H.J. and Vannimenus, J. (1983). J. Phys. (Paris) Lett. 44, L701.
Deutscher, G., Zallen, R. and Adler, J. (1983). Eds., Percolation Structures and Processes, Vol. 5 of Annals of Israel Physical Society. Adam Hilger, Bristol.
Fisch, R. and Harris, A.B. (1978). Phys. Rev. B18, 416-420.
Heermann, D.W. and Stauffer, D. (1981). Z. Phys. B44, 339-344.
Herrmann, H.J., Derrida, B. and Vannimenus, J. (1984). Phys. Rev. B30, 4080.
Kantor, Y. and Webman, I. (1984). Phys. Rev. Lett. 52, 1891.
Lobb, C.J. and Frank, D.J. (1984). Phys. Rev. B30, 4090.
Nienhuis, B. (1982). J. Phys. A15, 199-213.
Roux, S. (1986). J. Phys. A19, L351-L356.
Shante, V.K.S. and Kirkpatrick, S. (1971). Adv. Phys. 20, 325.
Skal, A.S. and Shklovskii, B.I. (1975). Fiz. Tekh. Poluprov. 8, 1029-1032.
Zabolitzky, J.G. (1984). Phys. Rev. B30, 4077.
Zabolitzky, J.G., Bergman, D.J. and Stauffer, D. (1986). J. Stat. Phys. 44, 211-223.

ADAPTIVE ANISOTROPY: AN EXAMPLE IN LIVING BONE

S. C. Cowin
Tulane University, New Orleans, Louisiana, USA

Abstract

Many natural materials adjust to their environmentally applied loads by changing their microstructure. These microstructural changes induce changes in the mechanical anisotropy of the material. Thus, the anisotropy of the material adapts to the applied loads. Materials that adapt their anisotropy by mechanical means include soft living tissue, rocks and all granular materials. Materials that adapt their anisotropy by chemical means include living bones and trees and saturated sandstones and limestones. In this paper an example of adaptive anisotropy in living bone is presented.

Introduction

Many natural materials possess mechanisms by which they adjust the degree of their anisotropy in order to carry, more efficiently, the load to which they are being subjected. By the degree of anisotropy we mean the relative stiffness or compliance of the material in different directions. For example, some fibrous composites have a ratio of Young's modulus in their fiber direction to Young's modulus in their transverse direction of 200. These materials are said to be strongly anisotropic. On the other hand bone can be described as mildly anisotropic because the ratio of Young's moduli in different directions generally exceeds two. However, bone has a mechanism by which it changes the degree of its anisotropy to adjust to its environmental load and manmade fibrous composites do not generally possess these adaptive mechanisms. As an example of the effect of this adaptive mechanism in bone one can note the fact that the ratio of Young's moduli near the mid-shaft of the human femur is about two, but it decreases to near one (almost isotropic) near the joints.

A review of these adaptive anisotropy mechanisms of natural materials was presented by Cowin (1985a). In that work these mechanisms were divided into two broad classes, mechanical and chemical. Examples of materials that adapt their degree of anisotropy by microstructural mechanical mechanisms include soft biological tissue, rock, ice and all granular materials. This paper deals with a particular chemical adaptation mechanism which requires that the material be perfused with a reacting fluid. Examples of materials that adapt their internal structure and shape by chemical means include living bones and trees and certain geological materials like saturated sandstone and limestone. In this work an example of adaptive anisotropy in living cancellous bone tissue is described. This topic is also discussed in Cowin (1986).

Adaptive Anisotropy in Cancellous Bone Tissue

Cancellous bone tissue is the type of bone tissue that exists inside the whole bone of an animal. It is a very highly porous material like a manufactured foam and is sometimes called "spongy" bone tissue.

In a work entitled "Die Architektur der Spongiosa", Von Meyer (1867) presented a line drawing of the cancellous bone structure he had observed in the proximal end of the human femur. An important structural engineer of the period, C. Culmann, was struck by the similarity between von Meyer's sketches of the cancellous bone in the proximal femur and the principal stress trajectories in a crane-like curved bar he was then in the process of designing. A sketch of the trabecular architecture of the type von Meyer published is shown in Figure 1b and a sketch of Culmann's crane is shown in Figure 1a. Culmann and von Meyer noted that the orientation of the trabeculae of cancellous bone of the upper femur appeared to be coincident with the principal stress trajectories in the loaded femur. Starting from this idea, Wolff (1872) stated that, when the environmental loads on a bone are changed by trauma, pathology or change in life pattern, functional remodeling reorients the trabeculae so that they align with the new principal stress trajectories. Wolff never attempted to prove this assertion. Roux (1885) produced a well known analysis of the structure of the cancellous tissue of an ankylosed knee. Roux's analysis was generally accepted as a proof of Wolff's law of the remodeling of bone.

Pauwels (1954) reexamined the same knee that Roux studied. The knee had been preserved at the Pathological Institute of the University of Wurzburg. Using two dimensional photoelastic models to determine principal stress trajectories, Pauwels showed that Roux's analysis did not support Wolff's hypothesis. The ankylosed knee studied by Roux had a discontinuous shelf at the knee that disrupted a smooth trabecular architecture. Pauwels obtained another ankylosed knee, one with smooth exterior surfaces, from the collection at Wurzburg and applied his methods of analyzing stress trajectories to this second knee. Figure 2a shows Pauwels' sketch of the cancellous architecture of the ankylosed knee he considered, and Figure 2b shows the stress trajectories in his photoelastic model of the situation. Pauwels indicates that this analysis is a proof of Wolff's hypothesis.

The work presented here relates a quantitative stereological measure of the microstructural arrangement of trabeculae and pores of cancellous bone, called the fabric tensor, to the elastic constants of the tissue and leads to a mathematical formulation of Wolff's Law at remodeling equilibrium. Quantitative stereology is the measurement of

Figure 1. On the left (a) is Culmann's crane with the principal stress trajectories indicated. On the right (b) is von Meyer's sketch of the trabecular architecture in a section through the proximal end of the human femur. Both the femur and Culmann's crane are loaded transversely at their cantilevered ends as illustrated on the little insert at the lower far right. Taken from Wolff (1872).

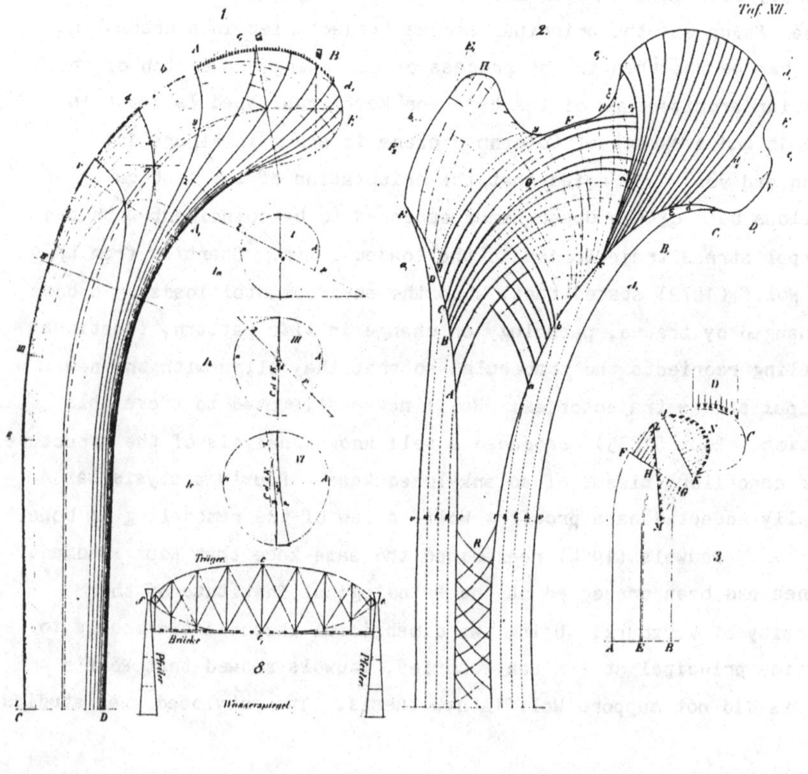

three dimensional geometric properties of materials from two dimensional images such as sections or projections. Based on work by Whitehouse (1974) and Whitehouse and Dyson (1974), Harrigan and Mann (1984) introduced a tensor valued measure of bone microstructure called the mean intercept length tensor. The fabric tensor employed in the present theory is the inverse square root of the mean intercept length tensor.

Figure 2. Upper: Pauwels' (1954) sketch of the cancellous architecture in the ankylosed knee he considered. Lower: Pauwels' sketch of the principal stress trajectories in the photoelastic model he constructed. The two transverse dotted lines in the photo-elastic model are the boundaries of the anatomically correct model of the knee.

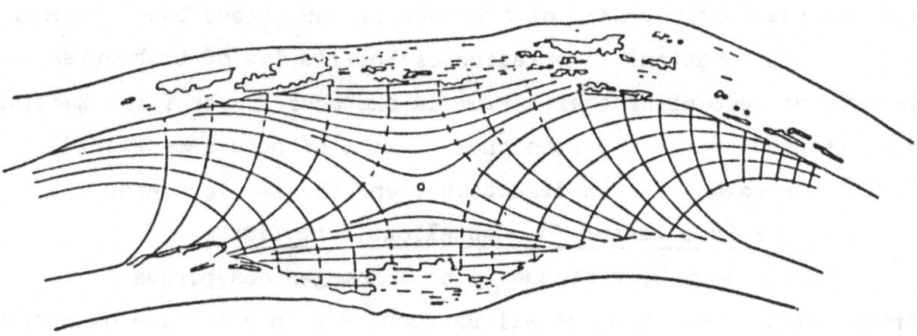

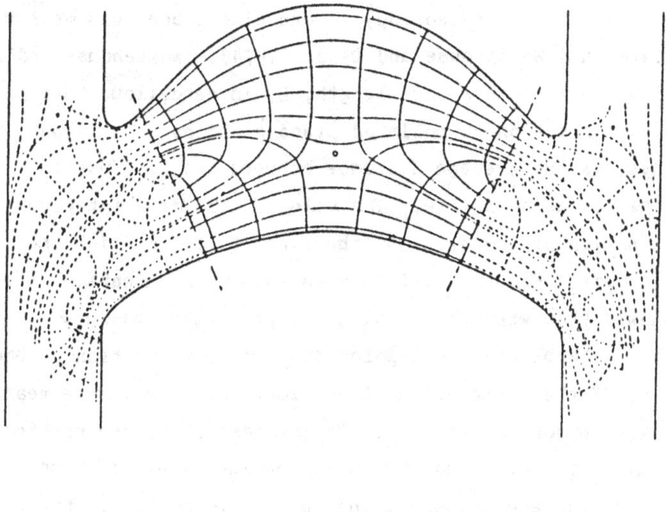

The fabric tensor and its experimental determination are discussed in the following section.

The relationship between the elastic constants and the fabric tensor in a local region of cancellous bone is described in a subsequent section. It is known that the porosity and orientation of the trabecular architecture as well as the elastic constants of cancellous bone vary with location in the bone tissue over a fairly wide range. This relationship permits the determination of the elastic constants from the fabric tensor if certain elastic parameters are known. These elastic parameters depend upon porosity but are independent of trabecular orientation of the tissue. They can be determined empirically from a knowledge of the elastic constants, porosity and the fabric tensor of a number of cancellous bone specimens.

A mathematical formulation of Wolff's law of trabecular architecture at remodeling equilibrium is sketched in the fifth section. The formulation depends only upon the concepts of bone remodeling equilibrium, principal stress directions, and the fabric tensor.

The Fabric Tensor for Cancellous Bone Tissue

Cancellous bone tissue is an inhomogeneous porous anisotropic structure. The cancellous structure is a lattice of narrow rods and plates (70 to 200 μm in thickness) of calcified bone tissue called trabeculae. The trabeculae are surrounded by marrow which is vašcular and provides nutrients and waste disposal for the bone cells. The distinctive structural anisotropy of cancellous bone was measured by Whitehouse (1974) and Whitehouse and Dyson (1974). Whitehouse and Dyson (1974) measured the mean intercept length, L, in cancellous bone tissue as a function of direction on polished plane sections. The mean intercept length is the average distance between two bone/marrow interfaces measured along a line. The value of the mean intercept length is a function of the slope of the line, θ, along which the measurement is made. Figure 3 illustrates such measurements. Whitehouse showed that when the mean intercept lengths measured in cancellous bone were plotted in a polar diagram as a function of the direction (i.e. of the slope of the line along which they were measured) the polar diagram produced ellipses. If the test lines are rotated through several values of θ and the corresponding values of mean intercept length L(θ) are measured, the data is found to fit the equation for an ellipse very closely:

$$\frac{1}{L^2(\theta)} = M_{11}\cos^2\theta + M_{22}\sin^2\theta + 2M_{12}\sin\theta\cos\theta, \tag{1}$$

where M_{11}, M_{22} and M_{12} are constants when the reference line from which the angle θ is measured is constant. The subscripts 1 and 2 indicate the axes of the x_1,x_2 coordinate system to which the measurements are referred. Harrigan and Mann (1984) observed that in three dimensions the mean intercept length would be represented by ellipsoids and would therefore be equivalent to a positive definite second rank tensor. The constants M_{ij} introduced above are then the components of a second rank tensor $\underset{\sim}{M}$ which are related to the mean intercept length $L(\underset{\sim}{n})$, where $\underset{\sim}{n}$ is a unit vector in the direction of the test line, by

$$\frac{1}{L^2(\underset{\sim}{n})} = M_{ij}n_in_j. \tag{2}$$

Figure 3. Test lines superimposed on a cancellous bone specimen. The test lines are oriented at the angle θ. The mean intercept length measured at this angle is denoted $L(\theta)$.

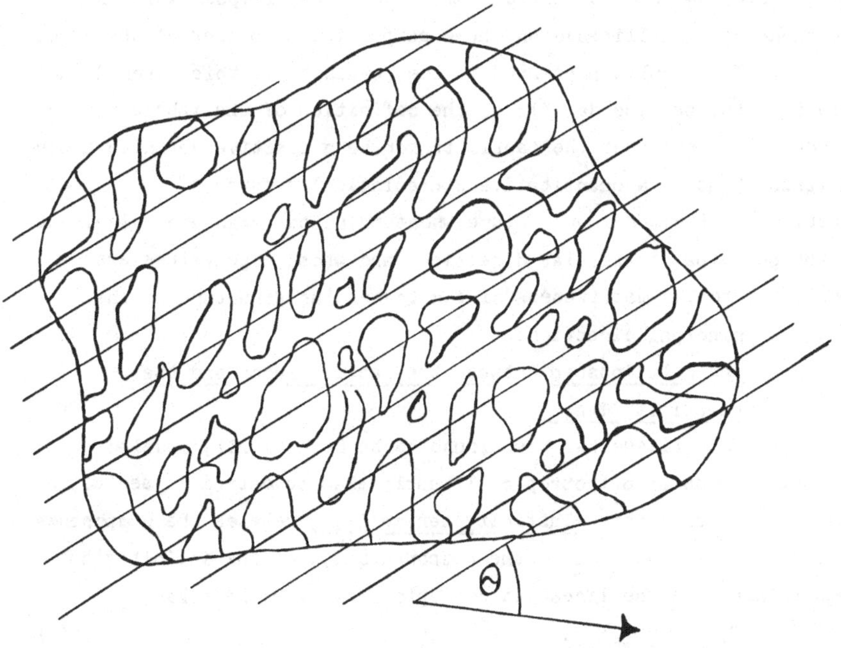

The work of Whitehouse (1974), Harrigan and Mann (1984), and others have shown that this second rank tensor is a good measure of the structural anisotropy in cancellous bone tissue. Cowin (1985b), noting the similarity to the structural ellipsoid/tensor employed by Oda (1976, see also Cowin, 1978, and Oda, et al. 1980) in a soil mechanics study, called an algebraically related tensor the fabric tensor of cancellous bone. $\underset{\sim}{H}$ is related to the mean intercept length tensor of Harrigan and Mann (1984) $\underset{\sim}{M}$ by

$$\underset{\sim}{H} = \underset{\sim}{M}^{-1/2} \tag{3}$$

The positive square root of the inverse of $\underset{\sim}{M}$ is well defined because $\underset{\sim}{M}$ is a positive definite symmetric tensor. $\underset{\sim}{M}$ is a positive definite symmetric tensor because it represents an ellipsoid. The motivation for defining a measure of fabric that depends inversely on the mean intercept length will be discussed later. The fabric tensor or mean intercept ellipsoid can be measured using the techniques described by Harrigan and Mann (1984) for a cubic specimen. On each of three orthogonal faces of a cubic specimen an ellipse will be determined from the directional variation of mean intercept length on that face. The fabric tensor can be constructed from these three ellipses which are the projections of the ellipsoid on three perpendicular planes of the cube.

The results presented in the remainder of this paper do not depend upon the precise details of the definition of the fabric tensor. It is only required that the fabric tensor be a positive definite second rank tensor that is a quantitative stereological measure of the trabecular architecture, a measure whose principal axes are coincident with the principal trabecular directions and whose eigenvalues are proportional to the massiveness of the trabecular structure in the associated principal direction.

The Relationship Between the Fabric Tensor and the Elasticity Tensor

Bone tissue has been found to be mechanically modeled satisfactorily as an orthotropic, linearly elastic material; see Cowin (1983). The fourth rank elasticity tensor C_{ijkm} relates the components T_{ij} of the stress tensor $\underset{\sim}{T}$ to the components E_{ij} of the infinitesimal strain tensor $\underset{\sim}{E}$ in the linear anisotropic form of Hooke's Law,

$$T_{ij} = C_{ijkm}E_{km}. \tag{4}$$

The elasticity tensor C_{ijkm} completely characterizes the linear elastic mechanical behavior of the cancellous bone tissue. If it is assumed that all the anisotropy of cancellous bone is due to the anisotropy of its trabecular structure, that is to say that the matrix material is itself isotropic, then a relationship between the components of the elasticity tensor C_{ijkm} and $\underset{\sim}{H}$ can be constructed. From previous studies of cancellous bone it is known that its elastic properties are strongly dependent upon its apparent density or, equivalently, the solid volume fraction of trabecular structure. This solid volume fraction is denoted by ν and is defined as the volume of solid trabecular struts per unit bulk volume of the tissue. Thus C_{ijkm} will be a function of ν as well as $\underset{\sim}{H}$. Cowin (1985b) has constructed the general representation of C_{ijkm} as a function of ν and $\underset{\sim}{H}$. The development of this representation was based on the notion that the matrix material of the bone tissue is isotropic and that the anisotropy of the whole bone tissue is due only to the geometry of the microstructure represented by the fabric tensor $\underset{\sim}{H}$. The mathematical statement of this notion is that the stress tensor $\underset{\sim}{T}$ is an isotropic function of the strain tensor $\underset{\sim}{E}$ and the fabric tensor $\underset{\sim}{H}$ as well as the solid volume fraction ν. Thus the tensor valued function

$$\underset{\sim}{T} = \underset{\sim}{T}(\nu, \underset{\sim}{E}, \underset{\sim}{H}) \tag{5}$$

has the property that

$$\underset{\sim}{Q}\underset{\sim}{T}\underset{\sim}{Q}^T = \underset{\sim}{T}(\nu, \underset{\sim}{Q}\underset{\sim}{E}\underset{\sim}{Q}^T, \underset{\sim}{Q}\underset{\sim}{H}\underset{\sim}{Q}^T) \tag{6}$$

for all orthogonal tensors $\underset{\sim}{Q}$. This definition of an isotropic tensor valued function is that given, for example, by Truesdell and Noll (1965).

It is shown by Cowin (1985b) that the most general form of the relationship between the elasticity tensor and the fabric tensor consistent with the isotropy assumption described above is

$$\begin{aligned} C_{ijkm} = {} & a_1\delta_{ij}\delta_{km} + a_2(H_{ij}\delta_{km} + \delta_{ij}H_{km}) + a_3(\delta_{ij}H_{kq}H_{qm} + \delta_{km}H_{iq}H_{qj}) + \\ & + b_1H_{ij}H_{km} + b_2(H_{ij}H_{kq}H_{qm} + H_{is}H_{sj}H_{km}) + b_3H_{is}H_{sj}H_{kq}H_{qm} + \\ & + c_1(\delta_{ki}\delta_{mj} + \delta_{mi}\delta_{kj}) + c_2(H_{ik}\delta_{mj} + H_{kj}\delta_{mi} + H_{im}\delta_{kj} + H_{mj}\delta_{ki}) + \\ & + c_3(H_{ir}H_{rk}\delta_{mj} + H_{kr}H_{rj}\delta_{mi} + H_{ir}H_{rm}\delta_{kj} + H_{mr}H_{rj}\delta_{ik}), \end{aligned} \tag{7}$$

where a_1, a_2, a_3, b_1, b_2, b_3, c_1, c_2 and c_3 are functions of ν and $\mathrm{tr}\underset{\sim}{H}$, $\mathrm{tr}\underset{\sim}{H}^2$ and $\mathrm{tr}\underset{\sim}{H}^3$. Substitution of (7) into (4) yields an explicit equation for the dependence of $\underset{\sim}{T}$ upon $\underset{\sim}{H}$ and $\underset{\sim}{E}$,

$$
\begin{aligned}
T_{ij} = {} & a_1\delta_{ij}E_{kk}+a_2(H_{ij}E_{kk}+\delta_{ij}H_{km}E_{km})+a_3(\delta_{ij}H_{kq}H_{qm}E_{km}+E_{kk}H_{iq}H_{qj})+ \\
& + b_1H_{ij}H_{km}E_{km}+b_2(H_{ij}H_{kq}H_{qm}E_{km}+H_{is}H_{sj}H_{km}E_{km})+ \\
& + b_3(H_{is}H_{sj}H_{kq}H_{qm}E_{km})+2c_1E_{ij}+2c_2(H_{ik}E_{kj}+H_{kj}E_{ki})+ \\
& + 2c_3(H_{ir}H_{rk}E_{kj}+H_{kr}H_{rj}E_{ki})
\end{aligned} \tag{8}
$$

The representation (7) for the fourth rank elasticity tensor is not capable of representing all possible elastic material symmetries. It cannot represent triclinic material symmetry, which is a total lack of symmetry, or monoclinic material symmetry which is characterized by a single plane of reflective material symmetry. The least material symmetry that can be represented by (7) is orthotropy. To see that this is the case we expand (7) in the indicial notation in the coordinate system that diagonalizes the fabric tensor. Thus the H_{12}, H_{13} and H_{23} components of $\underset{\sim}{H}$ vanish in this system and H_{11}, H_{22} and H_{33} are the three eigenvalues of $\underset{\sim}{H}$. The result of this expansion of (7) is the following nine non-zero components

$$
\begin{aligned}
C_{1111} &= a_1+2c_1+2(a_2+2c_2)H_{11}+(2a_3+b_1+4c_3)H_{11}^2+2b_2H_{11}^3+b_3H_{11}^4, \\
C_{2222} &= a_1+2c_1+2(a_2+2c_2)H_{22}+(2a_3+b_1+4c_3)H_{22}^2+2b_2H_{22}^3+b_3H_{22}^4, \\
C_{3333} &= a_1+2c_1+2(a_2+2c_2)H_{33}+(2a_3+b_1+4c_3)H_{33}^2+2b_2H_{33}^3+b_3H_{33}^4, \\
C_{1122} &= a_1+a_2(H_{11}+H_{22})+a_3(H_{22}^2+H_{11}^2)+b_1H_{11}H_{22}+b_2(H_{11}H_{22}^2+H_{11}^2H_{22})+b_3H_{11}^2H_{22}^2, \\
C_{1133} &= a_1+a_2(H_{11}+H_{33})+a_3(H_{33}^2+H_{11}^2)+b_1H_{11}H_{33}+b_2(H_{11}H_{33}^2+H_{11}^2H_{33})+b_3H_{11}^2H_{33}^2, \\
C_{2233} &= a_1+a_2(H_{22}+H_{33})+a_3(H_{22}^2+H_{33}^2)+b_1H_{22}H_{33}+b_2(H_{22}H_{33}^2+H_{33}H_{22}^2)+b_3H_{22}^2H_{33}^2, \\
C_{1212} &= c_1+c_2(H_{11}+H_{22})+c_3(H_{11}^2+H_{22}^2), \\
C_{1313} &= c_1+c_2(H_{11}+H_{33})+c_3(H_{11}^2+H_{33}^2), \\
C_{2323} &= c_1+c_2(H_{22}+H_{33})+c_3(H_{22}^2+H_{33}^2),
\end{aligned} \tag{9}
$$

and that all other components C_{ijkm} vanish. If an elastic material symmetry has one cartesian coordinate system in which only these nine components are non-zero and distinct from one another, then the material is orthotropic. Notice that these nine components of C_{ijkm} are distinct if and only if the eigenvalues of $\underset{\sim}{H}$ are distinct. It is easy to show that, if two eigenvalues of $\underset{\sim}{H}$ are coincident, then the symmetry reduces to transversely isotropic symmetry and, if all the eigenvalues of $\underset{\sim}{H}$ are equal, it reduces to isotropic material symmetry.

Adding together the relations (9) it can be seen that

$$\begin{aligned}C_{1111}+C_{2222}+C_{3333} = C_{1122}+C_{1133}+C_{2233} + 2(C_{1212}+C_{1313}+C_{2323}) +\\ +b_1(H_{11}(H_{11}-H_{22})+H_{22}(H_{22}-H_{33})+H_{33}(H_{33}-H_{11}))\\ +b_2(H_{11}^2(H_{11}-H_{22})+H_{22}^2(H_{22}-H_{11})+H_{33}^2(H_{33}-H_{22})\\ +H_{11}^2(H_{11}-H_{33})+H_{22}^2(H_{22}-H_{33})+H_{33}^2(H_{33}-H_{11}))\\ +b_3(H_{11}^2(H_{11}^2-H_{22}^2)+H_{22}^2(H_{22}^2-H_{33}^2)+H_{33}^2(H_{33}^2-H_{11}^2). \qquad (10)\end{aligned}$$

This result shows that if second order terms in $\underset{\sim}{H}$ are neglected the nine non-zero components of C_{ijkm} will no longer be independent. It follows that orthotropic symmetry could not be represented by (8) if all but the linear terms in $\underset{\sim}{H}$ were neglected.

Previous theoretical and experimental work suggests that the g's and h's should be proportional to the square or the cube of ν. It can be shown that the value of the tensor $\underset{\sim}{H}$ in a particular direction, say H_{11}, will increase with increasing elastic Young's modulus in that direction. Thus larger values of $\underset{\sim}{H}$ are associated with larger values of the Young's modulus. The reverse is true for the tensor $\underset{\sim}{M}$ considered in the previous section; larger values of $\underset{\sim}{M}$ are associated with smaller values of Young's modulus. This is part of the reason that $\underset{\sim}{H}$ and not $\underset{\sim}{M}$ is taken here as the measure of fabric.

The relationship $C_{ijkm}(\nu,\underset{\sim}{H})$ is the foundation stone of the model of the stress adaptation of trabecular architecture described in the section that follows. It relates the elastic constants to measures of the trabecular architecture, namely the solid volume fraction and the fabric tensor. The relationship $C_{ijkm}(\nu,\underset{\sim}{H})$ will determine the changes in the local elastic constants of cancellous bone as its trabecular architecture changes with stress adaptation.

Wolff's Law at Remodeling Equilibrium

A mathematical formulation of Wolff's Law at remodeling equilibrium is presented in this section. The relationship $C_{ijkm}(\nu,\underset{\sim}{H})$ given by (7) specifies the elastic constants of the cancellous bone tissue in terms of the solid volume fraction ν and the fabric tensor $\underset{\sim}{H}$. The elastic constants C_{ijkm}, in turn, relate the components T_{ij} of the stress $\underset{\sim}{T}$ to the components E_{km} of strain $\underset{\sim}{E}$, as illustrated in equation (4). The basic parameters characterizing the stress-strain-material microstructure relation of cancellous bone tissue are then ν, $\underset{\sim}{T}$, $\underset{\sim}{H}$, $\underset{\sim}{E}$ and C_{ijkm}.

The absence of bone remodeling is remodeling equilibrium. Specifically, remodeling equilibrium (RE) is the set of conditions under which there is no realignment of trabecular architecture and no net deposition or resorption of cancellous bone tissue. RE is thus characterized by a particular equilibrium architecture, denoted by ν^* and $\underset{\sim}{H}^*$, and a particular stress and strain state, denoted by $\underset{\sim}{T}^*$ and $\underset{\sim}{E}^*$, respectively. The RE stress $\underset{\sim}{T}^*$ and strain $\underset{\sim}{E}^*$ may actually be ranges of stress and strain, or some special measures of stress and strain history. This matter is discussed by Cowin (1984). It is reasonable to think of $\underset{\sim}{T}^*$ and $\underset{\sim}{E}^*$ as long time averages of the environmental stress and strain, although they may be more specific functions of the stress and strain history of the bone tissue.

In the introduction it was noted that Wolff's law of trabecular architecture states that the principal stress axes coincide with the principal trabecular directions in cancellous bone in RE. This means that, in RE, the principal axes of stress $\underset{\sim}{T}^*$ must coincide with the principal axes of fabric $\underset{\sim}{H}^*$. This coincidence of principal axes is assured if the matrix multiplication of $\underset{\sim}{T}^*$ and $\underset{\sim}{H}^*$ is commutative, i.e. if

$$\underset{\sim}{T}^*\underset{\sim}{H}^* = \underset{\sim}{H}^*\underset{\sim}{T}^*. \qquad (11)$$

This appears to be the first mathematical formulation of what is considered by many to be Wolff's law of trabecular architecture in remodeling equilibrium, namely that the principal stress directions are coincident with the principal directions of the trabecular architecture. It is also generally agreed that the direction associated with largest (smallest) Young's modulus in the cancellous bone should be associated with the direction in which the massiveness of the trabecular architecture is greatest (least). If the Young's moduli are ordered by $E_1 \geq E_2 \geq E_3$, then this condition is satisfied if $H_1 \geq H_2 \geq H_3$.

Since the principal axes of $\underset{\sim}{T}^*$ and $\underset{\sim}{H}^*$ are coincident by Wolff's Law, it can be shown using equations (4) and (9) that the principal axes of $\underset{\sim}{E}^*$ are also coincident with those of $\underset{\sim}{T}^*$ and $\underset{\sim}{H}^*$. To see this consider a principal coordinate system of $\underset{\sim}{H}^*$. This coordinate system is also a principal coordinate system of $\underset{\sim}{T}^*$. Equations (9) are applicable and may be substituted in equations (4). Since the shear stresses are all zero in this coordinate system, it follows that the shearing strains are also zero, hence the coordinate system is also a principal coordinate system of $\underset{\sim}{E}^{*}$. These results are summarized by any two of the following three relations:

$$\underset{\sim}{T}^{*}\underset{\sim}{H}^{*} = \underset{\sim}{H}^{*}\underset{\sim}{T}^{*},\ \underset{\sim}{T}^{*}\underset{\sim}{E}^{*} = \underset{\sim}{E}^{*}\underset{\sim}{T}^{*},\ \underset{\sim}{H}^{*}\underset{\sim}{E}^{*} = \underset{\sim}{E}^{*}\underset{\sim}{H}^{*}. \quad (12)$$

ACKNOWLEDGEMENT

This work has been supported by the Musculoskeletal Diseases Program of NIADDKA and NSF.

REFERENCES

Cowin, S. C. (1978), Microstructural Continuum Models for Granular Materials, In Continuum Mechanical and Statistical Approaches in the Mechanics of Granular Materials, S. C. Cowin and M. Satake, eds. pp. 162-170, Bunken Fukyu-Kai, Tokyo.

Cowin, S. C. (1983), The Mechanical and Stress Adaptive Properties of Bone, Ann. Biomed. Engr., 11, 263-295.

Cowin, S. C. (1984), Modeling of the Stress Adaptation Process in Bone, Cal. Tissue Int., 36, S99-S104.

Cowin, S. C. (1985a), Adaptive Anisotropy in Natural Materials, In Colloque International du CNRS No. 319, Compartement Plastique Des Solids Anisotropes, ed. J. P. Boehler, pp. 3-28.

Cowin, S. C. (1985b), The Relationship Between the Elasticity Tensor and the Fabric Tensor, Mechanics of Materials, 19, 137-147.

Cowin, S. C. (1986), Wolff's Law of Trabecular Architecture at Remodeling Equilibrium, J. Biomech. Engr., 108, 83-88.

Harrigan, T. and Mann, R. W. (1984), Characterization of Microstructural Anisotropy in Orthotropic Materials Using a Second Rank Tensor, J. Mat. Sci., 19, 761-767.

Meyer, G. H. (1867), Die Architektur der Spongiosa. Archiv fur Anatomie, Physiologie und wissenschaftliche Medizin (Reichert und DuBois-Reymonds Archiv), 34, 615-628.

Oda, M. (1976), Fabrics and Their Effects on the Deformation Behaviors of Sand, Dept. of Foundation Engr., Saitama University.

Oda, M., Konishi, J. and Nemat Nasser, S. (1980), Some Experimentally Based Fundamental Results on the Mechanical Behavior of Granular Materials, Geotechnique, 30, 479-495.

Pauwels, F. (1954), Kritische Uberprufung der Roux'schen Abhandlung Beschreibung und Erlauterung einer knochernen Kniegelenksankylose. Z. Anat. EntwGesch. 117, 528-532.

Roux, W. (1885), Gesammelte Abhandlungen uber die Entwicklungsmechanik der Organismen, W. Engelmann, Leipzig.

Truesdell, C. and Noll, W. (1965), The Non Linear Field Theories of Mechanics, In Handbuch der Physik, ed. S. Flugge, III/3.

Whitehouse, W. J. (1974), The Quantitative Morphology of Anisotropic Trabecular Bone, J. Microscopy, 101, 153-168.

Whitehouse, W. J. and Dyson, E. D. (1974), Scanning Electron Microscope Studies of Trabecular Bone in the Proximal End of the Human Femur, J. Anat., 118, 417-444.

Wolff, J. (1872), Uber die innere Architektur der Knochen und ihre Bedeutung fur die Frage vom Knochenwachstum. Archiv fur pathologische Anatomie und Physiologie und fur klinische Medizin (Virchovs Archiv), 50, 389-453.

STABILITY OF ELASTIC CRYSTALS

Irene Fonseca
CMAF
Ave Prof. Gama Pinto 2
1699 Lisboa Codex
Portugal

1. INTRODUCTION

Twinning, phase transitions and memory shape effects are frequently observed in solid materials with crystalline structure (cf. [3], [19]). As remarked by Wayman & Shimizu [19], it is of great theoretical and technological interest to investigate and predict the onset of such phenomena and to analyse equilibria and stability for this type of material.

The mathematical approach to these problems has recently been addressed by several authors, including Chipot & Kinderlehrer [4], Ericksen [7], [8], [9], Fonseca [10], [11], [12], Fonseca & Tartar [13], James [13], Kinderlehrer [15], Parry [17] and Pitteri [18], within the framework of a continuum theory based on nonlinear thermoelasticity proposed by Ericksen [7], [8]. The thermoelastic behaviour is quite noticeable in certain alloys for some ranges of temperature. As an example, Basinski & Christian [3] found that at room temperature Indium-Thallium alloys are perfectly plastic but for temperatures above 100°C or below 0°C they exhibit rubber like behaviour.

We summarize briefly the notion of an elastic crystal as described by Ericksen [9] and Kinderlehrer [15] (see also Ericksen [7], [8], and Fonseca [10]). In terms of molecular theory, a (pure) crystal is a countable set of identical atoms arranged in a periodic manner. By fixing a cartesian coordinate system with origin at one of the atoms and orthonormal basis $[e_1, e_2, e_3]$, the position vectors of the atoms of the lattice relative to some choice of linearly independent lattice vectors $\{a_1, a_2, a_3\}$ may be given by

$$x = A\begin{bmatrix} m_1 \\ m_2 \\ m_3 \end{bmatrix},$$

where $m_i \in Z$, $i = 1,2,3$, and A is the matrix with columns a_1, a_2, a_3. It is possible to choose a lattice basis in infinitely many ways: the same periodic structure is generated by $\{a_1', a_2', a_3'\}$ if and only if

$$A' = AM$$

where $M \in GL(Z^3) := \{M \in M^{3\times 3} | M_{ij} \in Z,\ i,j = 1,2,3, \text{ and } \det M = \pm 1,\}$ and $M^{3\times 3}$ denotes the set of 3×3 real matrices endowed with the topology of $\mathbb{R}^9$.

Due to frame indifference, the Helmholtz free energy Φ is given by

$$\Phi = \Phi\,(A^T A, \theta),$$

where θ is the absolute temperature. Since Φ does not distinguish between choices of lattice basis, we have then

$$\Phi(A^T A, \theta) = \Phi(M^T A^T A M, \theta) \qquad (1.1)$$

for all $M \in GL(Z^3)$.

At constant temperature θ, if in a reference configuration the crystal occupies an open bounded strongly Lipschitz domain $\Omega \subset \mathbb{R}^3$ and if $u : \Omega \to \mathbb{R}^3$ is a deformation of the lattice with deformation gradient

$$F = \nabla u,\ F_{ij} = \frac{\partial u_i}{\partial x_j},\ i,j, = 1,2,3$$

and

$$\det F > 0 \text{ a.e. in } \Omega,$$

then the energy density (per unit reference volume) is given by

$$W(F) = W(F,\theta) := \Phi((FA)^T(FA),\theta).$$

It follows from (1.1) that

$$W(RFM) = W(F) \qquad (1.2)$$

for all $R \in O^+(3)$, $F \in M_+^{3\times 3}$ and $M \in G_A^+$ where $O^+(3)$ is the proper orthogonal group, $M_+^{3\times 3} := \{F \in M^{3\times 3} \mid \det F > 0\}$ and $G_A^+ := \{AMA^{-1} | M \in GL(Z)^3 \text{ and } \det M = 1\}$.

As is customary in nonlinear elasticity, we suppose that W is bounded below and Borel measurable and, in order to prevent interpenetration of the matter, we further assume that

$$W(F) \to +\infty, \text{ when } \det F \to 0^+ \qquad (1.3)$$

with

$$W(F) = +\infty \text{ if and only if } \det F \leq 0.$$

Moreover, if W is differentiable in $M_+^{3\times 3}$, the first Piola-Kirchhoff stress tensor is given by

$$S(F) = \frac{\partial W}{\partial F}(F)$$

for all $F \in M_+^{3\times 3}$.

It is the behaviour of W with respect to θ together with the invariance (1.2) that help us to understand the onset of cubic-tetragonal transitions and the appearance of twinned states in the martensite phase at low temperature (cf. Basinski & Christian [3], Ericksen [8], Fonseca [10], Kinderlehrer [15], Wayman & Shimizu [19]). As discussed by James [14], a twinned configuration is composed of several morphologically different phases that are symmetry related and energetically equivalent, meeting at surfaces of composition (interfaces) where the deformation gradient suffers jump discontinuities.

However, due to the invariance of W under the infinite discrete group G_A^+, we are faced with serious difficulties when analyzing equilibria and stability problems. In fact, as shown by Ericksen [9] and Kinderlehrer [15], the presence of twinned co- existent metastable states leads to the breakdown of ellipticity of the nonlinear system of equilibrium equations

$$-\text{div } S(\nabla u) = f \quad \text{a.e. in } \Omega .$$

Also, the minimization of the total energy functionals cannot be achieved by direct methods of the calculus of variations (cf. Ball [1]) since, as W is bounded on several directions, there is failure of growth at infinity and the total stored energy functional

$$u \to \int_\Omega W(\nabla u(x))dx$$

is not sequentially weakly * lower semicontinuous (cf. Chipot & Kinderlehrer [4] and Fonseca [10]). Moreover, the usual relaxation techniques (cf. Dacorogna [5]) cannot be applied because of condition (1.3).

As we will show next, it turns out that these variational problems are very unstable (cf. Chipot & Kinderlehrer [4], Fonseca [11], [12] and Fonseca & Tartar [13]).

2. VARIATIONAL PROBLEMS AND RELAXATION

Throughout this section, we assume that (1.2) and (1.3) hold.

Firstly, we discuss briefly the stability properties of solutions of the dead loading traction boundary value problem.

If $f : \Omega \to \mathbb{R}^3$ is the body force per unit volume in the reference configuration and if $t : \partial\Omega \to \mathbb{R}^3$ is the surface traction per unit area of the undeformed configuration, we seek minimizers of the total energy functional $E(.)$ which is given by

$$E(u) := \int_\Omega W(\nabla u)dx - \int_\Omega f.u\, dx - \int_{\partial\Omega} t.u\, dS.$$

As proved in Fonseca [11], only residual stresses can provide minima, namely

<u>Theorem 2.1</u> ([11])

Let $\Omega \subset \mathbb{R}^3$ be a bounded C^1 domain, and for $1 < p < +\infty$, $\frac{1}{p} + \frac{1}{q} = 1$, let $f \in L^q(\Omega;\mathbb{R}^3)$ and $t \in W^{1-\frac{1}{q},q}(\partial\Omega;\mathbb{R}^3)$ satisfy the compatibility condition

$$\int_\Omega f\, dx + \int_{\partial\Omega} t\, dS = 0.$$

If W is continuous then

$$\inf_{u \in W^{1,p}(\Omega;\mathbb{R}^3)} E(u) > -\infty$$

if and only if $f = 0$ a.e. in Ω and $t = 0$ a.e. on $\partial\Omega$.

<u>Remark 2.2</u>

(1) The conclusion of Theorem 2.1 holds for the cases $p = 1$ or $p = +\infty$ whenever the problem

$$\begin{cases} -\text{div } T = f & \text{in } \Omega, \\ T\upsilon = t & \text{on } \partial\Omega, \end{cases}$$

admits a solution $T \in L^1(\Omega;M^{3\times 3})$, where υ is the outward unit normal to $\partial\Omega$.

(2) When there is no loading,

$$\inf_{u \in W^{1,p}(\Omega;\mathbb{R}^3)} \int_\Omega W(\nabla u)dx$$

is attained at some $u_0 \in W^{1,p}(\Omega;\mathbb{R}^3)$ if and only if

$$W(\nabla u_0) = \min_{F \in M_+^{3\times 3}} W(F) \text{ a.e. } x \in \Omega.$$

The nature and number of such natural states can be very complicated. In fact, if W is differentiable in $M_+^{3\times 3}$ it may be possible to construct infinite collections of twinned stable solutions of the homogeneous problem

$$\begin{cases} -\text{div } S(\nabla u) = 0 & \text{in } \Omega, \\ S(\nabla u).\upsilon = 0 & \text{on } \partial\Omega. \end{cases}$$

It can be shown also that metastable states are subjected to severe restrictions (cf. Fonseca [11]). In fact, relative minimizers of $E(.)$ with respect to the norm $\|.\|_\infty + \|.\|_{1,p}$, for $1 \leq p < +\infty$, must satisfy a balance law similar to the balance of angular momenta, where the deformation is replaced by its gradient. Several conclusions can be drawn from this necessary condition; in particular, it implies that no piecewise homogeneous deformation is a relative minimizer unless there is no loading.

As shown by Ball & Murat [2], $W^{1,\infty}$ - quasiconvexity is a necessary condition for the sequential weak * lower semicontinuity of a multiple integral. Hence, since

$$u \to \int_\Omega W(\nabla u)dx$$

is not sequentially weak * lower semicontinuous, it might be helpful to identify the lower $W^{1,\infty}$ - quasiconvex envelope of W, QW. In [12], we proved that

$$QW(F) = h(\det F) \tag{2.3}$$

for all $F \in M^{3\times 3}$, where h is the lower convex envelope of the sub-energy function ϕ given by

$$\phi(\delta) := \inf \{W(F) | \det F = \delta\}, \quad \delta \in \mathbb{R}.$$

Moreover, we showed that the characterization given by Dacorogna [6] of the quasiconvex envelope of a continuous bounded below and everywhere finite function is still valid for W satisfying (1.2) and (1.3), namely

$$QW(F) = Z(F) \tag{2.4}$$

for all $F \in M_+^{3\times 3}$, where

$$Z(F) := \inf \{\frac{1}{\text{meas } \Omega} \int_\Omega W(F + \nabla\xi(x))dx | \xi \in W_0^{1,\infty}(\Omega;\mathbb{R}^3)\}.$$

A similar result was obtained by Chipot & Kinderlehrer [4], who proved that

$$Z(F) = h(\det F)$$

for every $F \in M_+^{3\times 3}$.

In Fonseca & Tartar [13], the pure displacement problem was studied. On taking into account (2.3), necessary and sufficient conditions were sought for the existence of minimizers of

$$I(u) := \int_\Omega h(\det \nabla u)dx - \int_\Omega f.u \, dx$$

where $f \in L^1(\Omega;\mathbb{R}^3)$ and $u \in u_0 + W_0^{1,\infty}(\Omega;\mathbb{R}^3)$.

We must notice, however, that on setting

$$\bar{I}(u) := \int_\Omega W(\nabla u)dx - \int_\Omega f.u \, dx,$$

$$\alpha := \inf \{I(u) | u \in u_0 + W_0^{1,\infty}(\Omega;\mathbb{R}^3)\}$$

and

$$\beta := \inf \{\bar{I}(u) | u \in u_0 + W_0^{1,\infty}(\Omega;\mathbb{R}^3)\}.$$

it was not possible to show that $\alpha = \beta$. In fact on letting

$$\alpha' := \inf \{I(u) | u \in u_0 + W_0^{1,\infty}(\Omega;\mathbb{R}^3) \text{ and } u \text{ is piecewise affine}\}$$

and

$$\beta' := \inf \{\bar{I}(u) \mid u \in u_0 + W_0^{1,\infty}(\Omega;\mathbb{R}^3) \text{ and } u \text{ is piecewise affine}\}$$

it follows from (2.3) and (2.4) that

$$\alpha \leq \beta \leq \beta' = \alpha'.$$

However, due to (1.3) we have that

$$h(\delta) \to +\infty \quad \text{when} \quad \delta \to 0^+,$$

and so a density argument could not be devised allowing us to conclude that $\alpha' = \alpha$.

We define the set of admissible deformations

$$H = \{u \in W^{1,\infty}(\Omega;\mathbb{R}^3) \mid \det \nabla u > 0 \text{ a.e. in } \Omega \text{ and } u|_{\partial\Omega} = u_0|_{\partial\Omega}\}.$$

<u>Theorem 2.5</u> ([13])

Let $\Omega \subset \mathbb{R}^3$ be an open bounded strongly Lipschitz domain and let $h:(0,+\infty) \to \mathbb{R}$ be convex and bounded below with $h(\delta) \to +\infty$ when $\delta \to 0^+$. Assume that $f \in L^1(\Omega;\mathbb{R}^3)$ and that $u_0 \in C(\bar{\Omega};\mathbb{R}^3)$ is one-to-one in Ω.

(a) If $u \in H$ is such that $I(u) < +\infty$ and $I(u) \leq I(v)$ for all $v \in H$, then

(i) $\nabla u^T f$ is a gradient in $D'(\Omega;\mathbb{R}^3)$,

(ii) there exists $D \subset \Omega$ with meas $\partial D = 0$ such that for every finite collection $\{x_1, \ldots, x_n, x_{n+1} \equiv x_1\} \subset \Omega \setminus D$ one has

$$\sum_{i=1}^{n} f(x_i).(u(x_i) - u(x_{i+1})) \geq 0,$$

(iii) there exists a convex function G: convex hull $u_0(\Omega) \to \mathbb{R}$ such that

$$\int_\Omega G(u(x))dx \in \mathbb{R} \quad \text{and} \quad f(x) \in \partial G(u(x)) \quad \text{a.e. } x \in \Omega$$

and

(iv) if, in addition, there exists $\alpha \in (0,1)$ such that

$$\int_\Omega h'(\alpha \det \nabla u) \det \nabla u \, dx > -\infty$$

then

$$G(u(x)) = h(\det \nabla u(x)) - h'(\det \nabla u(x)) \det \nabla u(x) + C \text{ a.e. } x \in \Omega$$

for some constant $C \in \mathbb{R}$.

(b) Let $u \in H$ be such that $I(u) < +\infty$ with

$$\int_\Omega h'(\det \nabla u) \det \nabla u \, dx > -\infty.$$

If there exists a convex function G: convex hull $u_0(\Omega) \to \mathbb{R}$ such that

$$G(u(x)) = h(\det \nabla u(x)) - h'(\det \nabla u(x)) \det \nabla u(x) \text{ a.e. } x \in \Omega$$

with

$$f(x) \in \partial G(u(x)) \text{ a.e. } x \in \Omega$$

then

$$I(u) \leqq I(v) \text{ for all } v \in H.$$

Remarks 2.6

(1) The necessary conditions are obtained regardless of the boundary conditions. In fact, it is possible to generalise part (a) of Theorem 2.2 as follows:
Let $3 < p < +\infty$, $t \in L^1(\partial\Omega_2;\mathbb{R})$ where $\partial\Omega = \overline{\partial\Omega_1} \cup \overline{\partial\Omega_2}$ with $\partial\Omega_1 \cap \partial\Omega_2 = \emptyset$ and let $H_p := \{u \in W^{1,p}(\Omega;\mathbb{R}^3) | u|_{\partial\Omega_1} = u_0\}$. Assume that $u \in H_p$ and that there exists $u_1 \in C(\overline{\Omega};\mathbb{R}^3)$ one-to-one in Ω such that $u|_{\partial\Omega} = u_1|_{\partial\Omega}$. If

$$J(u) := \int_\Omega h(\det \nabla u) dx - \int_\Omega f.u \, dx - \int_{\partial\Omega_2} t.u \, dS < +\infty$$

and if there exists $\epsilon > 0$ such that

$$J(u) \leqq J(v) \text{ for every } v \in H_p \text{ with } \|v - u\|_{L^p} < \epsilon$$

then (i), (ii) and (iii) hold. Moreover,
(iv') if there exists $0 < \alpha < 1 < \beta$ for which

$$-\infty < \int_\Omega h'(\alpha \det \nabla u) \det \nabla u \, dx \quad \text{and} \quad \int_\Omega h'(\beta \det \nabla u) \det \nabla u \, dx < +\infty$$

then

$$G(u(x)) = h(\det \nabla u(x)) - h'(\det \nabla u(x)) \det \nabla u(x) + C$$

for some constant $C \in \mathbb{R}$.

(2) The conditions (i) and (ii) of Theorem 2.5 may be used to

detect body forces f for which $\inf_{u \in H} I(u)$ is not attained in H. As an example, let $f \in W^{1,1}_{loc}(\Omega;\mathbb{R}^3)$. Then from (i) it follows that

$$\nabla f^T \nabla u \text{ is a symmetric matrix.} \tag{2.7}$$

Also, if f is smooth enough (e.g. $f \in C^1(\Omega;\mathbb{R}^3)$) then (ii) and (2.7) yield

$$\nabla f^T \nabla u \text{ is a positive symmetric matrix}$$

and so,

$$\det \nabla f \geq 0 \quad \text{a.e. in } \Omega . \tag{2.8}$$

In particular, (2.8) excludes compressive body forces of the type $f(x) = -kx$, $k > 0$.

From the theorem above, we may conclude that $\inf_{u \in H} I(u)$ is always achieved in the case where u_0 = identity, $f = -ke_3$ with $k > 0$ (gravity field) and the material is "strong", i.e.

$$\frac{h(\delta)}{\delta} \to +\infty \quad \text{when} \quad \delta \to +\infty. \tag{2.9}$$

If (2.9) fails, it is possible to construct an example where the material breaks down for k bigger than a certain critical k_c (cf. Fonseca & Tartar [15]).

REFERENCES

[1] BALL, J.M. "Convexity conditions and existence theorems in nonlinear elastostatics", Arch. Rat. Mech. Anal. 63 (1977), 337-403.

[2] BALL, J.M. & MURAT, F. "$W^{1,p}$-quasiconvexity and variational problems for multiple integrals", J. Funct. Anal. 58 (1984), 225-253.

[3] BASINSKI, Z.S. & CHRISTIAN, J.W., "Crystallography of deformation by twin boundary movements in Indium-Thallium alloys", Acta Metallurgica 2 (1954), 101-116.

[4] CHIPOT, M. & KINDERLEHRER, D. "Equilibrium configurations of crystals", to appear

[5] DACOROGNA, B. "Quasiconvexity and relaxation of non convex problems in the calculus of variations", J. Funct. Anal 46 (1982), 102-118.

[6] DACOROGNA, B. "Remarques sur les notions de polyconvexite, quasi-convexité et convexité de rang 1", J. Math. Pures et Appl. 64 (1985), 403-438.

[7] ERICKSEN, J.L. "Special topics in elastostatics", Advances in Applied Mech. 17 (1977), 188-244.

[8] ERICKSEN, J.L. "Some phase transitions in crystals", Arch. Rat.

Mech. Anal. 73 (1980), 99-124.
[9] ERICKSEN, J.L. "Twinning of crystals", IMA Preprint #95, Univ of Minn., Minneapolis, 1984.
[10] FONSECA, I. Variational Methods for Elastic Crystals. Thesis, Univ. of Minn., Minneapolis, 1985.
[11] FONSECA, I. Variational methods for elastic crystals", Arch. Rat. Mech. Anal., to appear.
[12] FONSECA, I. "The lower quasiconvex envelope of the stored energy function for an elastic crystal", to apear.
[13] FONSECA, I. & TARTAR, L. "The displacement problem for elastic crystals", in preparation.
[14] JAMES, R.D. "Finite deformation by mechanical twinning", Arch. Rat. Mech. Anal. 77 (1981), 143-176.
[15] KINDERLEHRER, D. "Twinning of crystals II". IMA Preprint #106, Univ. of Minn., Minneapolis, 1984.
[16] MORREY, C.B. "Quasi-convexity and the lower semi-continuity of multiple integrals", Pacific J. Math. 2 (1952), 25-53.
[17] PARRY, G. "Coexistent austenitic and martensitic phases in elastic crystals", Proceedings of the 5th International Symp. on Continuum Models of Discrete Systems, (SPENCER, A.J.M. ed. to appear).
[18] PITTERI, M. "On the kinematics of mechanical twinning in crystals", Arch. Rat. Mech. Anal. 88 (1985), 25-57.
[19] WAYMAN, C.M. & SHIMIZU, K. "The shape memory ('Marmen') effect in alloys", Metal Science Journal 8 (1972), 175-183.

OPTIMAL BOUNDS FOR CONDUCTION IN TWO-DIMENSIONAL, TWO-PHASE, ANISOTROPIC MEDIA

G.A. Francfort
Laboratoire Central des Ponts & Chaussées,
58 bld Lefebvre - 75732 PARIS Cedex 15 - France.

F. Murat
Laboratoire d'Analyse Numérique, Université Pierre & Marie Curie
5, Place Jussieu - 75230 PARIS Cedex 05 - France.

INTRODUCTION

This paper is concerned with the determination of the set of all possible effective conductivities of a two-phase anisotropic material with arbitrary phase geometry. Since Hashin & Shtrikman's original bounds on the set of possible isotropic effective tensors of a two-phase material with isotropic phases due attention has been paid to the case of isotropic phases (*cf.* Hashin (1983), Tartar (1985), Kohn & Milton (1985), Francfort & Murat (1986), Ericksen, Kinderlehrer, Kohn & Lions (1986) and references therein).

The case of polycrystalline media has been considerably less investigated (*cf.* Schulgasser (1977)).

In a two-dimensional setting, Lurie & Cherkaev (1984) addressed the problem of characterizing the set of all anisotropic effective conductivity tensors of a two-phase material with anisotropically conducting phases in arbitrary volume fraction. In the present paper (which describes the results of Francfort & Murat (1987)), we revisit Lurie & Cherkaev's bounds and derive a complete characterization in the two-dimensional case.

We consider two homogeneous and anisotropic conducting materials. If they are positioned in a common reference configuration, there exists an orthonormal basis e_1, e_2 of $\mathbb{R}^2$ such that the conductivity tensors A_1 and A_2 of the two phases read as

$$\begin{cases} A_1 = \alpha_1 e_1 \otimes e_1 + \alpha_2 e_2 \otimes e_2, \\ A_2 = \beta_1 e_1 \otimes e_1 + \beta_2 e_2 \otimes e_2, \end{cases} \tag{1}$$

and we assume with no loss of generality that

$$\begin{cases} 0<\alpha_1 \leqslant \alpha_2 <+\infty, \\ 0<\beta_1 \leqslant \beta_2 <+\infty, \\ \alpha_1\alpha_2 \leqslant \beta_1\beta_2. \end{cases} \tag{2}$$

We seek the set of all possible anisotropic effective tensors corresponding to the mixture of the two phases with no restriction on the volume fractions.

When both materials are isotropic ($\alpha_1 = \alpha_2$, $\beta_1 = \beta_2$), the result has been known since Tartar (1974). When the materials are anisotropic, the investigated set is shown to depend only on the eigenvalues of the effective conductivity tensor. Specifically it coincides with the set of all bounded measurable symmetric mappings on $\mathbb{R}^2$ whose eigenvalues lie in a compact subset L of $[\mathbb{R}_+^\star]^2$. The associated eigenvectors are arbitrary.

The region L is uniquely determined in terms of α_1, α_2, β_1, β_2 (*cf.* Definition 5). In fact it is the outermost region bounded by the eigenvalues of the effective tensors corresponding to rank-1 lamination of both phases with each other or with themselves. The direction of the rank-1 lamination which produces the boundary of that region strongly depends on the ordering properties of the eigenvalues of the original phases. Three cases have to be considered in the analysis, namely,

$\alpha_1 \leqslant \beta_1$ and $\alpha_2 \leqslant \beta_2$,
$\alpha_1 \leqslant \beta_1$ and $\alpha_2 > \beta_2$,
$\alpha_1 > \beta_1$ and $\alpha_2 \leqslant \beta_2$.

In the three cases parts of the boundary of L are achieved by lamination of the tensor A_1 (resp. A_2) with its image by a rotation of angle $\frac{\pi}{2}$ in the direction e_1 or e_2. The other parts of the boundary of L are obtained through a layering of the tensor A_1 with the tensor A_2 in the direction e_1 when $\alpha_1 \leqslant \beta_1$ and $\alpha_2 \leqslant \beta_2$ and in the direction e_2 otherwise. The reader is referred to Francfort & Murat (1987) for a complete exposition of this lamination process.

Our results agree with those of Lurie & Cherkaev (1984) in the case when $\alpha_1 \leqslant \beta_1$ and $\alpha_2 \leqslant \beta_2$ but disagree with theirs in the case when $\alpha_1 \leqslant \beta_1$ and $\alpha_2 > \beta_2$. The third case was not investigated in Lurie & Cherkaev (1984).

In the first Section of the paper the problem is formulated in the mathematical setting of H-convergence. The characterization of the possible effective tensors is given in the second Section.

It relies on a stability criterion pertaining to the form of the sets of conductivity tensors which remain stable under H-convergence. This stability criterion is the object of the third Section. The fourth and last Section is devoted to a sketch of its proof.

A complete and detailed exposition of the results presented here will be found in Francfort & Murat (1987). Our analysis is based on the theories of H-convergence and compensated compactness (*cf. e.g.* Murat (1977), (1978), Tartar (1977), (1979), (1985)).

1. SETTING OF THE PROBLEM

An arbitrary mixture of the two phases defined in (1) is obtained by considering the characteristic function $\chi(x)$ of the first phase in $\mathbb{R}^2$ and the orientation matrix $R(x)$ which quantifies the rotation of the conductivity tensor at the point x with respect to its reference configuration. In other words a conductivity tensor associated with a mixture of the two phases is a conductivity tensor of the form

$$A(x) = \chi(x)\ {}^tR(x)A_1R(x) + (1-\chi(x))\ {}^tR(x)A_2R(x),$$

where χ is the characteristic function of a measurable subset of $\mathbb{R}^2$, R is a measurable orthogonal matrix on $\mathbb{R}^2$, and A_1 and A_2 are the tensors defined in (1).

We consider a family A_ε of such mixtures, *i.e.* a sequence of χ_ε and R_ε such that

$$A_\varepsilon(x) = \chi_\varepsilon(x)\ {}^tR_\varepsilon(x)A_1R_\varepsilon(x)+(1-\chi_\varepsilon(x))\ {}^tR_\varepsilon(x)A_2R_\varepsilon(x), \tag{3}$$

where ε is a small parameter which may be viewed as the typical size of the heterogeneities in the mixture. We propose to investigate its macroscopic behaviour, with the help of the theory of H-convergence (Murat (1977), Tartar (1977)).

DEFINITION 1. *If* K *is a compact subset of* $[\mathbb{R}_+^\star]^2$, $\mathcal{M}(K)$ *is the set of all symmetric tensors* A *with coefficients in* $L_\infty(\mathbb{R}^2)$ *whose eigenvalues* λ_1, λ_2 *satisfy*

$$(\lambda_1(x),\lambda_2(x)) \in K,\ (\lambda_2(x),\lambda_1(x)) \in K \ \textit{almost everywhere} \ \bullet$$

DEFINITION 2. *Let* α,β *be two elements of* $\mathbb{R}_+^\star$. *A sequence* A_ε *of elements of* $\mathcal{M}([\alpha,\beta]^2)$ *H-converges to a symmetric linear mapping* A_o *if and only if for any bounded domain* Ω *of* $\mathbb{R}^2$ *the relation*

$$q_o = A_o w_o \tag{4}$$

holds true for any sequence w_ε *in* $[L_2(\Omega)]^2$ *such that*

$$\begin{cases} w_\varepsilon \longrightarrow w_o, \\ q_\varepsilon = A_\varepsilon w_\varepsilon \longrightarrow q_o, \end{cases} \tag{5}$$

weakly in $[L_2(\Omega)]^2$ *as* ε *tends to zero while*

$$\begin{cases} \operatorname{curl} w_\varepsilon \overset{\text{def}}{=} \dfrac{\partial (w_\varepsilon)_1}{\partial x_2} - \dfrac{\partial (w_\varepsilon)_2}{\partial x_1}, \\ \operatorname{div} q_\varepsilon \overset{\text{def}}{=} \dfrac{\partial (w_\varepsilon)_1}{\partial x_1} + \dfrac{\partial (w_\varepsilon)_2}{\partial x_2}, \end{cases} \tag{6}$$

lie in a compact set of $H^{-1}_{loc}(\Omega)$ •

REMARK 1. The basic properties resulting from the above definition can be found in *e.g.* Murat (1977), Tartar (1977), (1985), Francfort & Murat (1986), (1987) •

The notion of H-limit is meaningful by virtue of the

THEOREM 1. *If* A_ε *is a family of elements of* $\mathcal{M}([\alpha,\beta]^2)$, *there exists a subsequence of* A_ε *which H-converges to an element* A_o *of* $\mathcal{M}([\alpha,\beta]^2)$ *and all possible H-limits belongs to* $\mathcal{M}([\alpha,\beta]^2)$ •

The above theorem was first proved by Spagnolo (67), then revisited by Tartar (*cf.* Murat (77), Tartar (77)). A similar proof may be found in *e.g.* Simon (79).

REMARK 2. The family A_ε considered in (3) belongs to $\mathcal{M}([\alpha,\beta]^2)$ with

$$\alpha = \min(\alpha_1,\beta_1) \ , \quad \beta = \max(\alpha_2,\beta_2),$$

and according to Theorem 1, a subsequence of A_ε H-converges to an element A_o of $\mathcal{M}([\alpha,\beta]^2)$ •

At the possible expense of extracting converging subsequences, we are left with the

DEFINITION 3. *Consider a sequence* A_ε *of the form* (3) *where* A_1, A_2 *are given by* (1), χ_ε *is a sequence of measurable characteristic functions and* R_ε *is a sequence of orthogonal matrices with measurable coefficients. Whenever* A_ε *H-converges to* A_o *as* ε *tends to zero, the tensor* A_o *is referred to as an effective tensor for the mixture of* A_1 *and* A_2 •

2. CHARACTERIZATION OF THE POSSIBLE EFFECTIVE TENSORS.

The complete characterization of all effective tensors for the mixture of A_1 and A_2 depends on the relative magnitudes of their eigenvalues.

DEFINITION 4. *The tensors* A_1 *and* A_2 *defined by* (1) *are said to be well ordered if and only if*

$$\alpha_1 \leqslant \beta_1 \quad \text{and} \quad \alpha_2 \leqslant \beta_2 .$$

Otherwise, i.e. *if*

$$\alpha_1 \leqslant \beta_1 \quad \text{and} \quad \alpha_2 > \beta_2$$

or if

$$\alpha_1 > \beta_1 \quad \text{and} \quad \alpha_2 \leqslant \beta_2$$

they are said to be badly ordered •

DEFINITION 5. *If* α_1, α_2, β_1, β_2 *satisfy* (2), *and if further* $\alpha_1\alpha_2 \neq \beta_1\beta_2$ *the sets* L_w *and* L_b *are defined as follows :*

L_w *is the set of all* $(\lambda_1,\lambda_2) \in [\mathbb{R}_+^*]^2$ *such that*

$$\alpha_1\alpha_2 \leqslant \lambda_1\lambda_2 \leqslant \beta_1\beta_2, \tag{7}$$

$$\frac{(\beta_1-\alpha_1)\lambda_1\lambda_2 + (\beta_2-\alpha_2)\alpha_1\beta_1}{\beta_1\beta_2 - \alpha_1\alpha_2} \leqslant \lambda_1,\lambda_2 \leqslant \frac{\lambda_1\lambda_2(\beta_1\beta_2-\alpha_1\alpha_2)}{(\beta_1-\alpha_1)\lambda_1\lambda_2 + (\beta_2-\alpha_2)\alpha_1\beta_1}. \quad (8)$$

L_b *is the set of all* $(\lambda_1,\lambda_2) \in [\mathbb{R}_+^\star]^2$ *such that*

$$\alpha_1\alpha_2 \leqslant \lambda_1\lambda_2 \leqslant \beta_1\beta_2, \quad (9)$$

$$\frac{\lambda_1\lambda_2(\beta_1\beta_2-\alpha_1\alpha_2)}{(\beta_2-\alpha_2)\lambda_1\lambda_2 + (\beta_1-\alpha_1)\alpha_2\beta_2} \leqslant \lambda_1,\lambda_2 \leqslant \frac{(\beta_2-\alpha_2)\lambda_1\lambda_2+(\beta_1-\alpha_1)\alpha_2\beta_2}{\beta_1\beta_2 - \alpha_1\alpha_2}. \quad (10)$$

If $\alpha_1\alpha_2 = \beta_1\beta_2$,

$$L_w = L_b = \{(\lambda_1,\lambda_2) \in [\mathbb{R}_+^\star]^2 \mid \lambda_1\lambda_2 = \alpha_1\alpha_2 = \beta_1\beta_2, \min(\alpha_1,\beta_1) \leqslant \lambda_1,\lambda_2 \leqslant \max(\alpha_2,\beta_2)\}$$ •

REMARK 3. The set L_w and L_b are compacts subsets of $[\mathbb{R}_+^\star]^2$. Their boundaries are hyperbolic segments in the (λ_1,λ_2) variables (see Figures 1 to 3 and Remark 5 below for a geometrical representation of L_w and L_b in an other set of variables) •

The set of all effective tensors for the mixture of A_1 and A_2 is characterized with the help of Definitions 4 and 5. Specifically we obtain the following

THEOREM 2. *In the context of Definition* 1 *to* 5, *a symmetric tensor* $A_o(x)$ *with coefficients in* $L_\infty(\mathbb{R}^2)$ *is an effective tensor for the mixture of* A_1 *and* A_2 *if and only if it belongs to* $\mathcal{M}(L_w)$ *when* A_1 *and* A_2 *are well-ordered and to* $\mathcal{M}(L_b)$ *when they are not* •

In the case where both A_1 and A_2 are isotropic ($\alpha_1 = \alpha_2$, $\beta_1 = \beta_2$) Theorem 2 is the result of Tartar (1974).

REMARK 4. Theorem 2 says that all elements of $\mathcal{M}(L_w)$ (respectively $\mathcal{M}(L_b)$) can be achieved as H-limits of a sequence of A_ε satisfying (3) (*cf.* Definition 3) *and* that they are the only ones. Note the absence of any restrictions on the eigenvectors of the possible effective tensors $A_o(x)$.

The proof of the achievability of all elements of $\mathcal{M}(L_w)$ and $\mathcal{M}(L_b)$ is performed through explicit construction using multiple layering.

It will not be given here and the reader is referred to Tartar (1985), Francfort & Murat (1987). In the present paper we focus our attention on the "only if" part of Theorem 2 •

Figure 1. (d-λ) representation of L_w in the well ordered case and L_b in the badly ordered cases.

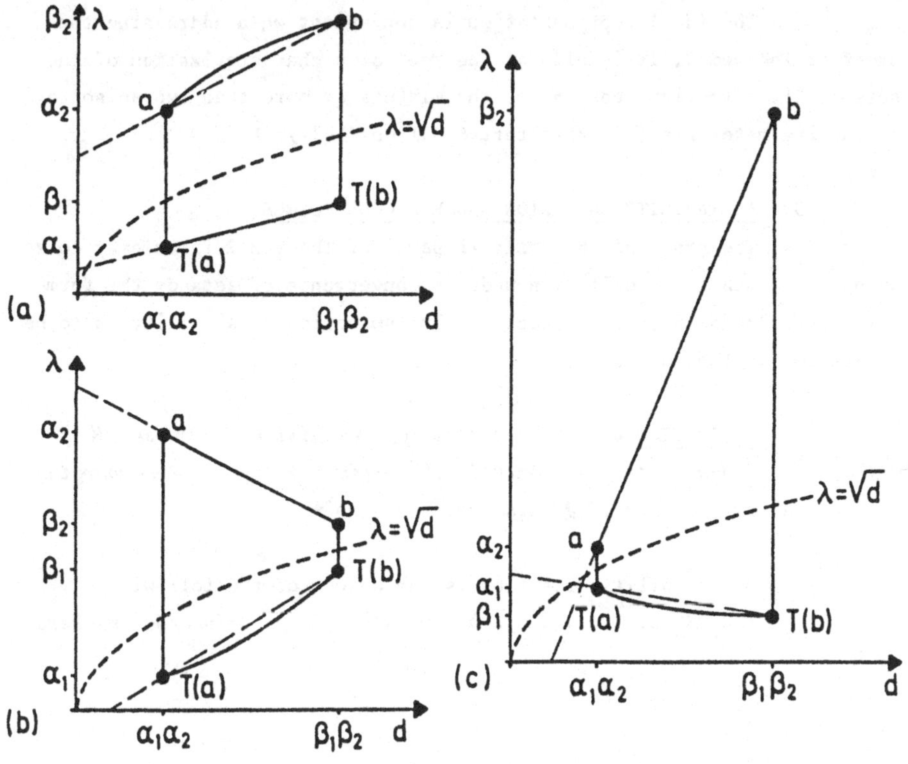

REMARK 5. Various geometrical representations of the sets L_w and L_b are presented in Francfort & Murat (1987). Figures 1(a), 1(b), 1(c) correspond to the so-called (d-λ)-representation of the sets L_w in the well-ordered case or L_b in the badly ordered case. Each point (λ_1,λ_2) of L_w (or L_b) is mapped onto two points p,T(p) whose coordinates are $(\lambda_1\lambda_2, \max(\lambda_1,\lambda_2))$ and $(\lambda_1\lambda_2, \min(\lambda_1,\lambda_2))$. Straight vertical line segments correspond to effective tensors with equal determinants. In the well ordered case the boundaries of the set L_w (for which inequalities (7), (8) become equalities) become the two vertical straight line segments ([a,T(a)],

$[b,T(b)])$ together with the concave hyperbolic segment $\widehat{ab}$ and the straight line segment $[T(a),T(b)]$ (*cf.* Fig. 1(a)). In the badly ordered cases the boundaries of the sets L_b (for which inequalities (9), (10) become equalities) are the two vertical line segments $([a,T(a)], [b,T(b)])$ together with the straight line segment $[a,b]$ and the convex hyperbolic segment $\widehat{T(a)T(b)}$ (*cf.* Fig. 1(b)-1(c)). When $\alpha_1\alpha_2 = \beta_1\beta_2$, the set $L_w (= L_b)$ reduces to a vertical straight line segment in its (d,λ)-representation.

The (d,λ)-representation is convenient when addressing the proof of Theorem 2. It is also at the root of a characterization of the sets of all effective tensors for the mixture of more than two anisotropic conducting materials (*cf.* Francfort & Milton (1987)) •

3. A STABILITY CRITERION UNDER H-CONVERGENCE

The proof of the "only if part" of Theorem 2 is an easy consequence of a stability criterion under H-convergence of sets of the form $\mathcal{M}(K)$ (*cf.* Remark 8 below). Specifically, our notion of stability is to be understood as the following

DEFINITION 6. *In the context of Definition 2 a subset $\mathcal{N}$ of $\mathcal{M}([\alpha,\beta]^2)$ is H-stable if and only if the H-limit A_0 of any H-converging sequence A_ε of elements of $\mathcal{N}$ also belongs to $\mathcal{N}$* •

Our stability criterion is the object of the following

THEOREM 3. *Let γ and δ be two strictly positive real numbers with*

$$\alpha^2 \leq \gamma \leq \delta \leq \beta^2.$$

Let φ and ψ be two real-valued functions defined on $[\gamma,\delta]$ with the following properties :

$$\left\{\begin{array}{l} \varphi \text{ and } \psi \text{ are } C^1\text{-functions with values in } \mathbb{R}_+^\star, \\ \varphi \text{ is concave,} \\ \psi \text{ is convex,} \\ \varphi(d)\psi(d) = d \text{ for any } d \text{ in } [\gamma,\delta]. \end{array}\right. \tag{11}$$

Define $K(\gamma,\delta,\varphi,\psi)$ as the set of all (λ_1,λ_2) in $[\alpha,\beta]^2$ such that

$$\gamma \leq \lambda_1\lambda_2 \leq \delta,$$
$$\psi(\lambda_1\lambda_2) \leq \lambda_1, \lambda_2 \leq \varphi(\lambda_1\lambda_2).$$

Then $\mathcal{M}(K(\gamma,\delta,\varphi,\psi))$ *is H-stable* •

REMARK 6. The last assumption of (11) is natural. Indeed whenever the boundary of a given set can be parametrized in the form $\lambda_2 = \varphi(\lambda_1\lambda_2)$, it can be equally parametrized in the form $\lambda_1 = \psi(\lambda_1\lambda_2)$, with $\psi(d) = \dfrac{d}{\varphi(d)}$ •

REMARK 7. Conditions (11) essentially characterize the sets of the form $\mathcal{M}(K)$ which are stable under H-convergence. The following converse of Theorem 3 holds true. Define

$$K = \{(\lambda_1,\lambda_2) \in [\alpha,\beta]^2 \text{ such that } \psi(\lambda_1\lambda_2) \leq \lambda_1, \lambda_2 \leq \varphi(\lambda_1\lambda_2)\},$$

where $\varphi \leq \psi$ are two C^1-function from $\mathbb{R}_+^\star$ into itself such that $\varphi(d)\psi(d) = d$ for any d in $\mathbb{R}_+^\star$ with

$$\min_{(\lambda_1,\lambda_2)\in K} \lambda_1\lambda_2 \leq d \leq \max_{(\lambda_1,\lambda_2)\in K} \lambda_1\lambda_2. \tag{12}$$

If K is stable under H-convergence, φ is concave and ψ is convex on the interval defined by inequalities (12).

The proof of this last result can be found in Francfort & Murat (1987) •

REMARK 8. The proof of the "only if" part of Theorem 2 is now straightforward. Since

$$\{(\alpha_1,\alpha_2)\} \cup \{(\beta_1,\beta_2)\} \subset L_w \cap L_b$$

(*cf.* Definition 5) all sequences A_ε satisfying (3) (*cf.* Definition 3) belong to $\mathcal{M}(L_w) \cap \mathcal{M}(L_b)$ and the "only if" part will result from the H-stability of $\mathcal{M}(L_w)$ when A_1 and A_2 are well-ordered and of $\mathcal{M}(L_b)$ when A_1 and A_2 are badly ordered (*cf.* Definition 4).

Let us set

$$\gamma = \alpha_1\alpha_2, \ \delta = \beta_1\beta_2.$$

The case where $\gamma = \delta$ is left to the reader and we are thus at liberty to assume that

$$\gamma < \delta.$$

In the well ordered case, we define, for any d in $[\gamma,\delta]$

$$\varphi_w(d) = \frac{d(\beta_1\beta_2-\alpha_1\alpha_2)}{(\beta_1-\alpha_1)d+(\beta_2-\alpha_2)\alpha_1\beta_1}, \ \psi_w(d) = \frac{d}{\varphi_w(d)},$$

whereas in the badly ordered case, we define,

$$\varphi_b(d) = \frac{(\beta_2-\alpha_2)d+(\beta_1-\alpha_1)\alpha_2\beta_2}{(\beta_1\beta_2-\alpha_1\alpha_2)}, \ \psi_b(d) = \frac{d}{\varphi_b(d)}.$$

By virtue of the ordering properties of $\alpha_1,\alpha_2,\beta_1,\beta_2$ the functions (φ_w,ψ_w) (respectively (φ_b,ψ_d)) are trivially seen to satisfy (11) in the well-ordered (respectively badly ordered) case.

Theorem 3 applies to the set $K(\gamma,\delta,\varphi_w,\psi_w)$ (respectively $K(\gamma,\delta,\varphi_b,\psi_b)$) and yields the H-stability of $\mathcal{M}(K(\gamma,\delta,\varphi_w,\psi_w))$ (respectively $\mathcal{M}(K(\gamma,\delta,\varphi_b,\psi_b))$). Since

$$\begin{cases} L_w = K(\gamma,\delta,\varphi_w,\psi_w), \\ L_b = K(\gamma,\delta,\varphi_b,\psi_b), \end{cases}$$

the desired result is proved •

4. SKETCH OF THE PROOF OF THEOREM 3.

A complete proof of Theorem 3 is presented in Francfort & Murat (1987). It relies on two main ingredients : a decomposition of the set $K(\gamma,\delta,\varphi,\psi)$ as a (countable) intersection of sets with specific properties (*cf.* (14) below) and a few lemmae pertaining to the theory of H-convergence.

4.1. A decomposition of $K(\gamma,\delta,\varphi,\psi)$.

It is firstly remarked that if φ and ψ satisfy (11) there exists a sequence of real numbers d_n in $[\gamma,\delta]$ such that, upon setting

$$\varphi_n = \varphi(d_n),\ \psi_n = \psi(d_n),$$
$$\varphi'_n = \varphi'(d_n),\ \psi'_n = \psi'(d_n),$$

then

$$\begin{cases} \psi'_n\varphi_n + \varphi'_n\psi_n = 1, \\ \varphi'_n \geqslant 0 \text{ and/or } \psi'_n \geqslant 0, \\ \varphi(d) = \inf_{n\geqslant 1} \{\varphi'_n d + \varphi_n^2\psi'_n\},\ \gamma \leqslant d \leqslant \delta, \\ \psi(d) = \sup_{n\geqslant 1} \{\psi'_n d + \psi_n^2\varphi'_n\},\ \gamma \leqslant d \leqslant \delta. \end{cases} \tag{13}$$

A simple computation implies that for any d in $[\gamma,\delta]$

$$(\varphi'_n d + \varphi_n^2\psi'_n)(\psi'_n d + \psi_n^2\varphi'_n) = d + \varphi'_n\psi'_n(d-d_n)^2.$$

Thus, if $\varphi'_n \geqslant 0$ and $\psi'_n \geqslant 0$, the inequality $\lambda_1,\lambda_2 \geqslant \psi'_n\lambda_1\lambda_2 + \psi_n^2\varphi'_n$ implies that $\lambda_1,\lambda_2 \leqslant \varphi'_n\lambda_1\lambda_2 + \varphi_n^2\psi'_n$, whereas, if $\varphi'_n\psi'_n \leqslant 0$, the inequality $\lambda_1,\lambda_2 \leqslant \varphi'_n\lambda_1\lambda_2 + \varphi_n^2\psi'_n$ implies that $\lambda_1,\lambda_2 \geqslant \psi'_n\lambda_1\lambda_2 + \psi_n^2\varphi'_n$.

We set

DEFINITION 7. *For any real numbers* a,b,ζ, $a \geqslant 0$ *and/or* $b \geqslant 0$, $\alpha^2 \leqslant \zeta \leqslant \beta^2$,

$$L_{\leqslant}(a,b) = \{(\lambda_1,\lambda_2) \in [\alpha,\beta]^2 \mid \lambda_1,\lambda_2 \leqslant a\lambda_1\lambda_2 + b\},$$
$$L_{\geqslant}(a,b) = \{(\lambda_1,\lambda_2) \in [\alpha,\beta]^2 \mid \lambda_1,\lambda_2 \geqslant a\lambda_1\lambda_2 + b\},$$
$$D_{\leqslant}(\zeta) = \{(\lambda_1,\lambda_2) \in [\alpha,\beta]^2 \mid \zeta \leqslant \lambda_1\lambda_2\},$$
$$D_{\geqslant}(\zeta) = \{(\lambda_1,\lambda_2) \in [\alpha,\beta]^2 \mid \zeta \geqslant \lambda_1\lambda_2\} \bullet$$

In view of the above computations, the set $K(\gamma,\delta,\varphi,\psi)$ decomposes as

$$K(\gamma,\delta,\varphi,\psi) = D_{\leqslant}(\gamma) \cap D_{\geqslant}(\delta) \bigcap_{\varphi'_n\psi'_n \leqslant 0} L_{\leqslant}(\varphi'_n,\varphi_n^2\psi'_n) \bigcap_{\substack{\varphi'_n \geqslant 0 \\ \psi'_n \geqslant 0}} L_{\geqslant}(\psi'_n,\psi_n^2\varphi'_n). \tag{14}$$

REMARK 9. The intersection of H-stable sets is H-stable. Thus, by virtue of (14), the set $\mathcal{M}(K(\gamma,\delta,\varphi,\psi))$ is H-stable if each of the sets $\mathcal{M}(D_{\leqslant}(\gamma))$, $\mathcal{M}(D_{\geqslant}(\delta))$, $\mathcal{M}(L_{\leqslant}(\varphi'_n,\varphi_n^2\psi'_n))$ $(\varphi'_n\psi'_n \leqslant 0)$, $\mathcal{M}(L_{\geqslant}(\psi'_n,\psi_n^2\varphi'_n))$ $(\varphi'_n \geqslant 0$ and $\psi'_n \geqslant 0)$ is H-stable, which is the object of the next subsection $\bullet$

4.2. A few results of the theory of H-convergence.

Our purpose here is not to give a detailed exposition of the theory of H-convergence, but rather to state and/or derive the few results needed in the proof of Theorem 3. Further details may be found in Francfort & Murat (1987).

The following theorem, *specific to the two-dimensional case,* is central to our analysis.

THEOREM 4. *Let* A_ε *belong to* $\mathcal{M}([\alpha,\beta]^2)$ *and H-converge to an element* A_0 *of* $\mathcal{M}([\alpha,\beta]^3)$. *Then*

$$\frac{A_\varepsilon}{\det A_\varepsilon} \text{ H-converges to } \frac{A_0}{\det A_0}\bullet$$

This result traces back to Keller (1964). A proof was presented by Kohler & Papanicolaou (1982) in a probabilistic setting. The proof presented below is due to L. Tartar.

Proof of Theorem 4. Let w_ε, q_ε be as in Definition 2 (*cf.* (4)-(6)). If R denotes the rotation of angle $-\pi/2$, we set

$$\begin{cases} \overline{w}_\varepsilon = Rq_\varepsilon, \\ \overline{q}_\varepsilon = Rw_\varepsilon, \\ B_\varepsilon = RA_\varepsilon^{-1}\,{}^tR. \end{cases}$$

Then, as ε tends to zero,

$$\begin{cases} \overline{w}_\varepsilon \longrightarrow \overline{w}_0 = Rq_0, \\ \overline{q}_\varepsilon \longrightarrow \overline{q}_0 = Rw_0, \end{cases} \tag{15}$$

weakly in $[L_2(\Omega)]^2$. Furthermore, recalling (6),

$$\begin{cases} \operatorname{curl} \overline{w}_\varepsilon = -\operatorname{div} q_\varepsilon, \\ \operatorname{div} \overline{q}_\varepsilon = \operatorname{curl} w_\varepsilon, \end{cases} \tag{16}$$

lie in a compact set of $H^{-1}_{loc}(\Omega)$.

Finally,

$$\overline{q}_\varepsilon = B_\varepsilon \overline{w}_\varepsilon . \qquad (17)$$

Theorem 1 applies to B_ε and yields the existence of a subsequence $B_{\varepsilon'}$ of B_ε and of an element B_0 of $\mathcal{M}([\alpha,\beta]^2)$ such that $B_{\varepsilon'}$ H-converges to B_0 as ε' tends to zero. In view of (15)-(17) and the very definition of H-convergence (*cf.* Definition 2), we conclude that

$$\overline{q}_0 = B_0 \overline{w}_0 ,$$

or still that

$$(I - {}^tRB_0RA_0)w_0 = 0,$$

where I stands for the identity mapping on $\mathbb{R}^2$.

A classical argument in the theory of H-convergence permits to choose w_0 as an arbitrary vector of $\mathbb{R}^2$ (at least locally), from which it is easily concluded that

$$B_0(x) = R A_0^{-1}(x)\ {}^tR$$

almost everywhere.

The identity

$$R\,C^{-1}\ {}^tR = \frac{{}^tC}{\det C}$$

which holds true for any invertible two by two matrix yields the desired result •

As announced in Remark 9, the H-stability of $\mathcal{M}(K(\gamma,\delta,\varphi,\psi))$ will result from the H-stability of four kinds of sets. Specifically, the following lemmae hold in the context of Definition 7 :

<u>LEMMA 1</u>. *The sets* $\mathcal{M}(D_{\leqslant}(\zeta))$ *and* $\mathcal{M}(D_{\geqslant}(\zeta))$ *are* H-*stable* •

<u>LEMMA 2</u>. *If* $a \geqslant 0$ *and* $b \geqslant 0$, *the set* $\mathcal{M}(L_{\geqslant}(a,b))$ *is* H-*stable*. *If* $ab \leqslant 0$, *the set* $\mathcal{M}(L_{\leqslant}(a,b))$ *is* H-*stable* •

Lemma 1 easily results from Theorem 4 together with the following comparison lemma (Tartar (1979a)) :

LEMMA 3. *Let* A_ε *and* B_ε *belong to* $\mathcal{M}([\alpha,\beta]^2)$ *and* H-*converge* A_o *and* B_o *respectively. If*

$$A_\varepsilon(x) \leqslant B_\varepsilon(x), \text{ almost everywhere on } \mathbb{R}^2,$$

then

$$A_o(x) \leqslant B_o(x), \text{ almost everywhere on } \mathbb{R}^2 \bullet$$

A complete proof of Lemma 2 is given in Francfort & Murat (1987). It is based on an adequate rewriting of the set $\mathcal{M}(L_{\leqslant}(a,b))$ (respectively $\mathcal{M}(L_{\geqslant}(a,b))$), namely

$$\mathcal{M}(L_{\leqslant(\geqslant)}(a,b)) = \{A \in \mathcal{M}([\alpha,\beta]^2) \mid I \leqslant (\geqslant) a\, A(x) + b\, \frac{A(x)}{\det A(x)} \text{ for almost any } x \text{ of } \mathbb{R}^2\}.$$

The actual proof uses Theorem 4 together with Lemma 4 (*cf.* Tartar (1979a)) and Lemma 5 (Murat (1976), Boccardo & Marcellini (1978)) stated below.

LEMMA 4. *Let* A_ε *belong to* $\mathcal{M}([\alpha,\beta]^2)$ *and* H-*converge to* A_o. *If*

$$A_\varepsilon \longrightarrow \bar{A}$$

weak-$\star$ *in* $[L_\infty(\mathbb{R}^2)]^4$ *as* ε *tends to zero, then*

$$A_o(x) \leqslant \bar{A}(x),$$

for almost any x *of* $\mathbb{R}^2$ $\bullet$

LEMMA 5. *Let* A_ε *belong to* $\mathcal{M}([\alpha,\beta]^2)$ *and* H-*converge to* A_o. *Let* w_ε *be a sequence of* $[L_2(\Omega)]^2$ *such that, as* ε *tends to zero,*

$$w_\varepsilon \longrightarrow w_o \text{ weakly in } [L_2(\Omega)]^2,$$

while curl w_ε *lies in a compact set of* $H^{-1}_{loc}(\Omega)$. *Then, for any positive* φ *in* $C^\infty_0(\Omega)$,

$$\int_\Omega \varphi\, A_0 w_0 \cdot w_0\, dx \leq \lim \int_\Omega \varphi\, A_\varepsilon w_\varepsilon \cdot w_\varepsilon\, dx \bullet$$

Acknowledgements. The results derived in the work were obtained in collaboration with Luc Tartar. It is a pleasure for the authors to acknowledge his friendly help.

REFERENCES

Boccardo, L. & Marcellini, P. (1976). Sulla convergenza delle soluzioni di disequazioni variazionali. Ann. Mat. Pura ed Appl. 110, 137-159.

Ericksen, J.L., Kinderlehrer, D., Kohn, R. & Lions, J.L. (editors) (1986) *Homogenization and Effective Moduli of Materials and Media*, Springer-Verlag, New York, The IMA Volumes in Mathematics and Its Applications 1.

Francfort, G.A. & Milton, G.W. (1987). Optimal Bounds for Conduction in Two-Dimensional, Multiphase, Polycrystalline Media. J. Stat. Phys. 46, 161-177.

Francfort, G.A. & Murat, F. (1986). Homogenization and Optimal Bounds in Linear Elasticity. Arch. Rat. Mec. Anal. 94, 307-334.

Francfort, G.A. & Murat, F. (1987). Optimal Bounds for Homogenization of Two Anisotropic Conducting Media in Two Dimensions. To appear.

Hashin, Z. (1983). Analysis of Composite Materials, A survey. J. Appl. Mech. 50, 481-505.

Keller, J.B. (1964). A Theorem on the Conductivity of a Composite Medium. J. Math. Phys. 5, 548-549.

Kohler, W. & Papanicolaou, G.C.(1982). Bounds for the Effective Conductivity of Random Media. *Macroscopic Properties of Disordered Media, Proceedings New York* 1981, ed. by R. Burridge, S. Childress, G.C. Papanicolaou, Springer-Verlag, Berlin, Lecture Notes in Physics 154, 111-132.

Lurie, K.A. & Cherkaev, A.V. (1984). G-Closure of a Set of Anisotropically Conducting Media in the Two-Dimensional Case. J. Opt. Th. Appl., 42, 283-304.

Murat, F. (1976). Sur l'homogénéisation d'inéquations elliptiques du 2ème ordre, relatives au convexe $K(\psi_1,\psi_2) = \{v \in H^1_0(\Omega) \mid \psi_1 \leq v \leq \psi_2$ p.p. dans $\Omega\}$. Thèse d'Etat, Université Paris VI.

Murat, F. (1977). H-Convergence. Séminaire d'Analyse Fonctionnelle et Numérique, 1977/1978, Université d'Alger, Multigraphed.

Murat, F. (1978). Compacité par compensation. Ann. Sc. Norm. Sup. Pisa 5, 489-507.

Murat, F. (1979). Compacité par compensation : condition nécessaire et suffisante de continuité faible sous une hypothèse de rang constant. Ann. Sc. Norm. Sup. Pisa 8, 69-102.

Simon, L. (1979). On G-Convergence of Elliptic Operators. Indiana Univ. Math. J. 28, 587-594.

Spagnolo, S. (1968). Sulla convergenza di soluzioni di equazioni paraboliche ed ellittiche. Ann. Sc. Norm. Sup. Pisa 22, 577-597.

Tartar, L. (1974). Problèmes de contrôle des coefficients dans des équations aux dérivées partielles. *Control Theory, Numerical Methods and Computer Systems Modelling, Proceedings* 1974, ed. by A. Bensoussan & J.L. Lions, Springer-Verlag, Berlin, Lecture Notes on Economics and Mathematical Systems 107, 420-426.

Tartar, L. (1977). Cours Peccot, Collège de France.

Tartar, L. (1979a). Estimation de coefficients homogeneisés. *Computing Methods in Applied Sciences and Engineering,* I. *Proceedings* 1977, ed. by R. Glowinski & J.L. Lions, Springer Verlag, Berlin, Lecture Notes in Mathematics 704, 364-373.

Tartar, L. (1979b). Compensated Compactness and Applications to Partial Differential Equations. *Nonlinear Analysis and Mechanics : Heriot-Watt Symposium* Vol. IV, ed. by R.J. Knops, Pitman, London, Research Notes in Mathematics 39, 136-212.

Tartar, L. (1985). Estimations fines des coefficients homogénéisés. *Ennio de Giorgi Colloquium,* ed. by P. Krée, Pitman, London, Research Notes in Mathematics 125, 168-187.

RAPID FLOWS OF GRANULAR MATERIALS

J.T. Jenkins
Department of Theoretical and Applied Mechanics,
Cornell University, Ithaca, NY 14853

Abstract. We outline the derivation of balance laws, constitutative relations, and boundary conditions for the rapid flow of a granular material consisting of identical, smooth, nearly elastic, spheres. As an illustration, we consider the steady shearing flow maintained by the relative motion of two parallel boundaries, and treat, in detail, the exceptional situation in which the shear rate, the solid volume fraction, and the energy of the velocity fluctuations are uniform across the gap.

INTRODUCTION

When a granular material is sheared at a sufficiently high rate, the shear stress and the normal stress required to maintain its motion are observed to vary with the square of the shear rate (Bagnold (1954), Savage (1972), Savage & McKeown (1983), Savage & Sayed (1984), Hanes & Inman (1985a)). The interpretation of these observations is that, at high shear rates, the dominant mechanism of momentum transfer is collisions between grains, with the interstitial liquid or gas playing a relatively minor role.

There is an obvious analogy between the colliding macroscopic grains of a sheared granular material and the agitated molecules of a dense gas. There are also several important differences: collisions between grains are inevitably inelastic; typical grain flows involve spatial variations over far fewer grains than their molecular counterparts; and, consequently, the influence of boundaries is far more pervasive.

Exploiting the analogy, methods from the kinetic theory of dense gases have been adopted and extended to derive balance laws, constitutive relations, and boundary conditions for systems of macroscopic, dissipative, grains. Jenkins (1987) provides a brief review of this activity. Here we outline the derivation of the theory for identical, smooth, nearly elastic, spheres (Lun, et al. (1984), Jenkins & Richman (1985)). It involves balance laws for the means of the mass density, velocity, and, essentially, the kinetic energy of the velocity fluctuations. To the order of the approximations used in determining the velocity distribution function, the expressions for the pressure tensor and the flux of fluctuation energy are identical to those for a dense gas of perfectly elastic spheres. The only nonclassical term is the collisional rate of dissipation per unit volume present in the energy balance.

Because differences in the boundaries that drive the shearing flows are almost certainly responsible for the quantitative differences in the shear flow experiments, we also outline a derivation of boundary conditions (Jenkins & Richman (1986)). These boundary conditions highlight the exceptional nature of steady, homogeneous, simple shear.

Finally, in the context of this exceptional flow, we obtain a simple scaling for the observed dependence of the ratio of the shear stress to the normal stress on the solid volume fraction.

BALANCE LAWS AND CONSTITUTIVE RELATIONS

We consider smooth spheres with a diameter σ and a mass m. The dynamics of a collision between two such spheres is determined in terms of their coefficient of restitution e. The collisional change of any property ψ that is a function of the particle velocity **c** may then be expressed in terms of e and the geometry of the collision. For example, with $\psi = mc^2/2$, where $c^2 \equiv \mathbf{c} \cdot \mathbf{c}$, the total change of kinetic energy in a collision is

$$\Delta(mc^2/2) = -(m/4)(1-e^2)(\mathbf{g} \cdot \mathbf{k})^2, \qquad (1)$$

where $\mathbf{g} \equiv \mathbf{c}_1 - \mathbf{c}_2$ is the relative velocity prior to the collision and $\sigma\mathbf{k}$ is the vector from the center of particle 1 to that of particle 2 at collision. When e=1, the collision is elastic.

The distribution of velocities is given by a function of c, position r, and time t, defined so that the number n of particles per unit volume at r and t is

$$n(\mathbf{r},t) = \int f^{(1)}(\mathbf{c},\mathbf{r},t)d\mathbf{c} \, , \tag{2}$$

where the integration is over the entire volume of velocity space. The mean $<\psi>$ of a particle property is defined in terms of n and the velocity distribution function by

$$<\psi> = \frac{1}{n} \int \psi(\mathbf{c}) f(\mathbf{c}) d\mathbf{c} \, , \tag{3}$$

where the dependence on r and t is understood. For example, the mean mass density ρ is mn, the mean velocity $\mathbf{u}$ is $<\mathbf{c}>$, the fluctuation velocity $\mathbf{C}$ is $\mathbf{c}-\mathbf{u}$, and the granular temperature T is $<C^2>/3$.

The balance laws for the mean fields ρ, $\mathbf{u}$, and T have the familiar forms:

$$\dot{\rho} + \rho u_{i,i} = 0 \, , \tag{4}$$

where the dot denotes a time derivative calculated with respect to the mean velocity;

$$\rho \dot{u}_i = -P_{ik,k} + \rho f_i \, , \tag{5}$$

where $\mathbf{P}$ is the symmetric pressure tensor and $\mathbf{f}$ is the external force per unit mass; and

$$\frac{3}{2}\rho\dot{T} = -q_{i,i} - P_{ik}D_{ik} - \gamma \, , \tag{6}$$

where $\mathbf{q}$ is the flux of fluctuation energy, $\mathbf{D}$ is the symmetric part of the velocity gradients, and γ is the collisional rate of dissipation of fluctuation energy per unit volume.

The fluxes of momentum and energy are due both to transport between collisions and transfer in collisions. The transport parts are $\rho\langle CC\rangle$ and $\rho\langle CC^2\rangle/2$, respectively. The collisional transfers, expected to dominate at relatively high number densities, depend upon the exchange of momentum and energy in a collision and the frequency of collision.

The collision frequency is given in terms of the complete pair distribution function $f^{(2)}(c,r,v,x,t)$ governing the liklihood that, at time t, spheres with velocities near c and v will be located near r and x, respectively. For dense systems, Enskog supposes that the $f^{(2)}$ for a colliding pair is the product of the $f^{(1)}$ of each sphere, evaluated at its center, and a factor g that incorporates the influence of the volume occupied by the spheres on their collision frequency. This factor is the equilibrium radial distribution function, evaluated at the point of contact. It is given as a function of solid volume fraction $\nu\equiv\pi n\sigma^3/6$ by Carnahan & Starling (1969)as

$$g(\nu) = (2-\nu)/2(1-\nu)^3 \ . \tag{7}$$

In dilute systems g is unity, the positions of the centers of the spheres are not distinguished, and the presumed absence of correlation in position and velocity is called the assumption of molecular chaos. In distinguishing between the positions of the centers of a pair of colliding spheres, the possibility of collisional transfers is incorporated into the dense theory for even the simplest of velocity distribution functions. For whatever such distribution function, g determines the density dependence of the resulting theory.

For elastic spheres in equilibrium, the velocity distribution function is Maxwellian:

$$f_0^{(1)} = [n/(2\pi T)^{3/2}] \exp(-C^2/2T) . \qquad (8)$$

For smooth but inelastic spheres, no such equilibrium is possible; but, provided that not too much energy is dissipated in collisions, $f^{(1)}$ will probably not differ much from $f_0^{(1)}$. Consequently, for nearly elastic spheres, $f^{(1)}$ is assumed to be a perturbation of the Maxwellian:

$$f^{(1)} = [1+A_{ik}C_iC_k/2T^2 - b_iC_i(5-C^2/T)/10T^2]f_0^{(1)} , \qquad (9)$$

where the coefficient **A** is the deviatoric part of the second moment <CC> of $f^{(1)}$ and **b** is the contraction $<CC^2>$ of its third moment. They are determined as functions of the mean fields ρ, **u**, and T and their spatial gradients as approximate solutions of the balance laws governing their evolution. These equations have essentially the same structure as that governing T, the isotropic part of the second moment.

Let L and U be, respectively, a characteristic length and velocity of a typical flow. Then, when it is assumed that A/T, $b/T^{3/2}$, σ/L, and $(1-e)^{1/2}$ are all small and that $U/T^{1/2}$ and $G \equiv \nu g$ are near one, these balance laws are satisified identically at lowest order provided that

$$A_{ik}/T = -(\sqrt{\pi}/6)(1+5G^{-1}/8)(\sigma/\sqrt{T})\hat{D}_{ik} , \qquad (10)$$

where the hat denotes the deviatoric part; and

$$b_i/T^{3/2} = -(15\sqrt{\pi}/16)(1+5G^{-1}/12)(\sigma/T)T,_i . \qquad (11)$$

Up to an error proportional to the square of quantities assumed to be small, these expressions are identical to those for perfectly elastic spheres (Chapman & Cowling (1970), Sec. 16.34). Consequently, so also are the expressions for the pressure tensor and the energy flux vector calculated by employing the velocity distribution function (9) and Enskog's extension of the assumption of molecular chaos.

The pressure tensor is (Chapman & Cowling (1970), Sec. 16.41)

$$P_{ik} = (p-\omega D_{jj})\delta_{ik} - 2\mu \hat{D}_{ik} , \tag{12}$$

where

$$p = \rho T(1+4G), \tag{13}$$

$$\omega = (8/3\sqrt{\pi})\rho\sigma\sqrt{T}\ G , \tag{14}$$

and

$$\mu = (\sqrt{\pi}/6)\rho\sigma\sqrt{T}[(5/16)G^{-1}+1 +(4/5)(1+12/\pi)G]. \tag{15}$$

The energy flux vector is (Chapman & Cowling (1970), Sec. 16.42)

$$q_i = -\kappa T,_i , \tag{16}$$

where

$$\kappa = (3\sqrt{\pi}/16)\rho\sigma\sqrt{T}[(5/24)G^{-1}+1 +(6/5)(1+32/9\pi)G] . \tag{17}$$

The only non-classical quantity is the rate of dissipation, whose form is independent of the perturbations **A** and **b** ,

$$\gamma = (24/\sqrt{\pi})(1-e)(\rho T^{3/2}/\sigma)G \ . \tag{18}$$

The presence of γ in the energy equation (6) permits it to have steady solutions for inelastic spheres in situations where none are possible in the classical kinetic theory.

BOUNDARY CONDITIONS

Assuming that the velocity distribution function (9) applies in the neighborhood of a boundary, boundary conditions on the pressure tensor and the energy flux vector may derived by considering the mean rate of transfer of momentum and energy in collisions between the flowing spheres and the boundary.

At a point on a rigid boundary translating with velocity **U**, the mean flow velocity will, in general, differ from **U** and slip will occur. The slip velocity **v** is **U**-**u**. Because the boundary is impenetrable, at a point with inward unit normal **N**, **v**·**N** = 0.

Over a unit area of the boundary, the rate **M** at which momentum is supplied by the boundary must be balance by the rate at which it is removed by the flow; so

$$M_i = P_{ik}N_k \ . \tag{19}$$

The corresponding rate at which energy is supplied by the boundary is **M**·**U**-D, where D is the rate of dissipation in collisions. This must balance the rate at which energy is removed by the flow; so

$$M_iU_i - D = u_iP_{ik}N_k + q_iN_i , \tag{20}$$

or, with (19) and the definition of the slip velocity,

$$M_i v_i - D = q_i N_i \ . \tag{21}$$

Note that the energy flux at the boundary may be positive or negative, depending upon the relative magnitudes of the slip working and the rate of dissipation.

Consider, for example, a boundary consisting of a flat plate with smooth hemispheres of diameter d fixed to it. The centers of the hemispheres are assumed to be positioned randomly in such a way that the average distance between their edges is s. In this case, a natural measure of the roughness of the boundary is $\sin\theta \equiv (d+s)/(d+\sigma)$. Neither d nor σ is assumed to differ much from their average, $\bar{\sigma} \equiv (d+\sigma)/2$. Also, the coefficient of restitution ϵ for a collsion between a sphere of the flow and a hemisphere of the wall is, like e, assumed to be near one.

If now the velocity distribution function (9) is used to calculate the rate at which collisions over a unit area of the wall supply momentum to the flow, the result is

$$M_i = \rho\chi T\{N_i + \sqrt{2/\pi}[Hv_i/\sqrt{T} + I_{ijk}\bar{\sigma}u_{k,j}/\sqrt{T}]\} \ , \tag{22}$$

where

$$H = (2/3)[2\csc^2\theta(1-\cos\theta) - \cos\theta] \ , \tag{23}$$

$$\begin{aligned} I_{ijk} = HN_j\delta_{ik} + (1/2)[&(5\sin^2\theta - 4)N_iN_jN_k \\ &-\sin^2\theta(N_i\delta_{jk} + N_j\delta_{ik} + N_k\delta_{ij})] \\ &\times[1 + (5\pi/96\sqrt{2})(\sigma/\bar{\sigma})(G^{-1} + 8/5)] \ , \end{aligned} \tag{24}$$

and χ is a factor, roughly corresponding to g in the flow, providing the influence of the size and spacing of the hemispheres on the frequency of collision at the wall. This expression for **M** differs from that given by Jenkins & Richman (1986) by the correction of a sign error, the inclusion of the perturbations **A** and **b** in the averaging, and the evaluation of v and νu at a distance $\bar{\sigma}$ from the flat wall rather than at the wall.

The corresponding expression for the rate of dissipation per unit area is

$$D = 2\sqrt{2/\pi}\ \rho\chi(1-\epsilon)T^{3/2}(1-\cos\theta)\csc^2\theta\ . \qquad (25)$$

SIMPLE SHEAR

Consider the flow of identical, smooth, nearly elastic spheres maintained in the absence of gravity by the relative motion of identical, parallel, bumpy walls a fixed distance $L+2\bar{\sigma}$ apart. The upper wall moves with constant velocity U in the x direction, the lower wall moves with the same speed in the opposite direction. The nonvanishing x components of the flow velocity and slip velocity are denoted by u and v, respectively. In general, the flow velocity, granular temperature, and solid volume fraction will be functions of the transverse coordinate y, measured from the centerline. Simple shear, in which u=ky, with k,T, and ν independent of y, is exceptional. It occurs when the diameter and solid volume fraction of the flow spheres, the diameter and separation of the wall hemispheres, and the two coefficients of restitution are such that, at the boundaries, the slip working balances the collisional rate of dissipation. For simplicity, we treat this flow, keeping in mind that, in it, the apparent rate of shear 2U/L differs from k by 2v/L.

In this steady, homogeneous flow the shear stress $S\equiv-P_{xy}$ and normal stress $N\equiv P_{yy}$ are constant and the energy equation (6) reduces to

$$Sk - \gamma = 0. \qquad (26)$$

with (12), (15), and (18), this determines the grannular temperature:

$$T = (\pi/144)[\sigma^2 k^2/(1-e)] \times [(5/16)G^{-2}+G^{-1}+(4/5)(1+12/\pi)]. \quad (27)$$

Then the tractions required to maintain the flow are

$$N = \rho T(1+4G) \quad (28)$$

and

$$S = (24/\sqrt{\pi})[(1-e)/\sigma k]\rho T^{3/2}G . \quad (29)$$

With T given by (27), the normal stress and shear stress vary with the square of the shear rate. The results of the numerical simulations of Walton, et al. (1986) are in excellent agreement with the shear stress predicted by (29) and (27) provided that the coefficient of restitution is greater than 0.8.

For flows that are not dilute, say when G is significantly greater than 5/16, the explicit dependence of S/N on G may be eliminated in favor of quantities that are more directly observable. In this event, (27) and (28) may be used together to obtain

$$G = [45/(\pi+12)](\xi-\pi/36) , \quad (30)$$

where $\xi \equiv (1-e)N/\rho\sigma^2 k^2$. Then

$$\frac{1}{(1-e)}\left[\frac{S}{N}\right]^2 = \frac{\pi+12}{5\pi}\,\frac{\xi}{(\xi-\pi/36)} . \quad (31)$$

This provides a scaling for the type of experimental data reported, for example, by Savage & Sayed (1984) and Hanes & Inman (1985b).

An improvement of this scaling for simple shear involves the relationship between k and 2U/L. Richman (1987) shows how the tangential compenent of the boundary condition (19) may be reduced to yield an expression for v/U. The appropriate simplification of his result for relatively dense flows is

$$v/U = \sqrt{\pi}(\bar{\sigma}/L)\{(2\sqrt{2}/5)(\sigma/\bar{\sigma})(1+\pi/12)+2H - [1+(\pi/12\sqrt{2})(\bar{\sigma}/\sigma)]\sin^2\theta\}/H \, , \qquad (32)$$

where H is given by (23). This correction is of order σ/L and involves only the diameter of the spheres and the boundary geometry.

Numerical simulations of the inhomogeneous equilibrium of elastic spheres contained between two flat, elastic walls (Henderson & van Swol (1984)) provide an indication of the deficiencies of the constitutive relations and boundary conditions. Near a wall, the radial distribution function is not that of the homogeneous fluid and the variation of the local number density normal to the wall is rapid and nonmonotonic. Perhaps the simplest correction to the constitutive relations near a solid boundary parallels van der Waals' treatment of the liquid-gas interface (Rowlinson (1980)). This involves the inclusion of higher spatial grandients of the local number density in the pressure tensor. The exact form of such a correction remains to be determined.

ACKNOWLEDGEMENT

I am grateful to F. Mancini for providing me with the terms contributed to the boundary conditions by the perturbations and to J. Dent for discussions that led to the scaling of the stress ratio. The preparation of this manuscript was supported by the U.S. Army Research Office through the Mathematical Sciences Institute at Cornell University.

REFERENCES

Bagnold, R.A. (1954). Experiments on a gravity-free dispersion of solid spheres in a Newtonian fluid under shear. Proc. Roy. Soc. London, A225, 49-63.

Champman, S. & Cowling, T.G. (1970). The Mathematical Theory of Non-Uniform Gases, Third edition. Cambridge: Cambridge University Press.

Hanes, D.M. & Inman, D.L. (1985a). Observations of rapidly flowing granular-fluid mixtures. J. Fluid Mech., 150, 357-80.

Hanes, D.M. & Inman, D.L. (1985b). Experimental evaluation of a dynamic yield criterion for granular fluid flows. J. Geophys. Res., 90, 3670-4.

Henderson, J.R. & van Swol, F. (1984). On the interface between a fluid & a planar wall. Theory & simulation of a hard sphere fluid at a hard wall. Mol. Phys., 51, 991-1010.

Jenkins, J.T. (1987). Balance laws and constitutive relations for rapid flows of granular materials. In Proc. Army Research Office Workshop on Constitutive Relations. ed. J.Chandra & R. Srivastav, Philadelphia: SIAM (pending publication).

Jenkins, J.T. & Richman, M.W. (1985). Grad's 13-moment system for a dense gas of inelastic spheres. Arch. Rat'l. Mech. Anal., 87, 355-77.

Jenkins, J.T. & Richman, M.W. (1986). Boundary conditions for plane flows of smooth, nearly elastic, circular disks. J. Fluid Mech., 171, 53-69.

Lun, C.K.K., Savage, S.B., Jeffrey, D.J., & Chepurniy, N. (1984). Kinetic theories for granular flow: inelastic particles in Couette flow and slightly inelastic particles in a general flow field. J. Fluid Mech., 140, 223-56.

Richman, M.W. (1987). Improved boundary condiitons for flows of nearly elastic spheres. In Proc. Eleventh Canadian Congress of Applied Mechanics. (pending publication).

Rowlinson, J.S. (1980). van der Waals revisited. Chem. Brit., 16, 32-5.

Savage, S.B. (1972). Experiments on shear flows of cohesionless granular materials. In Continuum Mechanical and Statistical Approaches in the Mechanics of Granular Materials, ed. S.C. Cowin & M. Satake, pp. 241-54, Tokyo: Gakujutsu Bunken Fukyu-kai.

Savage, S.B. & McKeown, S. (1983). Shear stresses developed during rapid shear of dense concentrations of large spherical particles between concentric rotating cylinders. J. Fluid Mech., 127, 453-472.

Savage, S.B. & Sayed, M. (1984). Stresses developed by dry cohesionless granular materials sheared in an annular shear cell. J. Fluid Mech., 142, 391-430.

Walton, O.R., Braun, R.L. & Cervelli, D.M. (1986). Flow of granular solids: 3-dimensional discrete-particle computer modeling of uniform shear. Preprint UCRL-95232, Lawrence Livermore Nat'l. Lab.

THE CONSTRAINED LEAST GRADIENT PROBLEM

Robert Kohn
Courant Institute
New York University
New York, NY 10012

Gilbert Strang
Department of Mathematics
Massachusetts Institute of Technology
Cambridge, MA 02139

ABSTRACT

Two model problems in optimal design, both initially nonconvex and in need of relaxation, lead to a variant of the least gradient problem in R^2:

$$\text{Minimize} \int_\Omega |\nabla\phi| dx \quad \text{subject to} \quad |\nabla\phi| \le 1 \quad \text{and} \quad \phi|_{\partial\Omega} = g \,.$$

Without the bound on $|\nabla\phi|$ the solution is easy to construct. It minimizes the lengths of the level curves γ_t on which $\phi = t$. The extra constraint $|\nabla\phi| \le 1$ requires the level curves γ_t and γ_s to be separated by at least the distance $|t-s|$. We prove that each level curve γ_t still solves a minimum problem, now constrained, and we illustrate by two examples the simplicity of the construction.

INTRODUCTION

In the classical minimum principles the integrand is often quadratic and always convex. In physical terms there is a "potential well." The problem is to minimize $\int[W(\nabla\varphi)-F\varphi]dx$ and the graph of W resembles Figure 1a. Its convexity assures that the minimum is attained (assuming that the class of admissible fields ϕ is suitably closed and convex) and the calculus of variations takes over: the Euler-Lagrange equations are elliptic and their solution is the minimizing φ^*. If W is not convex then all this is extremely likely to fail.

With a second potential well, convexity is lost. The graph in Figure 1b illustrates two regions of low energy, and the best φ, which is influenced by the linear term $F\varphi$ and the boundary conditions, may attempt to fall in between. The gradient $\nabla\varphi$ will jump from well to well. It is easily shown that the more rapid the oscillation, the lower the total energy (that is normal with nonconvexity), and therefore minimizing sequences φ_n have no strong limit. Only in the weak (averaged) sense is there a limit, and that weak limit is far from minimizing! Its gradient is at a point in between.

Fig. 1. (a) Single well (b) Double well (c) Steep well

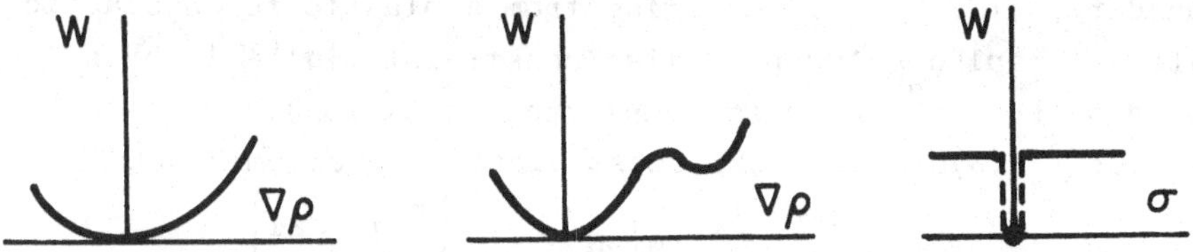

The same thing happens if the well is too steep. The graph in Figure 1c is the extreme case which we meet in optimal design, when forced to choose between using a given material or leaving a hole. The cost of the hole is $W(0) = 0$. The cost of the material is $W = 1$ and the graph jumps instantly to that value. In this case the argument in $W(\sigma)$ is the stress, and when it is zero we are free to create a hole. Where σ is nonzero we need material, and the difficulty is that the problem as it stands is strictly 0-1. The integral of W gives the volume of material, so that minimizing $\int W\,dx$ is equivalent to minimizing the weight. There will be constraints that modify the graph away from the well, but it is the steepness at that point that produces nonconvexity and makes the problem ill-posed. Our goal is to find a formulation that is well posed.

We mention three typical design problems with the integrand W:

1. Inf $\int W(\sigma)dx$ subject to $\frac{\partial \sigma_1}{\partial x} + \frac{\partial \sigma_2}{\partial y} = 0$, $\sigma.n = f$ on the boundary, and $|\sigma|^2 = \sigma_1^2 + \sigma_2^2 \leq 1$.

The pointwise bound on $|\sigma|$ comes from perfect plasticity. The equilibrium equation div $\sigma = 0$ is scalar for special geometries (see below), and the source term f is the surface traction.

2. Inf $\int W(\sigma)dx$ subject to div $\sigma = 0$, $\sigma.n = f$ on the boundary, $\int |\sigma|^2 \leq 1$. Changing from a plastic to an elastic material replaces the pointwise constraint $|\sigma| \leq 1$ by a bound on the ℓ^2 norm (a compliance condition).

3. Inf $\int W(\sigma)dx$ with the constraints of problem 2 but

$$\text{with} \quad \sigma = \begin{bmatrix} \sigma_{11} & \sigma_{21} \\ \sigma_{12} & \sigma_{22} \end{bmatrix}.$$

This is plane elasticity with a 2 by 2 symmetric stress tensor σ and a vector equation div $\sigma = 0$. Convexity of the energy density W is no longer the exact requirement for ellipticity of the Euler-Lagrange system. What we need in two dimensions is rank-one convexity, which only requires W to be convex between σ and σ' when those matrices differ by rank one. (In n dimensions that changes to rank $(\sigma-\sigma') < n$, while in the dual problem the Legendre-Hadamard condition on the strain energy density remains rank-one convexity.) This rank condition allows continuity, $\sigma.n = \sigma'.n$, of the normal stress while the tangential stress jumps. Therefore the stress can oscillate rapidly between σ and σ' in layers that are perpendicular to the null-vector n. It is this energy-reducing construction that leads us to the minimum of $\int W(\sigma)dx$.

In all three examples the principal question is the same. What happens if convexity (or rank-one convexity) is not present? The weight $\int W\,dx$ is bounded below, but the minimum weight is not attained. From the viewpoint of the Euler equations the problem is not elliptic. In the variational statement the functional is not weakly lower semicontinuous. Minimizing sequences converge weakly, but their limit does not minimize $\int W\,dx$. The question to answer is, what functional do the weak limits minimize?

That "relaxed" functional will be convex in the scalar case, and it is remarkably easy to find. It is the lower convex envelope of W, sketched in Figures 2a and 2b. We denote it by F; in the language of convex analysis it is W^{**}. For plasticity (example 1 above) the material can be absent $(W = 0)$ when there is no stress $(\sigma = 0)$. It is present $(W = 1)$ when there is an admissible stress. It is hopeless $(W = +\infty)$ when the stress exceeds $\sigma_{max} = 1$. Therefore

$$W(\sigma) = \begin{cases} 0 & \sigma = 0 \\ 1 & 0<|\sigma|\leq 1 \\ \infty & |\sigma|>1 \end{cases} \quad \text{and} \quad F(\sigma) = \begin{cases} |\sigma| & 0\leq|\sigma|\leq 1 \\ \infty & |\sigma|>1 \end{cases} .$$

For stresses between 0 and 1, the fraction of material required is proportional to the magnitude of the stress: $F = |\sigma|$. The details of the construction are given in [6]; what matters here is the convexity of the result. The relaxed problem minimizes $\int F(\sigma)dx$.

For elasticity the unrelaxed functional includes a term from the compliance constraint $\int |\sigma|^2 \leq C$ and its Lagrange multiplier λ :

$$W(\sigma) = \begin{cases} 0 & \sigma=0 \\ 1+\lambda|\sigma|^2 & \sigma\neq 0 \end{cases} \quad \text{and} \quad F(\sigma) = \begin{cases} 2\lambda^{1/2}|\sigma| & |\sigma|\leq\lambda^{-1/2} \\ 1+\lambda|\sigma|^2 & |\sigma|\geq\lambda^{-1/2} \end{cases} .$$

In this case the optimal design and the placement of material are described in [4]. The interest is in the "composite regime", where the stress is low and the design is a mixture of material and holes. In that region F is again proportional to $|\sigma|$.

Now we reach the problem whose solution is the object of this paper:

Fig. 2. Convexification $F(\sigma)$ of the functional $W(\sigma)$
(a) perfect plasticity (b) linear elasticity

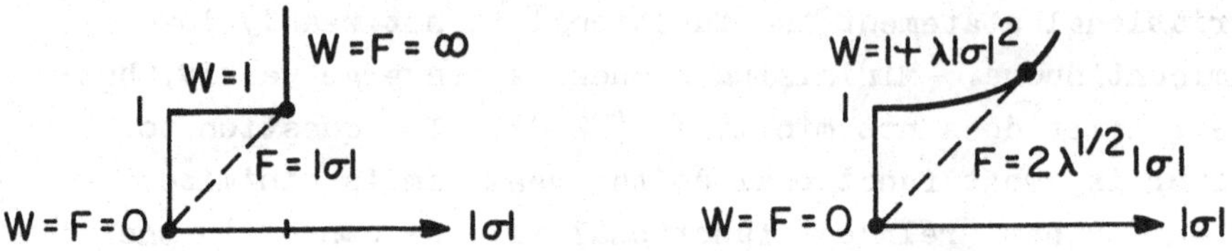

Fig. 3. Contradictions, to prove $\mathrm{dist}(\gamma_0, \gamma_t) \geq t$.
(a) case (4) (b) case (5)

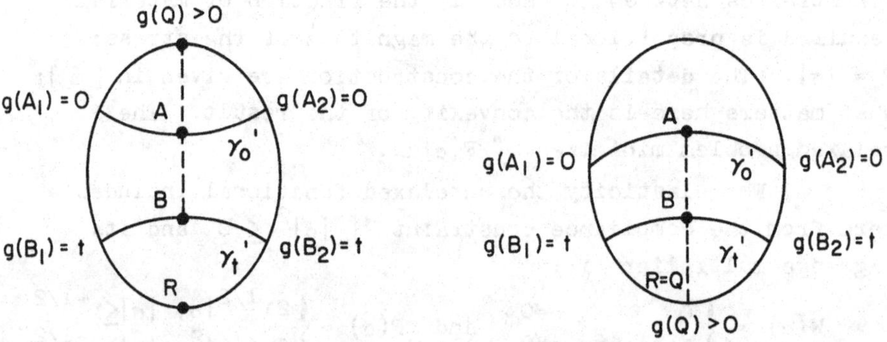

$$\underline{\text{Minimize}} \int_\Omega |\sigma| dx \quad \underline{\text{subject to}} \; |\sigma| \le 1, \; \frac{\partial \sigma_1}{\partial x} + \frac{\partial \sigma_2}{\partial y} = 0, \text{ and}$$

$$\sigma . n = f \quad \text{on} \quad \partial\Omega .$$

It comes directly from plastic design and indirectly from elastic design; those applications serve here as motivation. The problem is of natural interest in itself, and the first step will be to simplify its statement.

THE CONSTRAINED LEAST GRADIENT PROBLEM

In place of the stress (σ_1, σ_2) we introduce a scalar unknown, the stress function $\phi(x,y)$. They will be related by $\sigma = (\phi_y, -\phi_x)$, so that σ automatically satisfies the constraint of zero divergence: $\partial\sigma_1/\partial x + \partial\sigma_2/\partial y = \phi_{yx} - \phi_{xy} = 0$. In a simply-connected domain each divergence-free vector field comes from a stress function ϕ.

At the boundary, $\sigma \cdot n$ is the tangential derivative of ϕ, so that integrating along $\partial\Omega$ gives

$$\phi(Q) = \int_P^Q f \, ds = g(Q) \quad \text{on} \quad \partial\Omega .$$

Thus the indefinite integral of f yields the boundary values of ϕ which are defined up to a constant depending on P. The constraint $|\sigma| \le 1$ becomes

$$|\nabla\phi| = (\phi_x^2 + \phi_y^2)^{1/2} = (\sigma_2^2 + \sigma_1^2)^{1/2} = |\sigma| \le 1 .$$

Therefore the problem to solve is

(CLG) Minimize $\int |\nabla\phi| dx$ subject to $|\nabla\phi| \le 1$ and $\phi|_{\partial\Omega} = g$.

The inequality $|\nabla\phi| \le 1$ makes each admissible ϕ a contraction:

$$|\phi(x) - \phi(y)| \le d(x,y) \quad \text{for all} \quad x,y \quad \text{in} \quad \Omega ,$$

where distance is measured in the domain. Along the shortest path in Ω we have

$$|\phi(x) - \phi(y)| = \left|\int_x^y \frac{\partial\phi}{\partial s} ds\right| \le \left|\int_x^y ds\right| = d(x,y) .$$

Replacing $|\nabla\phi| \le 1$ by this infinite family of linear inequalities exhibits (CLG) as a "transportation problem" in infinite-dimensional mathematical programming.

Note that the constraints $|\nabla\psi| \leq 1$ and $\psi|_{\partial\Omega} = g$ are not automatically compatible: <u>The function</u> g <u>must be a contraction</u>. In plasticity this means that the load f must not exceed the limit load (or the structure will collapse). Beyond that limit no admissible stress can be in equilibrium with f. In elasticity the transition is less precipitous, between a composite material (with holes) and a solid material (with full density). We assume the load f to be small enough so that $g = \int f\, ds$ is a contraction. The MacShane-Kirzsbraun lemma assures that g extends to a contraction over all of Ω, which can be found explicitly from $\psi(x) = \max[g(Q)-d(Q,x)]$.

The key to the problem is the "coarea formula" from geometric measure theory. It expresses $\int |\nabla\psi| dx$ in terms of the lengths of the <u>level curves</u> of ψ. The curve on which $\psi = t$ is γ_t, its length is $|\gamma_t|$, and the formula is simply an integral over those lengths:

$$\int_\Omega |\nabla\psi| dx = \int_{-\infty}^{\infty} |\gamma_t|\, dt\ .$$

For proof we refer to [2,3] but a rough justification is possible here. The strip between the level curves γ_t and γ_{t+dt} has length $|\gamma_t|$ and varying width $dt/|\nabla\psi|$, since the gradient is normal to the curves and ψ changes by dt. The integral of $|\nabla\psi|$ over the strip is therefore $|\gamma_t| dt$, and integrating over all strips yields the coarea formula.

<u>Remark</u>: When ψ is constant on an open set the correct γ_t is the boundary of the set $E_t = \{x: \psi(x) \geq t\}$. In one dimension the length becomes the cardinality of γ_t and the formula is

$$\int_a^b |\psi'(x)| dx = \int_{-\infty}^{\infty} \mathrm{card}(\gamma_t)\, dt\ .$$

If ψ is strictly increasing, the level sets contain only a single point (for $\psi(a) \leq t \leq \psi(b)$) and both sides equal $\psi(b) - \psi(a)$. The formula extends to all functions of bounded variation, for which $\int |\nabla\psi| dx < \infty$. This is the admissible class for the least gradient problem of Bombieri, DeGiorgi, and Giusti [1], without the constraint $|\nabla\phi| \leq 1$. There the level sets are minimal surfaces; here they are modified by the constraint.

<u>Example</u>: Suppose Ω is the unit square $0 \leq x,y \leq 1$ and $\sigma.n = f = +1$ on the bottom, $\sigma.n = -1$ on the left side, and $\sigma.n = 0$ on the right side and top. The boundary values of ϕ are

$$g = \int_0 f \, ds = \begin{cases} x & \text{on the bottom} \\ y & \text{on the left side} \\ 1 & \text{on the right side and top} \end{cases}$$

With no constraint on $|\nabla\phi|$, the minimizer of $\iint |\nabla\phi| dx\, dy$ has straight lines as level sets. The line γ_t must connect the points $(t,0)$ and $(0,t)$ at which the boundary conditions yield $\phi = t$; thus γ_t is the line $x + y = t$. In the upper triangle $x + y \geq 1$, the solution is constant: $\phi \equiv 1$. In the lower triangle it can be reconstructed from its level lines, and it is $\phi = x+y$.

That solution has $|\nabla\phi| = (1+1)^{1/2}$ and violates $|\nabla\phi| \leq 1$. In fact the constraint $|\sigma| \leq 1$ and the boundary condition $\sigma.n = 1$ require that $\sigma = n$ along the bottom; the level lines (which are the streamlines in a flow problem) must exit <u>vertically</u>. The minimizing function becomes $\phi = r$, with $|\nabla\phi| \equiv 1$ in the quarter-disk $r \leq 1$. In the remainder of Ω the solution is again $\phi \equiv 1$. The level curves become circular arcs instead of straight lines.

In this example, where $\phi = g = 0$ at the origin, the level curve γ_t had to be <u>at least a distance</u> t <u>away</u>. It could not come closer to the origin, because $|\nabla\phi| \leq 1$ forbids it. If $\phi = 0$ at one point and $\phi = t$ at another, then the distance between those points is not less than t. This constraint on γ_t, reflecting the constraint $|\nabla\phi| \leq 1$ on ϕ itself, is the key to its construction.

THE CONSTRUCTION OF THE LEVEL CURVES

Our problem is to minimize $\int |\nabla\phi| dx = \int |\gamma_t| dt$. We approach it by minimizing each individual length $|\gamma_t|$. Then we show that the curves γ_t which solve this family of one-dimensional problems, with t as parameter, fit together as the level curves of a function $\phi(x,y)$ which is minimizing.

The boundary points at which $u = g = t$ must lie on γ_t. Without a constraint on $|\nabla\phi|$, γ_t would connect those points by straight lines of minimum total length. The condition $|\nabla\phi| \leq 1$ imposes an extra requirement: γ_t <u>cannot enter the disk of radius</u> $|t-g(Q)|$ <u>around any boundary point</u> Q. Each level curve avoids a set which is a union of open disks. If it entered that set we would have $\phi = t$ and $\phi = g(Q)$ at points whose distance apart is less than $|t-g(Q)|$, contradicting $|\nabla\phi| \leq 1$.

The key is that it suffices to consider only pairs of points, <u>one of which lies on the boundary</u>. The construction of γ_t is independent of the construction of γ_s (we give a further example below). The problem is to show that the distance between γ_t and γ_s is at least $|t-s|$.

This construction can be restated, more precisely but more awkwardly, in terms of $E_t = \{x: \phi(x) \geq t\}$. That set must contain the union

$$D_t = \bigcup_{g(Q)\geq t} \overline{B}(Q, g(Q)-t)$$

and it must not intersect the union

$$G_t = \bigcup_{g(Q)<t} B(Q, t-g(Q)) .$$

Here $B(Q,r)$ is the open ball of points in $\overline{\Omega}$ within r of the boundary point Q. The conditions $E_t \supset D_t$ and $E_t \cap G_t = \emptyset$ require the boundary γ_t of E_t to avoid all the disks $B(Q, |t-g(Q)|)$ as before. Now we show that the construction succeeds if the domain is simply connected.

THEOREM. The distance between γ_0 and γ_t, $t > 0$, is at least t.

Informal proof. Suppose $A \in \gamma_0$ and $B \in \gamma_t$ are as close as possible: $d(A,B) = d(\gamma_0, \gamma_t)$.

(1) If A or B is a boundary point (or if the shortest path from A to B touches the boundary, when Ω is not convex) then from the construction $d(A,B) \geq t$.

(2) If not (1), the tangent to γ_0 at A is parallel to the tangent to γ_t at B. In case both A and B are on straight pieces of γ_0 and γ_t, we can translate A and B until (1) occurs or until (say) $d(Q,A) = |g(Q)|$ for some boundary point Q. Then A is on one of the circles which γ_0 cannot enter, and γ_0 is tangent to it. The points A,B,Q are collinear because the tangents are parallel (and Q is not between A and B or (1) occurs).

(3) If $g(Q) < 0$ we have the expected situation

$$d(A,B) \geq d(B,Q) - d(A,Q) \geq t + |g(Q)| - |g(Q)| = t.$$

Similarly if B is on the circle of radius $g(R) - t > 0$ around a boundary point R, we conclude that $d(A,B) \geq d(A,R) - d(B,r) \geq t$. For the remaining cases we recognize that γ_0 and γ_t may consist of unions of disjoint curves (as in the example below) and we pick out the components γ_0' and γ_t' which contain A and B. They connect boundary points at which $g(A_1) = g(A_2) = 0$ and $g(B_1) = g(B_2) = t$.

(4) Suppose $g(Q) > 0$ and the points occur in the order QAB (Fig. 3a). Between B_1 and Q, g has an even number of zeros. Therefore in addition to A_1 there is a point A_3 from which a different component γ_0'' crosses to the other side QB_2. If γ_0'' is below γ_0' this contradicts $d(A,B) = \text{dist}(\gamma_0, \gamma_t)$. If γ_0'' is above γ_0' this contradicts $d(Q,\gamma_0) \geq g(Q) = d(Q,A)$.

Similarly there is no point, collinear in the order ABR, with $d(B,R) = t - g(R) > 0$.

Note: We assume the generic case in which g and $g - t$ have an even number of (simple) zeros on the boundary and γ_0 and γ_t are disjoint unions of smooth curves; otherwise perturb to $\gamma_\varepsilon, \gamma_{t+\varepsilon}$.

(5) Suppose finally that the points occur in the order ABQ. A still lies on the circle of radius $g(Q) > 0$ around Q. Because B is the point on γ_t' closest to A, it must lie on a circle of radius $|g(R)-t|$ around some boundary point R, collinear with AB and below γ_t'. (R = Q is likely but in principle $\partial\Omega$ could yield another candidate.) These properties of R lead to the same conclusion, for $g(R) > t$ in case (3) and $g(R) < t$ in case (4), that $d(A,B) \geq t$.

This completes the proof (not minimal!) that $d(\gamma_0,\gamma_t) \geq t$. Similarly $d(\gamma_s,\gamma_t) \geq |s-t|$. Therefore there is a function $\phi(x)$ with these as its level sets and with $|\nabla\phi| \leq 1$. That function solves the constrained least gradient problem.

Example 2: Suppose Ω is the unit circle and $\sigma.n = f = 2xy = \sin 2\theta$. Its integral is $g = -\frac{1}{2}\cos 2\theta = \frac{1}{2}(y^2-x^2)$, the boundary value for ϕ.

We look first at the minimization of $\int|\sigma|^2dx = \int|\nabla\phi|^2dx$. The solution is the harmonic function $\phi = \frac{1}{2}(y^2-x^2)$, or $\sigma=(\phi_y,-\phi_x) = (y,x)$. Its support is the whole of Ω, and its level curves are hyperbolas.

Next we minimize $\int|\sigma|dx$; the level curves are straight lines. For any value in the range $-\frac{1}{2} < t < \frac{1}{2}$ there are four boundary points at which $g = t$. The shortest γ_t connects the two closest pairs (AB and CD in Fig. 4). The case $t = 0$, with the four points equally spaced around the circle, is the transition from vertical lines to horizontal lines. The inscribed square caught in this transition is the set where $\phi \equiv 0$.

Fig. 4. Least gradient example: unconstrained and constrained

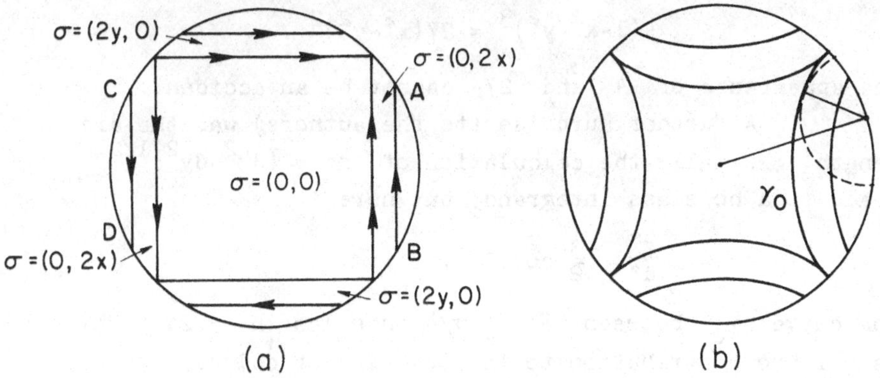

In the right hand sector of the circle, the function ϕ with vertical level lines depends only on x. To agree with the boundary condition at A, it is $\phi = \frac{1}{2}(1-2x^2)$. In that part of the circle $\sigma = (0,2x)$ and $\operatorname{div} \sigma = 0$ and $\sigma.n = 2xy$. Since $|\sigma| = |\nabla\phi|$ reaches 2, the constrained least gradient problem must look for another solution.

To satisfy $|\nabla\phi| \leq 1$, the level curves will have to "bow inwards" as in Fig. 4b. We concentrate on the right-hand quarter of the unit circle, and we need the envelope of a family of disks. They are centered at the boundary points $(\cos\theta, \sin\theta)$, their radius is $t - g = t + \frac{1}{2}\cos 2\theta$, and a calculation gives the points on γ_t:

$$x = \cos\theta(1-t-\tfrac{1}{2}\cos 2\theta),\ y = \sin\theta(1+t+\tfrac{1}{2}\cos 2\theta).$$

Those points satisfy $(x-\cos\theta)^2 + (y-\sin\theta)^2 = (t+\frac{1}{2}\cos 2\theta)^2$ and also the θ-derivative of that equation, which gives the envelope. Another approach to the envelope, based on duality, is given below.

The special curve γ_0 (with $t=0$) connects the boundary points $(1/\sqrt{2}, \pm 1/\sqrt{2})$. To the left of that curve, at the center of the unit circle, is the set where $\phi = 0$ and $\sigma = 0$. It is a hole where the optimal design places no material, but the hole is not square as in the unconstrained case. The curve itself can be described by eliminating θ from x and y above, to yield a remarkable equation for the envelope γ_0 :

$$4(1-x^2-y^2)^3 = 27(x^2-y^2)^2 .$$

The appearance of 4 and 27 cannot be an accident.

A further surprise (to the authors) was the arc length. Normally the calculation of $ds = (dx^2+dy^2)^{1/2}$ leads to a hopeless integrand, but here

$$\frac{ds}{d\theta} = \frac{3}{2}\cos 2\theta + t .$$

The curve γ_0 between $\theta = \pm \pi/4$ has length 3/2. This is our one contribution to the teaching of calculus, the discovery of a curve whose arc length is computable.

THE DUAL PROBLEM

It is unusual in a constrained problem to be able to verify directly that a proposed ϕ or σ is minimizing. Here that seemed possible; each level curve γ_t had minimum length. That was a one-dimensional problem, quickly solved. But that problem also had constraints, and strictly speaking it is not trivial to verify that <u>its</u> proposed solution is correct. The normal way to admit constraints is through <u>Lagrange multipliers</u>, leading to an optimality condition that involves both ϕ (or σ) and the multiplier. The latter is the solution to the dual problem, and we need to identify that problem for constrained least gradients.

Our approach to duality is the old-fashioned one, through the minimax theorem. We introduce a Lagrange multiplier u(x,y) for the constraint div $\sigma = 0$:

$$M = \min_{\substack{\operatorname{div}\sigma=0\\ \sigma\cdot n=f\\ |\sigma|\le 1}} \int_\Omega |\sigma|\,dx = \min_{\substack{\sigma\cdot n=f\\ |\sigma|\le 1}} \max_u \int_\Omega (|\sigma| + u \operatorname{div}\sigma)\,dx .$$

Applying the minimax theorem and Green's theorem, this is

$$M = \max_u \min_{|\sigma|\le 1} \int_\Omega (|\sigma| - \sigma\cdot\nabla u)\,dx + \int_{\partial\Omega} uf\,ds .$$

At each point of Ω, we minimize $|\sigma|-\sigma\cdot\nabla u$ subject to $|\sigma| \leq 1$. There are three possibilities. In the case of $|\nabla u| < 1$, the minimum is zero and it is attained at $\sigma = 0$. In the case of $|\nabla u| = 1$, the minimum is still zero but there is a range of minimizing vectors σ; any positive multiple $\sigma = \alpha\nabla u$ gives $|\sigma| - \sigma.\nabla u = \alpha - \alpha = 0$. The only restriction is $\alpha = |\sigma| \leq 1$. In the remaining case, $|\nabla u| > 1$, the minimum occurs when σ is the largest possible multiple of ∇u,

$$\sigma = \frac{\nabla u}{|\nabla u|} \quad \text{and} \quad |\sigma| - \sigma\cdot\nabla u = 1 - |\nabla u| .$$

We summarize these optimality conditions: σ <u>is a non-negative multiple of</u> ∇u, <u>and</u>

$$|\nabla u| < 1 \iff \sigma = 0$$

$$|\nabla u| = 1 \iff 0 \leq |\sigma| \leq 1$$

$$|\nabla u| > 1 \iff |\sigma| = 1 .$$

Under these optimality conditions we have:

$$|\sigma| - \sigma\cdot\nabla u = \min\{0, 1 - |\nabla u|\} = (1 - |\nabla u|)_- .$$

Then substituting into M, the <u>dual problem</u> is the maximization of a concave function:

(CLG*) Maximize $\Phi^*(u) = \iint_\Omega (1 - |\nabla u|)_- \, dxdy + \int_{\partial\Omega} ufds.$

The admissible functions u are those for which Φ^* is well defined. We assume that $|f| \leq 1$, or the constraints $\sigma.n = f$ and $|\sigma| \leq 1$ are incompatible. Then $\int uf\, ds$ is defined if at the boundary we have $u \in L^1(\partial\Omega)$, and the integral of $(1 - |\nabla u|)_-$ is finite over Ω if $\iint |\nabla u| < \infty$. The natural set of functions with these properties is BV, the space of functions of bounded variation. The gradient of any $u \in BV$ is a finite measure, and the trace on $\partial\Omega$ is in L^1. Therefore $\Phi^*(u) < \infty$. The maximum of Φ^* is finite, and it equals the minimum of $\int |\sigma|\, dx$, exactly when g is a contraction.

Remark: The least gradient problem without the constraint $|\sigma| \le 1$ (or $|\nabla\psi| \le 1$) has the dual:

$$\text{Maximize} \int_{\partial\Omega} uf\,ds \quad \text{subject to} \quad |\nabla u| \le 1 \text{ in } \Omega.$$

The primal involved the L^1 norm of $|\nabla\psi|$ and the dual (see [5]) constrains the L^∞ norm of $|\nabla u|$.

We return to interpret the optimality conditions. If $\sigma = (\psi_y, -\psi_x)$ is parallel to ∇u, then the gradients of ψ and u are everywhere perpendicular. Whenever the optimal ψ and u have smooth level curves ψ = constant and u = constant, these curves form an orthogonal family. The three subregions of Ω in the optimal design are

(i) holes: $\sigma = 0$ and $|\nabla u| < 1$.

(ii) fibers: $0 \le |\sigma| = |\nabla\psi| \le 1$ and $|\nabla u| = 1$; the level curves of ψ are straight lines orthogonal to the level curves of u.

(iii) solid regions: $|\sigma| = |\nabla\psi| = 1$ and $|\nabla u| > 1$; the level curves of u are straight lines orthogonal to the level curves of ψ.

This description includes a striking fact about the orthogonal level curves: one or the other is straight! The geometry of the design is very special. Where there is material, it is formed either of straight but not necessarily parallel fibers, or it has full density and the fibers are orthogonal to a family of straight but not necessarily parallel lines. To see this we assume sufficient smoothness and look at the consequences of $|\sigma| = |\nabla\psi| = 1$ or $|\nabla u| = 1$. With σ parallel to ∇u several approaches are possible:

1. In a region where $|\sigma| = 1$, of type (iii) above, σ is the vector field of unit normals to the curves u = constant. Then their curvature κ is the divergence of the unit normal. Therefore $\kappa = \operatorname{div}\sigma = 0$ and the curves u = constant are straight lines.

2. In a region where $|\nabla u| = 1$, of type (ii) above, differentiation of $u_x^2 + u_y^2 = 1$ gives

$$u_x u_{xx} + u_y u_{xy} = 0, \quad u_x u_{xy} + u_y u_{yy} = 0 .$$

The slope dy/dx, along u = constant, is $-u_x/u_y$. On the orthogonal family it is therefore u_y/u_x, and then

$$\frac{d^2y}{dx^2} = \frac{u_{yx} + u_{yy}y_x}{u_x} - \frac{u_y(u_{xx}+u_{yx}y_x)}{u_x^2} = 0 .$$

We owe to Richard Dudley this proof that the orthogonal families are straight lines.

3. We may identify the level curves of u as a family of involutes and the envelope of the straight fibers as the corresponding evolute. Or, if we fix one involute as the curve $u = 0$, the function u is simply the <u>distance from that curve</u>. Polar coordinates are typical, with $|\nabla r| = 1$ and straight lines θ = constant; r is then the distance from a degenerate curve, the point at the origin.

4. One could establish the condition for a smooth function ψ to have straight level sets; it seems to be

$$\psi_y^2\psi_{xx} + \psi_x^2\psi_{yy} = 2\psi_x\psi_y\psi_{xy} .$$

This must yield the linear behavior of the reciprocal $|\sigma|^{-1} = |\nabla\psi|^{-1}$ noted elsewhere [6] along the fibers.

<u>Question</u>: It would be of interest to remove the assumption of smoothness, and to describe all Lipschitz functions for which $|\nabla u| \equiv 1$. For finitely many constants c_i, closed sets S_i, and signs $\varepsilon_i = \pm 1$,

$$u(P) = \sup(c_i + \varepsilon_i \operatorname{dist}(P,S_i)) \quad \text{has} \quad |\nabla u| = 1.$$

We close by returning to the example in the unit circle and recomputing the "bowed" level curves of ψ. Since they are curved, the lines u = constant are straight. At the boundary point $Q = (\cos\theta, \sin\theta)$, the stress is $\sigma(Q) = (\sin\theta, \cos\theta)$. (This satisfies $|\sigma| = 1$ and $\sigma\cdot n = \sin 2\theta = f$.) The straight line u = constant is orthogonal to σ, so $P = (x,y)$ is on the line if $(P-Q)\cdot\sigma(Q) = 0$:

$$x \sin \theta + y \cos \theta = Q\cdot\sigma(Q) = \sin 2\theta .$$

Orthogonal to those lines are the level curves γ_t of ψ. To find γ_t we go along the line a distance $g(Q) - t$:

$$P = Q + (g(Q)-t)(\cos \theta,-\sin \theta)$$

or

$$x = \cos \theta(1-\frac{1}{2} \cos 2\theta-t), \quad y = \sin \theta(1+\frac{1}{2}\cos 2\theta +t).$$

Thus we recover the envelope of disks. It is this solution of the constrained least gradient problem, through the curves γ_t rather than the values $\psi(x,y)$, which was the main object of the paper.

ACKNOWLEDGEMENT

The work of the first author was supported in part by NSF Grants MCS82-01308, DMS-8312229 and ONR Grant N00014-83-K-0536 and that of the second author was supported in part by NSF Grant DMS-8403222 and ARO Grant DAAG29-83-K-0025.

REFERENCES

1. Bombieri, E., De Giorgi, E. and Giusti, E. (1969). Minimal cones and the Bernstein problem. Inventiones Math., 7, 243-268.

2. Federer, H. (1969). Geometric Measure Theory. New York: Springer-Verlag.

3. Fleming, W. and Rishel, R. (1960). An integral formula for total variation. Archiv der Mathematik, 11, 218-222.

4. Kohn, R. V. and Strang, G. (1986). Optimal design and relaxation of variational problems, I-III. Comm. Pure and Applied Mathematics, 39, 113-137, 139-182, 353-377.

5. Strang, G. (1982). L^1 and L^∞ approximation of vector fields in the plane. Lecture Notes in Num. Appl. Anal., Nonlinear PDE in Applied Science, 5, 273-288.

6. Strang, G. and Kohn, R. V. (1986). Fibered structures in optimal design. In Proceedings of the Dundee Conference on Differential Equations, ed. B. Sleeman. To appear.

7. Strang, G. and Kohn, R. V. (1982). Optimal design of cylinders in shear. In Mathematics of Finite Elements and Its Applications, ed. J. Whiteman. New York: Academic Press.

THE FUSION OF PHYSICAL AND CONTINUUM-MECHANICAL CONCEPTS IN THE FORMULATION OF CONSTITUTIVE RELATIONS FOR ELASTIC-PLASTIC MATERIALS

E.H. Lee
Department of Mechanical Engineering, Aeronautical Engineering, & Mechanics, Rensselaer Polytechnic Institute, Troy, New York

A. Agah-Tehrani
Department of Mechanical Engineering, Aeronautical Engineering, & Mechanics, Rensselaer Polytechnic Institute, Troy, New York

Abstract. In the formulation of constitutive relations for materials, which express physical characteristics in mathematical form, it is important to ensure that a physically motivated concept be expressed by a valid continuum-mechanics structure and conversely that a mathematical construct express a relevant physical phenomenon. Examples are presented of the application and failure to apply these precepts drawn from the development of elastic-plastic theory valid for finite deformation in the presence of strain induced anisotropy.

On the foundation of nonlinear kinematics, in contrast to the classical linear assumption that the rate of total deformation is the sum of the elastic and plastic deformation rates, strict uncoupling of elastic and plastic deformation rate terms according to their physical origins is achieved.

Rate variables $\hat{D}^p$ and $\hat{D}^e$ associated with the current configuration are devised which respectively express the influence of plasticity modified by the elastic strain and of elasticity modified by the deforming unstressed reference state needed for the elastic law. In terms of these variables an elastic-plastic theory is generated valid for finite deformation in the presence of strain-induced anisotropy modeled as combined isotropic-kinematic hardening. The resulting formulation provides elastic-plastic moduli having the symmetries necessary for generating a rate potential function which can be incorporated into Hill's variational principle valid for solving problems involving finite deformation and convenient for finite-element exploitation.

1 INTRODUCTION

The formulation of constitutive relations, which express physical characteristics of materials in mathematical form, can pose a very demanding challenge. Eddington has stated in describing the attack on a physical problem, see Woods (1973), "the initial formulation of the problem was the most difficult part, as it was necessary to use one's brains all the time; afterwards one could use mathematics instead!" The formulation of constitutive relations for stress analysis of elastic-plastic deformation of ductile metals falls squarely into this initial stage of problem solving. As will be demonstrated, enormous errors have been introduced through incorrect constitutive relations, particularly when the problem involves finite strain and rotation of material elements.

The accurate solution of such problems can play a crucial role in the success of the design of a structure, of a component in a machine, or of a forming process. At the present time the formulation of constitutive relations to express the influence of strain-induced anisotropy has generated quite disparate theories. It is important to clarify this situation because strain induced anisotropy is commonly generated by plastic flow and can have a major influence on stress and strain distributions, particularly when non-proportional loading or straining occurs, i.e. when the stress-increment or strain-increment components suffer rapid changes in the relative magnitudes of the various tensor components. For example, such situations arise in a particularly severe form in the plastic buckling of columns having cruciform sections. If torsional buckling occurs, compressive straining can change suddenly to shear straining. A similar situation arises as wave fronts of combined compression and shear stress traverse a material element.

In formulating constitutive relations it is necessary that the introduction of the influence of a physically motivated phenomenon be expressed by a valid continuum-mechanics structure and conversely that a mathematical construct should comprise a valid representation of an associated physical phenomenon. Whereas this may seem to be a truism, it is not incorporated into some current formulations.

For some materials a simple mathematical statement encompasses their response to stress or deformation. For example, the deformation characteristics of an elastic body can be expressed by the statement that the stress generated is a function of the current deformation gradient, which incorporates both the strain and rotation. The stress does not depend on the history of the deformation. This requirement, in conjunction with the objectivity principle (that the superposition of an arbitrary time-dependent rigid-body rotation must simply rotate the current stress by the current superposed rotation of the body) and the existence of a strain energy function, results in the elastic constitutive relation

$$\sigma = \frac{2}{\det(F)} F \frac{\partial \psi}{\partial C} F^T, \quad C = F^T F \tag{1.1}$$

where σ is the Cauchy or true stress, F the current deformation gradient and ψ the strain energy function. The superscript T denotes the matrix transpose and det( ) indicates the determinant. Experiments by Rivlin & Saunders (1951) yielded the deduction of the form of the function ψ which provided an accurate constitutive relation of the form (1.1) to express the stress response of rubbers to deformation.

An analogous formulation of nonlinear viscosity that the stress is a function of the density and rate of deformation (the symmetric part of the velocity gradient) generates a constitutive relation which, for straining in simple shear, predicts a relationship between the normal stresses generated that is in disagreement with experimental measurements. The normal stresses arise in addition to the shear stress because of the nonlinearity. It was later shown that to correct the normal-stress prediction in simple shear for a nonlinearly viscous fluid it is necessary to incorporate a term involving a higher order velocity derivative which in effect introduces a viscoelastic contribution to the response.

Thus in the case of elasticity the simple concept that the stress is a function of the displacement gradient yielded a satisfactory constitutive relation, whereas in the case of nonlinear viscous liquids an analogous simple mathematical model failed to reproduce experimental results. In the case of elastic-plastic materials, the main concern of this paper, no adequate simple model presents itself and a thorough investigation of the interaction between elastic and plastic deformation is needed in order to develop a satisfactory constitutive relation.

2 THE KINEMATICS OF ELASTIC-PLASTIC DEFORMATION

The measurement of stress and strain in a tension test provides a convenient vehicle to discuss the coupling between elastic and plastic deformation and thus to select appropriate variables to express these components in a constitutive equation relating stress and total deformation.

Fig. 1 The tension test for a ductile metal.

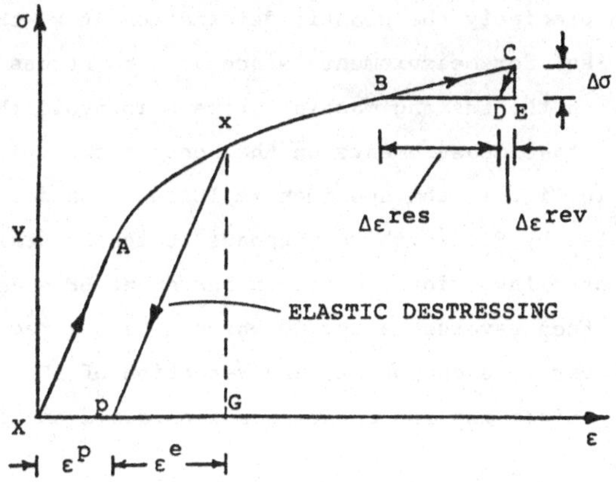

Figure 1 depicts the stress-strain relation in a tension test for a ductile polycrystalline metal. The undeformed specimen corresponds to the position X. On loading elastic deformation occurs along XA. Plastic flow commences at the yield point A followed by combined plastic and elastic deformation along Ax. If unloading then takes place elastic recovery occurs along xp to zero stress at p. In the regions of pure elastic deformation, XA and px, the deformation is reversible. Since the stress is zero at p, the elastic strain is zero as it is also in the initial undeformed state X. Thus ε^p, the strain at p, is purely plastic and will correspond directly to the distribution of plastic-flow micro-mechanisms such as dislocation migration and will be governed by the physical theory of plastic flow. Since purely elastic deformation takes place along xp, with dislocations and other plastic micro-mechanisms locked in place, the state at p, and hence the strain ε^p, also expresses the plastic flow at x as well as at p. Since the elastic strain is zero at p, pG expresses the elastic strain ε^e at x.

In the case of infinitesimal strains, all strains can be defined from the bases of the configurations X (initially undeformed), p (unstressed) or x(current) with negligible difference between them. Thus

$$\varepsilon = \varepsilon^e + \varepsilon^p \tag{2.1}$$

For finite strains, X is the natural base configuration for the total strain ε and the plastic strain ε^p, and the unstressed state p is the appropriate base for defining the elastic strain ε^e. Thus in this case

$$\varepsilon \neq \varepsilon^{e} + \varepsilon^{p} \tag{2.2}$$

Although, as explained, the deformation in the destressed configuration p expressed precisely the plastic deformation at x, this configuration is inconvenient for measurements since at zero stress the specimen is not held tightly in the testing machine grips. To avoid this predicament it is common to take measurements on the monotonically loaded specimen. As indicated in Fig. 1, the specimen is loaded in a series of stress increments $\Delta\sigma$ depicted by EC, which corresponds to the excursion BC along the stress-strain curve involving the strain increment BE = $\Delta\varepsilon$. The stress increment is then reversed along CD which causes a recovery of strain $\Delta\varepsilon^{rev}$, the reversible component, and retention of BD = $\Delta\varepsilon^{res}$, the residual component. Since the increments are considered small, as for infinitesimal theory

$$\Delta\varepsilon = \Delta\varepsilon^{rev} + \Delta\varepsilon^{res} \tag{2.3}$$

where the basis for strain increment measurement is the configuration at B.

Classical plasticity theory is based on the assumptions that

$$\Delta\varepsilon^{e} = \Delta\varepsilon^{rev} \ ; \ \Delta\varepsilon^{p} = \Delta\varepsilon^{res} \tag{2.4}$$

and the total, elastic and plastic components of strain are determined by adding the increments from the initial state.

In considering the incremental stressing and destressing along BCD, it must be borne in mind that the specimen is continuously at the yield stress and hence subject to elastic strain at the yield limit. Thus there is likely to be coupling of the elastic and plastic strains in the $\Delta\varepsilon^{rev}$ and $\Delta\varepsilon^{res}$ and (2.4) is not likely to be strictly correct.

In order to assess these considerations for general stressing, an undeformed body with the reference coordinates of material particles X is deformed according to the motion

$$x = x(X,t) \tag{2.5}$$

as shown in Fig. 2. Deformation is analyzed through the deformation gradient

$$F = \partial x/\partial X \quad \text{or} \quad F_{ij} = \partial x_i/\partial X_j \tag{2.6}$$

and the differential relation

$$dx = F\,dX \tag{2.7}$$

which expresses the strain and rotation of material elements defined by dx and dX, column vectors in the current and initial configurations respectively.

Fig. 2 Elastic-plastic Deformation in 3-D.

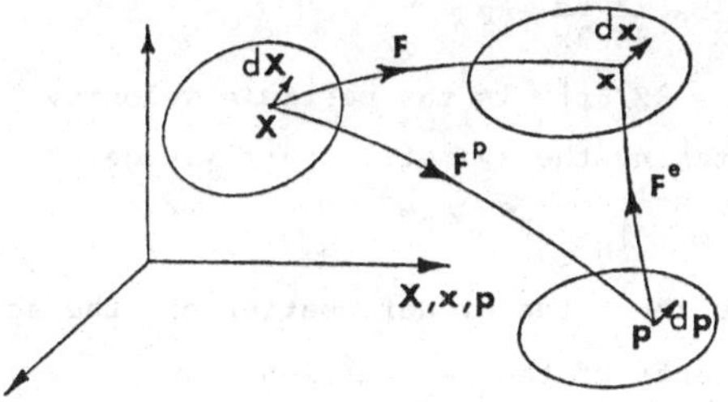

If the body in the deformed configuration x is destressed by the superimposed elastic destressing deformation gradient $F^{e^{-1}}$, the inverse of the elastic deformation gradient F^e, the resulting deformation gradient due to the sequential deformations F followed by $F^{e^{-1}}$ determines the plastic deformation

$$F^p = F^{e^{-1}} F \tag{2.8}$$

so that pre-multiplying by F^e gives

$$F = F^e F^p \tag{2.9}$$

Lee (1969). Since destressing from x leaves the body unstressed in the state p, the stress is zero there, and so also is the elastic strain. Thus F^p expresses the plastic deformation. Equation (2.9) uncouples precisely the elastic and plastic components of the total deformation gradient F. Because (2.9) is a matrix product which is in general non-commutative it will clearly not result in the summation law (2.1) for strains nor in a similar additive law for strain rates. This is an example of the much deeper physical content which the complete finite-deformation-valid law embodies compared with that associated with the classical linearized laws of summation of elastic and plastic strains or of strain rates.

The finite-deformation valid Lagrange strain E is given by

$$E = (F^T F - I)/2 \tag{2.10}$$

where I is the unit matrix. Substitution of (2.9) gives (Lee 1981)

$$E = F^{p^T} E^e F^p + E^p \tag{2.11}$$

where E^e and E^p are the Lagrange strains associated with the elastic and plastic deformations respectively.

The rate of deformation D is given by the symmetric part of the velocity gradient in the current state, the latter being given by

$$L = D + W = \frac{\partial v}{\partial x} = \frac{\partial v}{\partial X}\frac{\partial X}{\partial x} = \dot{F}F^{-1} \tag{2.12}$$

where W is the spin and $v = \partial x/\partial t|_X$ is the particle velocity. Substitution of (2.9) into (2.12) and taking the symmetric part yields

$$D = D^e + F^e D^p F^{e^{-1}}\Big|_S + F^e W^p F^{e^{-1}}\Big|_S \tag{2.13}$$

where D^e, W^e and D^p,W^p are the rates of deformation and the spins for the elastic and plastic components of the deformation:

$$L^e = D^e + W^e \text{ and } L^p = D^p + W^p \tag{2.14}$$

Some simplification can be achieved by noting that the unstressed configuration p remains an unstressed configuration if an arbitrary rotation is considered to be superposed on the destressing process. Hence taking note of the polar decomposition theorem F = RU = VR, where U and V are symmetric, without loss of generality F^e can be replaced by $V^e = V^{e^T}$, i.e. destressing without rotation is considered, and (2.13) becomes, Lee (1981)

$$D = D^e + V^e D^p V^{e^{-1}}\Big|_S + V^e W^p V^{e^{-1}}\Big|_S \tag{2.15}$$

The choice of $F^e = V^e$ will also have affected F^p since, according to (2.9), $F = F^e F^p$ and F is not affected.

Equation (2.15) can be used to extend the interpretation of the incremental stressing and destressing experiment along BCD in Fig. 1 to include arbitrary stressing. This is needed to investigate general elastic-plastic analysis. Writing

$$\Delta\varepsilon = D\,\Delta t \tag{2.16}$$

and substituting from (2.15) demonstrates that the classical assumption (2.4)which is equivalent to

$$D = D^e + D^p \tag{2.17}$$

is not valid. For arbitrary stressing the coupling of the elastic deformation with the rate terms prevents direct measurement of the elastic and plastic strain increments by means of the incremental stressing and destressing experiment. However, since elastic strains are usually small, V^e is close to the unit matrix, so that the second term on the right-hand side of (2.15) approximates D^p and the third term approximates zero since W^p is anti-symmetric. Thus (2.17) is approximately satisfied. In contrast additivity of elastic and plastic Lagrange strains is not even approximately valid since in (2.11) F^p is far from the unit matrix in the case of finite plastic strains.

It has been demonstrated that (2.9) precisely uncouples elastic and plastic deformation and hence provides a basis for elastic-plastic theory not restricted by the classical assumption of linear kinematics of the summation of elastic and plastic rates of deformation. In the case of isotropic hardening such a theory was presented by Lubarda and Lee (1981) and recently for anisotropic hardening by Agah Tehrani et al. (1986) which is summarized in this paper.

The expression (2.15) for the rate of total deformation involves complicated contributions from the rate terms D^p and W^p coupled with the elastic deformation. It would thus be difficult to use this in combination with the elastic and plastic laws to obtain the desired relation between stress rate and the rate of total deformation. Since the stress acts on the body in the current configuration, study of the kinetics of elastic-plastic deformation, which is needed to investigate normality and stability in the formulation of a constitutive relation, should take place in that configuration. Rate variables defined in the current configuration would be advantageous, in contrast, for example, to D^p which refers to deformation in the purely plastically deformed unstressed state. A further advantage of rate type variables involving elastic and plastic straining defined in the current state is that they would each involve a contribution to the velocity field in the current state and would thus be additive. A pair of such variables termed $\hat{D}^e$ and $\hat{D}^p$, which comprise D^e and D^p modified by the elastic deformation to express the contribution of elastic and plastic deformation to the velocity field in the current configuration has been adopted by Agah Tehrani et al. (1986). The rate variable $\hat{D}^p$ was first suggested by Creus et al. (1984). Although $\hat{D}^e$ does involve elastic strain rate in the current configuration, it was shown by Lubarda and Lee (1981) that the last term in (2.15) also comprises a rate of elastic strain term due to the rotation of the unstressed reference configuration p involved in the elasticity law. This does not appear in elasticity theory since the unstressed reference state is considered to be at rest throughout the history of elastic deformation.

The resultant contributions of elastic and plastic rate of straining to the velocity field in the current configuration were evaluated by Agah Tehrani et al. (1986) by considering the consistency of the sequence of deformations from the unstressed configuration p to the current configuration x at times t and t + dt. This study resulted in a simple algebraic manipulation of the expression (2.12), with $F = F^e F^p$ from (2.9), for the velocity gradient L in the current state of which D in (2.15) is the symmetric part.

$$L = D + W = \dot{V}^e V^{e^{-1}} + V^e D^p V^{e^{-1}} + V^e W^p V^{e^{-1}} \tag{2.18}$$

Since $\hat{D}^e$ and $\hat{D}^p$ are the resultant rate terms in the current state associated with elastic and plastic rates of deformation respectively and since the associated velocity contributions to the velocity field in the current configuration are additive, then

$$D = \hat{D}^e + \hat{D}^p \tag{2.19}$$

To obtain the velocity gradient in the current configuration the spin W must be added, giving

$$L = \hat{D}^e + \hat{D}^p + W \tag{2.20}$$

Substituting for L from (2.18) and solving for $L^p = D^p + W^p$ gives

$$D^p + W^p = - V^{e^{-1}} \dot{V}^e + V^{e^{-1}} (\hat{D}^e + \hat{D}^p + W) V^e \tag{2.21}$$

Taking the symmetric part of (2.21) in order to eliminate W^p, which complicated (2.15), gives

$$D^p = -V^{e^{-1}} \dot{V}^e\big|_S + V^{e^{-1}} [(\hat{D}^e + \hat{D}^p + W) C^e]_S V^{e^{-1}} \tag{2.22}$$

where $C^e = V^e V^e$ is the Cauchy Green Tensor appearing in (1.1) specialized for $F = V^e$. In an attempt to eliminate the dependence of D^p on elastic rate terms and spin, we try equating the terms in (2.22) involving D^p and $\hat{D}^p$. This turns out to be successful since the remainder provides a convenient vehicle for expressing the elastic law. Thus we have

$$D^p = V^{e^{-1}} (\hat{D}^p C^e)_S V^{e^{-1}} = V^{e^{-1}} \hat{D}^p V^e\big|_S \tag{2.23}$$

hence

$$C^e \hat{D}^p + \hat{D}^p C^e = 2 V^e D^p V^e \tag{2.24}$$

Thus $2 V^e D^p V^e$ is given by a linear operator Λ on $\hat{D}^p$

$$\Lambda : \hat{D}^p = 2 V^e D^p V^e \tag{2.25}$$

where Λ is a fourth order tensor with components

$$\Lambda_{ijkl} = C^e_{ik} \delta_{jl} + C^e_{jl} \delta_{ik} \tag{2.26}$$

Inverting the operator gives

$$\hat{D}^p = \Lambda^{-1} : 2 V^e D^p V^e \tag{2.27}$$

Having dealt with the plastic part of (2.22), the remainder takes the following form after pre-multiplying and post-multiplying it by V^e and utilizing $\dot{C}^e = \dot{V}^e V^e + V^e \dot{V}^e$.

$$\dot{C}^e - WC^e + C^eW = \overset{\circ}{C}^e = C^e\hat{D}^e + \hat{D}^eC^e = \Lambda : \hat{D}^e \tag{2.28}$$

where $\overset{\circ}{C}^e$ is the Jaumann derivative of C^e. These terms are all associated with elastic deformation. Thus the plastic relation (2.23) has produced a separation of elastic and plastic rate terms, $\hat{D}^e$ and $\hat{D}^p$, with coupling only between D^p and the elastic strain because this modifies the plastic contribution to the strain rate in the current configuration. The spin W^p does not appear at all and the spin W plays an independent role in the constitutive relation through the Jaumann derivative and also in the variational principle for stress analysis.

By taking the anti-symmetric part of (2.12) a linear operator relation for $(W - W^p)$ first given by Creuss et al. (1984) is obtained

$$V^e(W - W^p) + (W - W^p)V^e = V^e(D + D^p) - (D + D^p)V^e \tag{2.29}$$

Thus W^p is determined from the strain by a kinematic relation which throws new light on a recent concept that the plastic spin was involved as a constitutive property of the material.

There are two aspects of the kinematics of the unstressed state which do not in fact modify the theory presented but which are often commented on. When a body is unloaded following non-homogeneous plastic flow it is generally left in a state of residual stress. To achieve an unstressed state such a body must be considered cut into sufficiently small elements so that the residual stresses approach zero. This configuration is not then globally differentiable to determine a displacement gradient but for each unstressed particle a local linear mapping will determine F^p. This circumstance does not affect the stress analysis solution since as observed by Lee (1982) the unstressed configuration is the product of a thought experiment to determine the variables F^e and F^p which are point matrix functions. While the continuum mechanics aspect of the study is limited to the current state which is achieved physically and must satisfy compatibility relations, the unstressed state only plays a role in the formulation of the elastic-plastic constitutive relation. The second topic concerns the development of a severe Bauschinger effect, common at large strains, causing reverse plastic flow on destressing so that the unstressed state cannot be achieved without modifying the plastic strain generated in the sequence of current states x. In effect the origin of the stress space lies outside the yield surface associated with the state x. As discussed by Agah Tehrani et al. (1986), by extrapolating the application of the elasticity law to zero stress a formal plastic strain tensor can be obtained. This process is equivalent to considering the plastic micro-mechanisms, such as dislocations, locked in place. Because of the reversibility of the elastic law, adoption of this formal plastic strain will exactly reproduce the achievable conditions within and on the yield surface so that the theory presented will apply to stress and deformation analysis without modification.

3 KINETICS

Plasticity is an incremental or flow type theory involving a strain-rate type variable, rather than the total plastic strain, related to a stress-rate term for a strain hardening material. Equation (2.19) provides the kinematic coupling needed between elastic and plastic rates. The usual elasticity law relating stress and deformation must therefore be differentiated and (2.28) provides the means of connecting the kinematics to the elasticity law through the Jaumann derivative $\overset{\circ}{C}{}^{e}$.

Elastic strains are usually small in elastic-plastic problems because there are traction free surfaces and plastic flow will prevent the stress increasing beyond the order of the yield stress. However it is not appropriate to adopt Hooke's law as in classical elasticity because it is not objective so that taking a time derivative in the presence of finite rotation, which occurs because of plastic flow, is not legitimate. Use of (1.1), the elasticity law valid for finite deformation, eliminates this problem. In general the strain-induced plastic anisotropy overwhelms the effect of strain on elastic properties so that we consider the elastic law to be isotropic and invariant. Since V^e is always symmetric it is convenient to take (1.1) in the form

$$\tau = 2\,C^e\,\partial\psi/\partial C \tag{3.1}$$

where τ is the Kirchhoff stress $(\det V^e)\sigma$. Taking a Jaumann derivative gives

$$\overset{\circ}{\tau} = L : \overset{\circ}{C}{}^{e} \tag{3.2}$$

where

$$L_{ijk\ell} = 2\,\delta_{ik}\left(\frac{\partial\psi}{\partial C^e}\right)_{\ell j} + C^e_{im}\left(\frac{\partial^2\psi}{\partial {C^e}^2}\right)_{mjk\ell} \tag{3.3}$$

Using (2.28)

$$\overset{\circ}{\tau} = (L : \Lambda) : \hat{D}^e \tag{3.4}$$

Inversion of the operator gives

$$\hat{D}^e = (L : \Lambda)^{-1} : \overset{\circ}{\tau} \tag{3.5}$$

Lubarda and Lee (1981) indicated that the operator $(L:\Lambda)_{ijk\ell}$ has the necessary symmetries for $\hat{D}^e$ to be derivable from an elastic rate potential which means that $(\hat{D}^e dt)$ can be recovered upon unloading of the Jaumann stress increment $(\overset{\circ}{\tau} dt)$. The Jaumann stress increment is used since it eliminates the effect on the stress increment of a possible spin difference in the stressing and destressing processes. Since the unloading is elastic, $\hat{D}^e dt$ is thus the reversible part $\Delta\varepsilon^{rev}$ associated with incremental loading and $\hat{D}^p dt$ the residual part $\Delta\varepsilon^{res}$ since no plastic flow occurs on destressing. This conclusion provides a precise interpretation of the reversible and residual strain increments in the usual test for "elastic" and "plastic" increments of strain. It also shows that $\hat{D}^p$ is the candidate for normality in the elastic-plastic law.

4 STRAIN INDUCED PLASTIC ANISOTROPY

Strain induced anisotropy arises from two basic phenomena. One is the residual type stresses which result from the heterogeneous nature of plastic deformation in single crystals as well as in polycrystalline material. This can be modeled by kinematic hardening, a shift of the yield surface in stress space. The other is a change in the shape of the yield surface during deformation which is a result of the differential hardening of various slip systems and texture formation. The present discussion involves only the first kind of anisotropy although the approach adopted can include the latter as stressed by Creus et al. (1984).

Combined isotropic kinematic hardening is considered. A Mises type yield surface shifted in stress space by the back stress α takes the form

$$(\tau' - \alpha) : (\tau' - \alpha) = 2\,\bar{\tau}^{2/3} \tag{4.1}$$

in which the colon denotes the scalar product, τ' is the stress deviator, and $\bar{\tau}$ is the isotropic component of the yield stress. The evolution equation for α is based on the Prager-Ziegler rule for infinitesimal deformation

$$\dot{\alpha} = C(\bar{\varepsilon}^p)\, D^p \tag{4.2}$$

where C is the kinematic hardening modulus, a function of the generalized plastic-strain magnitude $\bar{\varepsilon}^p$. The superposed dot denotes the material derivative. For finite deformation analysis the material derivative is not objective and Nagtegaal and de Jong (1982) removed this difficulty by replacing the material derivative by the Jaumann derivative, an approach, which was commonly considered appropriate at the time to generalize such a relation for application in the case of finite deformation. The evolution equation then becomes

$$\overset{\circ}{\alpha} = \dot{\alpha} - W\alpha + \alpha W = C(\bar{\varepsilon}^p)\, D^p \tag{4.3}$$

However, this led to the deduction of spurious oscillations of the shear stress generated in analysing monotonically increasing straining in simple shear.

Lee et al. (1983) explained this anomaly by considering the physical significance of the spin terms in the Jaumann derivative through consideration of (4.3) in the form

$$\dot{\alpha} = C(\bar{\varepsilon}^p)D^p + W\alpha - \alpha W \tag{4.4}$$

The back stress is a type of residual stress embedded in the deforming material and the spin terms in (4.4) express the contribution of the consequent rotation of α to the material derivative with respect to fixed axes. Thus, applied to straining in simple shear, the spin terms in (4.4) imply that the embedded back stress is carried around with angular velocity $W = \dot{\gamma}/2$ where γ is the shear strain. But for constant shearing rate $\dot{\gamma}$, this implies that α rotates with constant angular velocity and yet no lines of material elements ever rotate across the planes on which shearing is taking place and hence it seems plausible that neither can the embedded stress α. Lee et al. (1983) suggested replacing W in (4.4) with W^* a constitutively defined function and suggested a particular case which eliminated the anomalous oscillations but generated a normal stress effect which was too large.

This example was a special case of a theorem given by Fardshisheh and Onat (1974) which limited possible general forms of α to satisfy

$$\dot{\alpha} = h(\alpha, D^p) + W\alpha - \alpha W \tag{4.5}$$

where h is an isotropic tensor function of α and D^p. Since W^*, the deformation induced spin of the residual type back stress α is not equal to W, part of h must comprise a spin term determined by α and D^p. This arises naturally since differently oriented lines of material elements, which carry components of the back stress, rotate with different speeds due to the deformation. Agah Tehrani et al. (1986) show that, based on (4.5), the resulting spin would be a function of $(\alpha D^p - D^p \alpha)$ and invariants of α. When principal directions of strain are fixed in the body, α and D^p are parallel tensors, $(\alpha D^p - D^p \alpha)$ is therefore zero and W^* equals W. Such considerations permit the strain rate generated spin terms in h to be separated from the direct strain-rate contribution by, for example, comparing the results of tension and torsion tests. A general form-invariant tensor function expressing (4.5) was presented in a form appropriate for planning experiments to determine an analytical representation of the function h.

As already noted, Nagtegaal and de Jong (1982) showed that in general the back stress α does not rotate with the spin W. Other spins determined solely by the deformation have been adopted for W^*, for example the spin associated with the polar-decomposition rotation of the deformation. But such laws cannot be satisfactory for general deformation. Lee (1984) considered the example of a situation analogous but simpler than the kinematic hardening anisotropy, a filamentary composite with filaments parallel to the x_1 axis. Consider simple shearing in the x_1 direction given by the velocity $v_1 = \dot{\gamma} x_2$, where $\dot{\gamma}$ is the rate of shear strain. The velocity of the filaments is directed along their lengths so that the anisotropy they cause is not changed by the shearing. However, with the same deformation field but the filaments parallel to the x_2 direction, the filaments and hence the anisotropy characteristics would rotate. Thus both variables α and D^p must be present in the function h of (4.5) which expresses the change in the anisotropy due to deformation.

Determination of the mismatch residual type stress generated by the heterogeneous nature of plastic flow is an extremely difficult task. In this section several examples have demonstrated the need for care in ensuring correct physical modeling and continuum mechanics. For example, the use of the Jaumann derivative in (4.3) ensured objectivity but ignored the physical significance of the spin terms added to the equation. And yet there is an infinite number of objective derivatives available.

5 DISCUSSION

In formulating elastic-plastic theory it is important to adopt continuum mechanics valid for finite deformation. Nonlinear terms can be significant even at small strains because the tangent modulus can be of the same order as the stress. The elastic strains are usually small but it is important to use the "rubber" elasticity law valid for finite deformation in place of Hooke's law because finite rotations associated with the plastic strain prevent objective differentiation of the latter, and this is needed for coupling with the incremental plasticity law. Using continuum mechanics valid for finite deformation generates a deductive structure in contrast to the use of Hooke's law for which a choice of objective derivatives must be made. The theory developed in this way is similar in structure to classical theory but with modified variables and material characteristics.

Strain induced plastic anisotropy can have a major influence on computed stress distributions. There is as yet no generally accepted experimentally validated constitutive relation to model elastic-plastic deformation at finite strain involving strain-induced anisotropy. As pointed out by Hughes (1984) it is extremely important to settle this question since computer programs involving strain induced anisotropy are currently being applied in connection with important technological projects. The formulation discussed in this paper results in a stress rate - total rate of deformation, rate-potential, operator relation appropriate for inclusion in finite-element programs designed for finite deformation applications. It is general enough to accommodate a wide range of evolution equations for the back stress, presenting possible choices which avoid the anomalies previously encountered.

ACKNOWLEDGMENT

This paper was written while the first author was visiting the Ruhr-Universität, Bochum under the sponsorship of a Humboldt-Award from the Alexander von Humboldt-Foundation which is gratefully appreciated. He also wishes to thank his host, Professor Dr.-Ing. Th. Lehmann, for his many kindnesses in making the visit to Bochum enjoyable and fruitful. The paper is based on work developed in the course of research carried out under a grant to RPI from the Division of Mechanical Engineering and Mechanics of the National Science Foundation, for which the authors are grateful.

REFERENCES

Agah-Tehrani, A., Lee, E.H., Mallett, R.L., and Onat, E.T. (1986). The theory of elastic-plastic deformation at finite strain with induced anisotropy modeled as combined isotropic-kinematic hardening. Metal Forming Report RPI-86/1. To appear in J. Mech. Phys. Solids.

Creus, G.J., Groehs, A.G. & Onat, E.T. (1984). Constitutive equations for finite deformations of elastic-plastic solids, Yale Technical Report.

Fardshisheh, F. & Onat, E.T. (1974). Representation of elastoplastic behavior by means of state variables. In Problems of Plasticity, ed. A. Sawczuk, pp. 89-114. Leyden: Noordhoff Int. Publishing.

Hughes, T.J.R. (1984). Statements by Organizing Committee. In Theoretical Foundations for Large-Scale Computations of Nonlinear Material Behavior, ed. S. Nemat-Nasser, pp. 400-401. Dordrecht: Martinus Nijhoff Publishers.

Lee, E.H. (1969). Elastic-plastic deformation at finite strains. J. Appl. Mech., 36, 1-6.

(1981). Some comments on elastic-plastic analysis. Int. J. Solids Structures, 17, 859-872.

(1982). Finite deformation theory with nonlinear kinematics. In Plasticity of Metals at Finite Strain, eds. E.H. Lee and R.L. Mallett, pp. 107-120. Stanford : Stanford University and Rensselaer Polytechnic Inst.

(1984). The structure of elastic-plastic constitutive relations for finite deformation. In Constitutive Equations : Macro and Computational Aspects, ed. K.J. Willam, pp. 103-110, New York : ASME.

Lee, E.H., Mallett, R.L. & Wertheimer, T.B. (1983). Stress analysis for anisotropic hardening in finite-deformation plasticity. J. Appl. Mech., 50, 554-560.

Lubarda, V.A., & Lee, E.H. (1981). A correct definition of elastic and plastic deformation and its computational significance. J. Appl. Mech., 48, 35-40.

Nagtegaal, J.C., & de Jong, J.E. (1982). Some aspects of non-isotropic work-hardening in finite strain plasticity. In Plasticity of Metals at Finite Strain, eds. E.H. Lee & R.L. Mallett, pp. 65-102. Stanford : Stanford University and Rensselaer Polytechnic Inst.

Rivlin, R.S. & Saunders, D.W. (1951). Large elastic deformations of isotropic materials VII. Experiments on the deformation of rubber. Phil. Trans. Roy. Soc., 243, A 865, pp. 251-288.

Woods, L.C. (1973). Beward of axiomatics in applied mathematics. Bull. Inst. Math. and its Applications, 40-44.

SINGULARITIES IN ELLIPTIC NON-SMOOTH PROBLEMS. APPLICATIONS TO ELASTICITY

D. Leguillon
Laboratoire de Mécanique Théorique, Univ. P. & M. Curie
4 place Jussieu, 75252 Paris Cedex 05, France.

E. Sanchez-Palencia
Laboratoire de Mécanique Théorique, Univ. P. & M. Curie
4 place Jussieu, 75252 Paris Cedex 05, France.

1. INTRODUCTION

Most problems in mathematical physics are solved by second order elliptic equations or systems. Accordingly, the corresponding solutions are well defined functions of the class H^1 of Sobolev, i.e., they are square-integrable functions with square-integrable first order derivatives. This is a very poor smoothness, and such functions u may be singular, i.e., **grad** u may tend to infinity at some points.

From a physical point of view, such solutions are meaningless in a neighbourhood of such singularities (for instance, the smallness hypotheses of linearization are not fulfilled). *The presence of singularities shows that new phenomena* (non linearities, qualitative modification of the medium, etc) *may appear*. An example is the *lightning rod* (Fig. 1). The electric potential u in air is a harmonic function exhibiting a singularity at 0 (which is a non-convex point of the boundary) ; **grad** u becomes infinite and this provokes ionization of the air : it becomes conducting and the lightning starts (but this phenomena is not described by harmonic functions !).

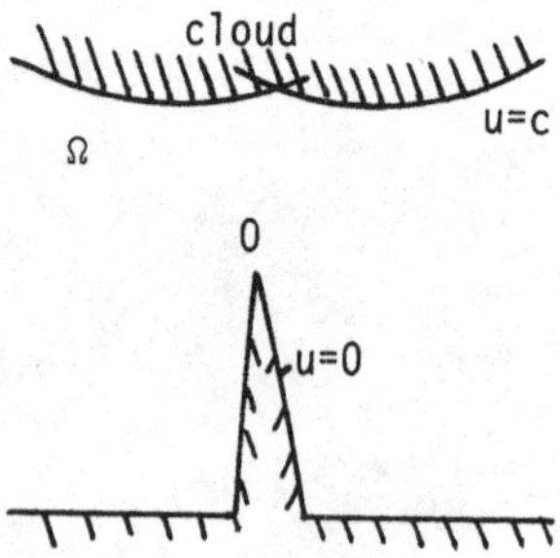

Fig. 1

Another classical example is the *rear edge of an airfoil*. Under standard circumstances, the rear edge introduces a singularity and the velocity (which is the gradient of a harmonic potential) tends to ∞ ; accordingly the pressure $\rightarrow -\infty$ and this produces a circulation of the velocity field. Actually the singularity desappears, but the lift is a consequence of the circulation.

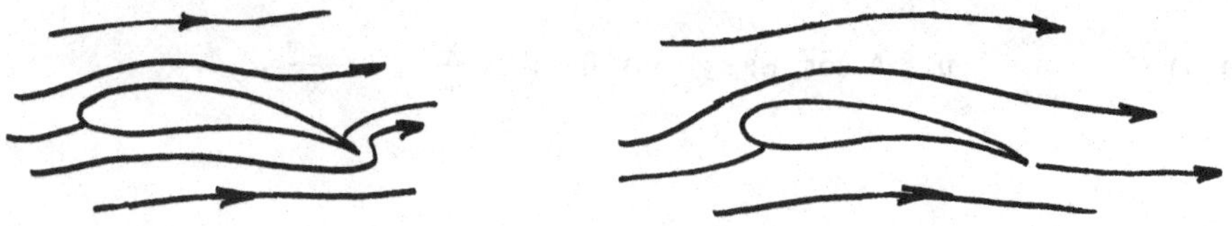

(a) (b)

Fig. 2

In linear *elasticity theory* it is well known that infinite stress usually appears at non- convex corners. According to the nature of the material, it may become plastic, or brittle fracture or other form of damage may appear.

For the mathematical theory of singularities the reader is referred to Grisvard (1985) and Kondratiev (1967). This communication is devoted to two numerical methods for computing singularities (Section 2) and some numerical experiments (Section 3). Only some generalities and results are given here ; a more explicit exposure along with other problems may be found in the booklet by Leguillon and Sanchez (1987). Other numerical results and methods are, for instance in Dempsey and Sinclair (1979, 1981) and Sonoratna and Ting (1986). In fact, the method of these last authors is near our first method, which was announced in Leguillon and Sanchez (1985).

As a first elementary example, let us consider the Laplace equation with Neumann boundary conditions :

(1.1) $$- \Delta U = 0 \quad , \quad \frac{\partial U}{\partial n} = 0$$

in the vicinity of an angular point of the boundary Σ of a domain $\Omega \subset \mathbf{R}^2$. Searching for solutions of the form

(1.2) $$U(x_1,x_2) = r^{\alpha}u(\theta)$$

in polar coordinates, we obtain for α, $u(\theta)$ the eigenvalue problem

(1.3) $-u'' - \alpha^2 u = 0$ for $\theta \in (0,\omega)$, $u' = 0$ for $\theta = 0$, $\theta = \omega$

where ω denotes the opening of the angular point (see Fig. 3). We have

(1.4) $$u = A \cos \alpha\theta \;;\; \alpha = 0 \;;\; \pm\frac{\pi}{\omega} \;,\; \pm\frac{2\pi}{\omega} \;,\; \ldots$$

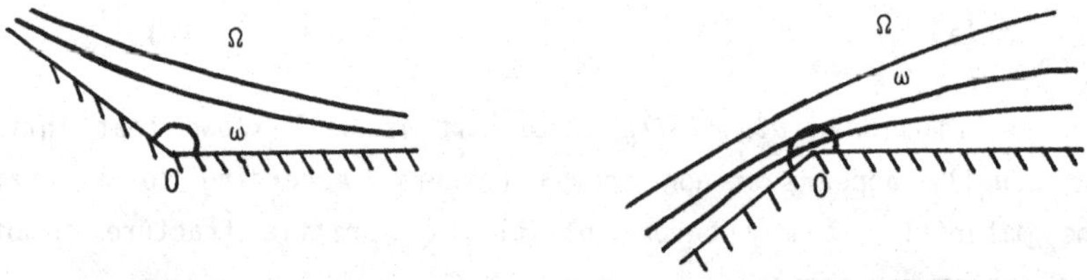

Fig. 3

As the gradient behaves as $r^{\alpha-1}$ the singularity appears for $\mathrm{Re}\,\alpha < 1$. On the other hand, we know in advance that $U \in H^1(\Omega)$ and this implies that grad U is square integrable in the vicinity of $r = 0$, i.e. $\mathrm{Re}\,\alpha > 0$. As a result, the existence of singularities amounts to the existence of solutions of the form (1.2) with

(1.5) $$0 < \mathrm{Re}\,\alpha < 1$$

According to (1.4) this happens for $\omega > \pi$, i.e. for a non-convex point. It is not difficult to obtain a physical interpretation of this result. We consider the flux lines tangent to **grad** u. At the interior of Ω, U is an analytic function and consequently it cannot vanish in a subdomain of Ω. In other words, the flux lines must go all over Ω ; then in the non-convex corner, the boundary "disturbs" the flux lines which push to each other. Moreover

(1.6) $$0 = -\Delta U = -\mathrm{div}\,(\mathbf{grad}\,U)$$

i.e. **grad** U behaves as an incompressible fluid, and |**grad** U| becomes infinite at points where the flux lines touch to each other.

In the case of an elliptic equation for a composite medium (i.e. containing an interface Γ) the criterion for the existence of a singularity becomes the *non-convexity of the boundary* Σ *with respect to the refracted fluxes*. Indeed, let us consider the elliptic equation in divergence form

(1.7) $$-\frac{\partial}{\partial x_j}\left(a_{ij}(x)\,\frac{\partial U}{\partial x_i}\right) = 0$$

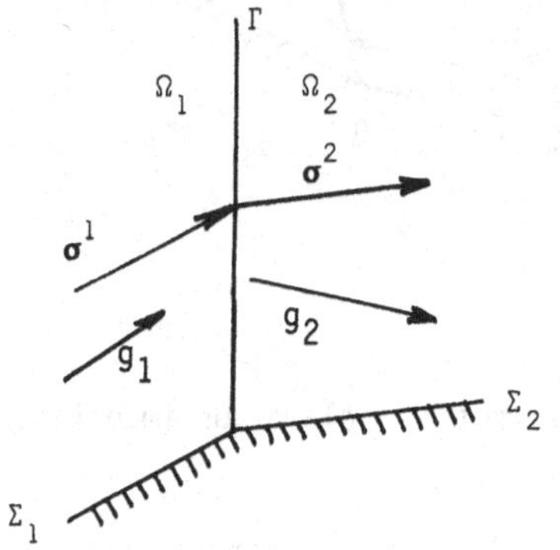

Fig. 4

with piecewise constant coefficients, taking constant values in each of the domains Ω_1, Ω_2 (Fig. 4). We define the gradient **g** and the flux σ by :

$$g_i = \frac{\partial U}{\partial x_i} \quad ; \quad \sigma_i = a_{ij}\,\frac{\partial U}{\partial x_j}$$

then, equation (1.7) in the distribution sense for U belonging locally to H^1 amounts to (1.7) in the classical sense in Ω_1 and Ω_2 and the transmission conditions

(1.8) $$(U) = 0 \iff (g_t) = 0 \quad ; \quad (\sigma_n) = 0$$

where the bracket denote the jump across Γ and the indices t and n denote the tangential and normal components.

If we consider solutions with σ and **g** constant in each region Ω_1, Ω_2, (1.8) implies a refraction law across Γ. Now, we add to (1.7) the Neumann boundary condition

(1.9) $$\sigma_n = 0 \quad \text{on } \partial\Omega$$

and we see that the above mentioned piecewise constant solution is consistent with a boundary with sides Σ_1, Σ_2 parallel to σ^1, σ^2 (Fig. 4). Then, the refracted flux lines are analogous in this problem to the straight lines of the Laplacian. The singularity criterion is the following :

Send a flux σ^1 in Ω_1 parallel to Σ_1 and compute the refracted flux σ^2 across Γ. Then, if Σ_2 is convex (resp. non-convex) with respect to σ^2 we have no singularity (resp. singularity) Fig. 5.

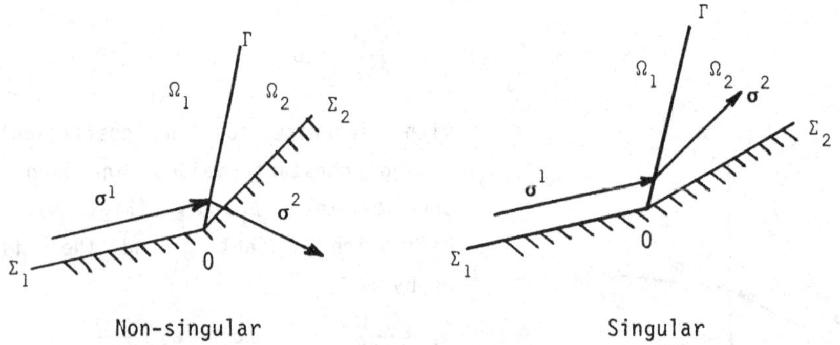

Fig. 5

We see in particular that the anisotropy plays an important role in the existence of singularities.

For Dirichlet (instead of Neumann) boundary conditions the criterion is the same with σ replaced by the normal to **g** (as $u = 0$ at the boundary implies that the gradient is normal to it).

2. NUMERICAL COMPUTATION OF SINGULARITIES

In elasticity problems a simple criterion (analogous to the refracted fluxes one) for the existence of singularities is not known. It is then worthwhile having computational methods to find the characteristic exponents α, and in particular the existence of such α satisfying

(2.1) $$0 < \text{Re}\,\alpha < 1$$

In fact, the existence of solutions of the form

(2.2) $$U(x_1,x_2) = r^\alpha u(\theta)$$

amounts in general to some kind of eigenvalue problem, where α and $u(\theta)$ are the eigenvalue and eigenvector, respectively. Our method starts with some variational formulation of the eigenvalue problem without using the equation and boundary conditions in polar coordinates. The variational formulation for the eigenvalue problem is obtained directly from the variational formulation for the problem in x_1, x_2. Of course, the boundary and transmission conditions are not explicitly handled. In fact, this may be done by two methods : the first one is easier and leads to an implicit eigenvalue problem. The second is more powerful and leads to an explicit eigenvalue problem for a matrix operator.

First method - Let $U(x_1 x_2)$ be a function (scalar, or vector with any number of components) of the two Cartesian variables (x_1, x_2). Let it be a solution of some elliptic problem and let

$$a(U,V) = 0 \quad (\text{or} = \int_\Omega f\ V\ dx) \tag{2.3}$$

be the variational formulation of the elliptic problem in a bounded domain Ω of $\mathbf{R}^2$. The solution is sought in $H^1(\Omega)$ and the test function is any function of the same space. This variational formulation includes the boundary conditions and the transmission conditions at the interfaces.

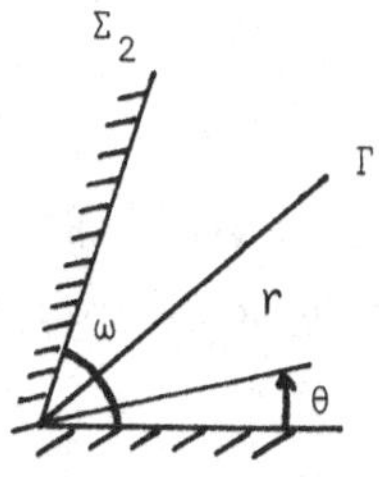

Fig. 6

Now, if Ω is a sector (Fig. 6), the variational formulation does not make sense for solutions of the form (2.2) because $r^\alpha u(\theta)$ does not belong to H^1(sector) : the integral diverges either at 0 or at ∞ according to the values of α. Instead of this, we take in (2.3)

$$(2.4) \quad \begin{cases} U = r^\alpha u(\theta) \quad , \quad u \in H^1(0,\omega) \\ V = \phi(r)\ v(\theta) \quad , \quad \phi \in \mathcal{D}(0,\infty) \quad , \quad v \in H^1(0,\omega) \end{cases}$$

where $\mathcal{D}(0,\infty)$ is the space of the smooth functions defined on R_+ and vanishing for small and large r. The formulation (2.3) makes sense, and contains the boundary conditions and transmission conditions on Σ and Γ ; otherwise the behavior with respect to r is explicitly written in (2.4). Changing to polar coordinates, (2.3) becomes (for an elliptic scalar equation, for instance) :

$$0 = \int_0^\infty r\, dr \int_0^\omega a_{ij}(\theta) \frac{\partial(r^\alpha u)}{\partial x_j} \frac{\partial(\phi v)}{\partial x_i} d\theta$$

or

$$0 = \int_0^\infty \{r^\alpha \phi' (\alpha b_{11}(u,v) + b_{12}(u,v)) + r^{\alpha-1} \phi (\alpha\, b_{21}(u,v) + b_{22}(u,v))\} dr \tag{2.5}$$

where b_{ij} are 4 sesquilinear forms for u,v in $H^1(0,\omega)$ which have nothing to do with r (we performed some kind of "separation of variables"). Integrating by parts (2.5) with respect to r we have :

$$0 = \int_0^\infty r^{\alpha-1} \phi(r)\, a(\alpha;u,v) dr \qquad \forall\, \phi \in \mathcal{D}(0,\infty) \tag{2.6}$$

where

$$a(\alpha;u,v) = -\alpha^2 b_{11} + \alpha(b_{21} - b_{12}) + b_{22} \tag{2.7}$$

and (2.6) is equivalent to

$$a(\alpha;u,v) = 0 \qquad \forall\, v \in H^1(0,\omega) \tag{2.8}$$

Defining the operator $A(\alpha)$ associated with the form a (which is an operator with compact resolvent depending holomorphically on the parameter α), we have

$$A(\alpha)u = 0 \tag{2.9}$$

and the existence of solutions of the form (2.2) amounts to saying that 0 is an eigenvalue of $A(\alpha)$. Finally, our problem amounts to find the values of α such that 0 is an eigenvalue of $A(\alpha)$; $u(\theta)$ is the corresponding eigenvector. Performing a finite-element discretization of (2.9) (which only contains the variable θ), it becomes a matrix problem, and we search for the values of α such that its determinant vanishes.

Second method - Coming back to Fig. 6, we now think about $U(x_1,x_2)$ as a function of the only variable r with values in $H^1(0,\omega)$. Performing the same transformations as in (2.5), we arrive at an expression analogous to (2.9) but with U, $\partial U/\partial r$ instead of $r^\alpha u$, $\alpha r^{\alpha-1}u$, and defining the operators B_{ij} associated with the forms b_{ij} we have the second order differential equation with operator coefficients

$$(2.10) \qquad -\left(r\frac{\partial}{\partial r}\right)^{(2)} B_{11}\,U + \left(r\frac{\partial}{\partial r}\right)(B_{21}-B_{12})U + B_{22}\,U = 0$$

We may transform this in a first order differential equation for a matrix operator (the formal operations are analogous to that for hyperbolic second order equations in t in order to apply semi group theory)

$$\mathcal{U} = \begin{pmatrix} U \\ r\dfrac{\partial U}{\partial r} \end{pmatrix} = \begin{pmatrix} U_1 \\ U_2 \end{pmatrix} \quad ; \quad A \equiv \begin{pmatrix} 0 & I \\ B_{11}^{-1}B_{22} & B_{11}^{-1}(B_{21}-B_{12}) \end{pmatrix}$$

$$(2.11) \qquad r\frac{\partial \vec{U}}{\partial r} = A\vec{U}$$

where A is an unbounded operator in $H^1(0,\omega) \times L^2(0,\omega)$ with compact resolvent. We see that solutions of the form (2.2) are associated with the eigenvalues :

$$(2.12) \qquad (A-\alpha)\vec{U} = 0$$

It should be noticed that A is not selfadjoint, and consequently Jordan blocks may appear. In this case, we have also logarithmic solutions. For instance, for a block of the form (2.13) we have the solution (2.14)

$$(2.13) \qquad \begin{pmatrix} \alpha & 1 \\ 0 & \alpha \end{pmatrix} \quad ; \quad (A-\alpha)\vec{u} = 0 \quad , \quad (A-\alpha)\vec{v} = \vec{u}$$

$$(2.14) \qquad \vec{U} = r^\alpha\,(\log r\,\vec{u}(\theta) + \vec{v}(\theta))\ .$$

Stress concentration factor - In fact, when the solution of a boundary value problem is singular, it appears under the form :

(2.15) $$u = c\, r^{\alpha} u(\theta) + u^{reg}$$

where u^{reg} is a regular function at $r = 0$. The coefficient c is called the "stress concentration factor". The eigenvalue problem (2.12) enjoys the property that if α is an eigenvalue, $-\alpha$ is too. Let u and v be the corresponding eigenvectors. $r^{-\alpha} v$ is a solution which does not satisfy (2.1) because it is much too singular. But we may use it in order to compute the stress concentration factor. It suffices (see the details in Leguillon and Sanchez (1987)) to write the Green formula for the two-dimensional problem in Ω_ρ (Fig. 7) and to take the limit as $\rho \to 0$. We then obtain a formula giving c as a function of u for $r = R$.

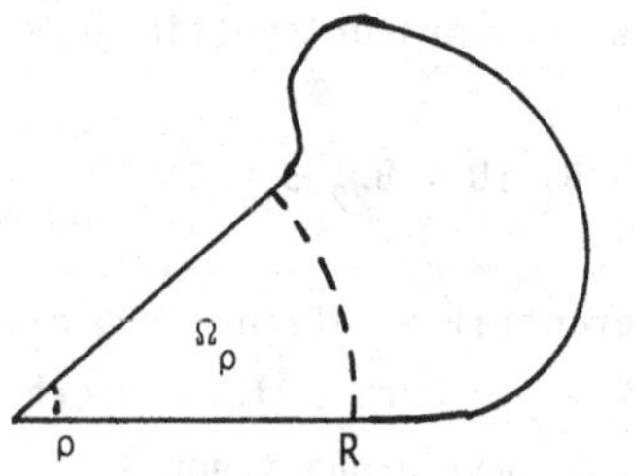

Fig. 7

Three-dimensional edge Let us now consider edge singularities for functions of (x_1,x_2,x_3). Under the hypothesis that the geometric and material properties vary slowly with respect to the tangential variable z (Fig. 8), we search for solutions of the form :

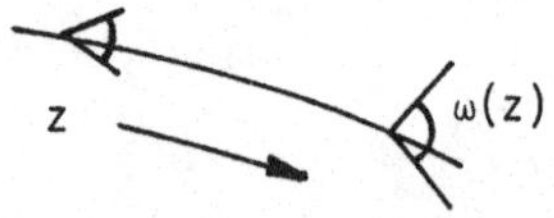

Fig. 8

(2.16) $$U(x_1,x_2,x_3) = r^{\alpha(z)}\, u(\theta,z) + U^{reg}$$

where r,θ are local polar coordinates in the normal plane to the edge. An asymptotic reasoning shows that in this case the tangential derivative $\partial/\partial z$ is negligibly small with respect to the normal derivatives. The problem of finding $\alpha(z)$ and $u(\theta,z)$ becomes the problem of the preceding sections, with parameter z and genuine variables x_1,x_2. This is the case of layered plates, for instance.

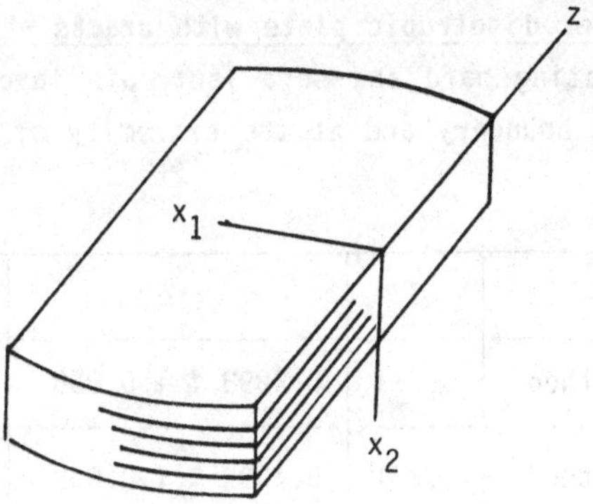

Fig. 9

3. NUMERICAL EXPERIMENTS

Layered anisotropic plate with a hole - We consider a plate made of orthotropic layers of a carbon-epoxy mixture, with elastic coefficients (fibers in direction x_1)

$$a_{1111} = 14.5 \quad , \quad a_{2222} = a_{3333} = 1.27$$

$$a_{2233} = 0.622 \quad , \quad a_{1122} = a_{1133} = 0.672$$

$$a_{2323} = 0.324 \quad , \quad a_{1212} = a_{1313} = 0.485 \quad (\text{X } 10^4 \text{ M pa})$$

The layers are disposed with orthogonal fibers (Fig. 10). We consider a hole in such a plate. To study the singularities, according to the preceding considerations, we must consider tangent and normal coordinates (and then the elastic coefficients in these coordinates). The exponent α depends on the curvilinear abscissa :

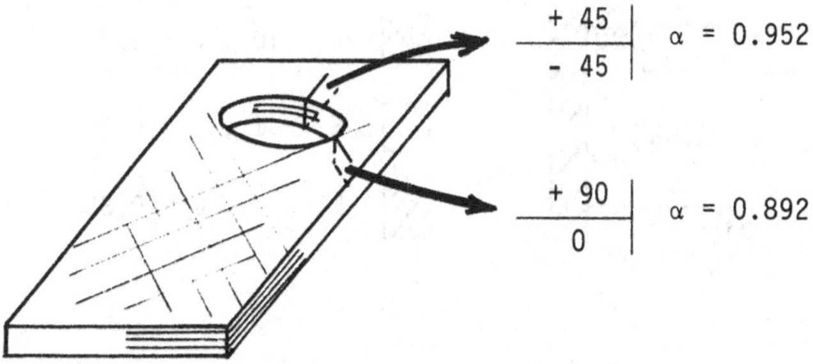

Fig. 10

Layeredisotropic plate with cracks - Now we consider a plate made of alternating hard and soft isotropic layers (Fig. 11). The values of α for the boundary and at the extremity of several types of cracks are :

Case	1	2	3	4	5
Compt. 1st method	0.771	0.4998 ± i 0.058	0.647	0.243	0.573
Compt. 2nd method	0.771	0.4999 ± i 0.059	0.647	0.245	0.573

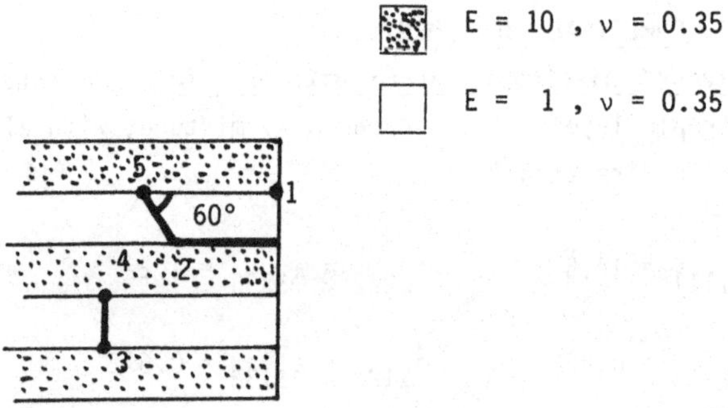

Fig. 11

Layered isotropic plate with covering - In the situation of the point 1 of Fig. 11, we put a covering of another material (Fig. 12). It is then seen that the exponent of the singularity changes, but it is contained between 0 and 1 (i.e. the singularity does not desappear) even for a very hard covering :

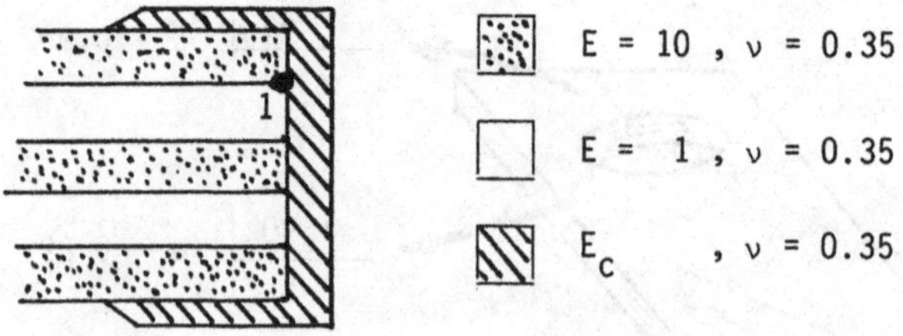

Fig. 12

E_c	-	0.5	5	20	50	100	1000	∞
α	0.771	0.755	0.678	0.671	0.695	0.711	0.731	0.734

In the present case, the singularity is at an interior point, and the method is slightly modified : the boundary conditions become 2π-periodicity conditions.

REFERENCES

Dempsey, J.P. and Sinclair, G.B. (1979). On the stress singularities in the plane elasticity of the composite wedge. Jour. of Elasticity, 9, 373-391.

Dempsey, J.P. and Sinclair, G.B. (1981). On the singular behavior at the vertex of a bi-material wedge. Jour. of Elasticity, 11, 317-327.

Grisvard, P. (1985). Elliptic equations in non smooth domains, London : Pitman.

Kondratiev, V.A. (1967). Boundary value problems for elliptic equations in domains with conical or angular points. Trudy Moskovs. Mat. Obs. 16, 209-292.

Leguillon, D. et Sanchez-Palencia, E. (1985). Une méthode numérique pour l'étude des singularités de bord dans les composites. Comptes Rendus Acad. Sci. Paris, 301, sér. II, 1277-1280.

Leguillon, D. and Sanchez-Palencia, E. (1987). Computation of singular solutions in elliptic problems. Application to elasticity. Paris : Masson (To be published).

Somaratna, N. and Ting, T.C.T. (1986). Three-dimensional stress singularities in anisotropic materials and composites. Intern. Jour. Engng. Sci., 24, 1115-1134.

SOLITONS IN ELASTIC SOLIDS EXHIBITING PHASE TRANSITIONS

G.A.Maugin
Université Pierre-et-Marie Curie, Laboratoire de Mécanique Théorique, Tour 66, 4 Place Jussieu, 75230 Paris Cédex 05, France

Abstract. Among elastic crystals which are subjected to phase transitions are those which represent so-called ferroic states. These are elastic ferroelectrics, elastic ferromagnets and ferroelastic crystals such as twinned shape-memory materials. In these three wide classes studied by the author and co-workers, the solitary waves which can be shown to exist and represent either moving domain walls or nuclei of transformations, are not true solitons since the interaction of two such waves always is accompanied by some linear radiation. This is a consequence of the very form of the governing systems of equations which may be of different types (e.g., sine-Gordon equation coupled to wave equations, modified Boussinesq equation) and are usually obtained either from a discrete lattice model or a rotationally invariant continuum model. For the sake of illustration the case of shape-memory materials is presented in greater detail through the first approach.

INTRODUCTION

Nearby, but below, the order-disorder phase-transition temperature, the dynamical equations that govern elastic crystals with a microstructure exhibit all the necessary, if not sufficient, properties (essentially , nonlinearity and dispersion with a possible compensation between the two effects) to allow for the propagation of so-called solitary waves. Whether these are true solitons or not is a question that can be answered only in each case through analysis and/or numerical simulations. These waves, however, are supposed to represent domain walls in motion . The latter are layers of relatively small thickness which carry a strong nonuniformity in a relevant parameter between two adjacent phases or degenerate ground states (see,e.g.,Maugin and Pouget ,1986). Simple models derived from lattice dynamics (e.g., Pouget et al,1986) or continuum models derived from rotationally-invariant continuum physics (e.g., Maugin and Pouget,1980) allow one to deduce such sets of governing dynamical equations for elastic crystals that exhibit a phase transition. As a first (phenomenological) approach sufficient for our purpose , the latter can be envisaged within the Ginzburg-Landau framework (Boccara,1976; Tolédano and Tolédano,1986) . Essentially two large classes have been considered in the case of solid elastic crystals , although some results also apply to liquid crystals of the nematic type (Magyari,1984) , to macromolecules exhibiting a twist (Yomosa, 1985) and, to some extent, to Korteweg fluids. With electric and magnetic properties of the ferroelectric and ferromagnetic type taken into account , we find that the primary order parameter may be either a nonmechanical entity (such as the electric polarization emanating from electric dipoles or the magnetization density resulting from magnetic spins) while elasticity remains of the classical linear type, or a strain component as is the case in ferroelasticity in general (Pouget J.P.,1981; Wadhawan ,1982) and martensitic phase transitions for crystals exhibiting twinning and shape-memory in particular (Klassen-Neklyudova, 1964) . The first situation has been studied in great detail elsewhere (Pouget and Maugin,1984,1985a,b;Maugin and Miled,1986a) and a general framework emerges for the description of the propagation of domain walls in ferroelectric and ferromagnetic elastic crystals as also in elastic crystals with a molecular group represented by oriented materials (Maugin and Pouget,1986 , Maugin and Miled, 1986b, Pouget,1986). This consists in a nonlinear hyperbolic system of equations which is shown to exhibit , both analytically and numerically , solitary-wave solutions with an almost soliton-like behaviour (i.e., solitary waves interact almost elastically for small amplitudes , but not exactly, being accompanied by linear radiation) . This is briefly recalled here by giving the general framework , but the emphasis is placed on the second method that corresponds to ferroelasticity where a similar phenomenon is exhibited, being described by a nearly integrable equation.

ELEMENTS OF PHASE-TRANSITION THEORY

In a phase transition the primary order parameter is that parameter which best describes the reduction in symmetry from the disordered high-temperature phase (with symmetry group G_o) to the ordered low-temperature phase (of symmetry group G) . Other parameters which can be deduced from the former through a coupling are called secondary order parameters. In Landau's classification the "order"of a transition refers to the order of the derivative of the free energy which presents a discontinuity at the transition. Therefore, a first-order phase transition is accompanied by a jump in the primary order parameter while the latter is continuous , but not smooth, for a second-order transition . If Γ is the symmetry group of the primary order parameter then $G = G_o \cap \Gamma$ is the maximal subgroup of G_o which leaves this order parameter invariant. Phase transitions in ferroelectric and ferromagnetic elastic crystals of many types are described phenomenologically by the first two columns in Table 1 where Ψ denotes the free energy. Ferroelastic bodies (which, by definition, exhibit a change of crystal system) may present coupled electric and magnetic properties and are described by the third column . Martensitic phase transitions are first-order phase transitions with constraint of an "invariant plane strain" (in the so-called"habit plane" , e.g., (110) plane in austenite-martensite alloys with a BCC structure). They do not necessarily correspond to a change of crystal system and, therefore, all martensitic phase transitions are not ferroelastic ones, but many are. The search of minimizers for the free energy must yield several distinct solutions which are more or less stable, and this requires nonconvex energy functions. In addition, the order that prevails at low temperature indicates a strong spatial correlation. Phenomenologically, this requires accounting for a length scale related to this correlation and introducing gradients of the primary order parameter in agreement with Ginzburg . With such gradient terms jumps become energetically unprofitable and there may exist spatial regions of rapidly varying order parameter . This yields domain walls and boundary layers. The nonlinearity accompanying the nonconvexity of energy functions and the dispersion due to the introduction of gradient terms, through possible compensation, may contribute to the existence of solitary waves in the dynamical case. Localized configurations , static or dynamic,thus arise fairly naturally and in fact abound in physics (e.g., dislocations, necking, dimples on shells , vapor bubbles in a liquid,etc; see Berdichevskii and Truskinovskii,1985) . Aizu (1970) has given a classification of all "ferroic" states described in Table 1.

VARIOUS WAYS TO CONSTRUCT MODELS

Several methods can be used to arrive at the local dynamical partial differential equations that can describe solitary waves in various elastic crystals . Three of these are schematized on the flow chart of Figure 1. They must all rely on some microscopic view or some of the

Table 1 - Ferroic states

	ferroelectrics	ferromagnets	ferroelastics
primary order parameter	polarization P -displacive -molecular	magnetization M	strain e
conjugate force	E	H	σ
work	- P E	- M H	- σ e
secondary order parameter	e	e	P (M)
coupling	-piezoelectricity -electrostriction	-piezomagnetism (rare) -magnetostriction	-piezoelectricity -electrostriction -(magnetostriction)
potential energy ϕ =	$\Psi(P,T;e) - PE$	$\Psi(M,T;e) - MH$	$\Psi'(e,T;P) - \sigma e$
localization (correlation length)	$\tilde{\phi} = \phi + \frac{1}{2}\delta^2(\nabla P)^2$ Ginzburg	$\tilde{\phi} = \phi + \frac{1}{2}\alpha(\nabla M)^2$ Heisenberg	$\tilde{\phi} = \phi + \frac{1}{2}\delta^2(\nabla e)^2$ Mindlin
effective field	$E^{eff} = -\delta\tilde{\phi}/\delta P$	$H^{eff} = -\delta\tilde{\phi}/\delta M$	$\sigma^{eff} = -\delta\tilde{\phi}/\delta e$

elements of Table 1 . The most "physical" method consists in constructing a simple lattice-dynamics model , obtaining a set of discrete equations and taking in some sense their continuum limit . The second method corresponds to by-passing the discrete equations of the first one by taking a continuum limit of the Lagrangian density of lattice-dynamics and then writing down the Euler-Lagrange equations of the continuum Lagrangian density. Both of these methods do not involve the second principle of thermodynamics which in any case should not intervene since Landau's phase-transition theory does not involve dissipative phenomena. The nonlinearity will come from the type of interactions assumed between neighbours while gradient terms will emerge from second-neighbour interactions . The third method, more in the tradition of rational mechanics is fully continuous and uses the principle of virtual power for finite fields. The notion of gradient will come into the picture through the notion of gradient theories . This is developed at length for all theories considered in the paper in Maugin (1980). Here the establishment of constitutive equations necessary to close the differential system of field equations requires a separate thermodynamical study involving the basic principles of thermodynamics. This method can accommodate dissipative processes in addition to the thermodynamically recoverable ones included in the other two methods. Method 1 is illustrated by Pouget et al (1986) and Pouget and Maugin (1984,1985a,b) for elastic

Figure 1. Construction of models of elastic crystals with phase transition

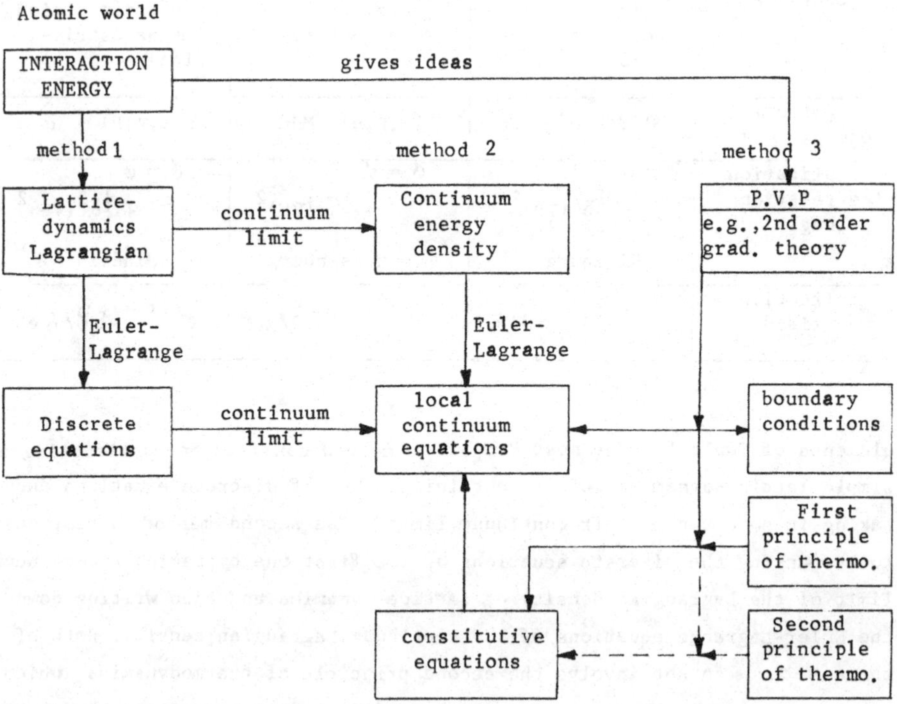

ferroelectrics of the molecular-group type such as KNO_3 and $NaNO_2$. Avenue 2 is illustrated by Maugin and Pouget (1980) for elastic ferroelectrics of the displacive type such as $BaTiO_3$, by Maugin and Miled (1986a) for elastic ferromagnets and by Maugin and Miled (1986b) and Pouget (1986) for elastic crystals with a molecular group. Remarkably enough, whenever the primary order parameter (e.g., electric polarization, magnetization, etc.) gives rise to a new rotational degree of freedom (this is not the case for ferroelectrics of the displacive type) while the elastic behaviour remains of a totally classical type (linear, but anisotropic and coupled to the rotational degree of freedom) one obtains governing dynamical continuum equations which form a nonlinear hyperbolic dispersive system given typically by (nondimensional units; we have discarded the coupling with the longitudinal elastic displacement without loss of generality)

$$\frac{\partial^2 v}{\partial t^2} - V_T^2 \frac{\partial^2 v}{\partial X^2} = - a \frac{\partial}{\partial X}(\sin \phi) , \tag{1}$$

$$\frac{\partial^2 \phi}{\partial t^2} - \frac{\partial^2 \phi}{\partial X^2} - \sin \phi = a \frac{\partial v}{\partial X} \cos \phi , \tag{2}$$

with Lagrangian ($\theta = \phi/2$)

$$L = \int \left\{ \frac{1}{2}[(\frac{\partial v}{\partial t})^2 + (\frac{\partial \theta}{\partial t})^2] - \frac{1}{2}[V_T^2(\frac{\partial v}{\partial X})^2 + (\frac{\partial \theta}{\partial X})^2] - (1+ \cos \phi) + a(\frac{\partial v}{\partial X})\sin \phi \right\} dX , \tag{3}$$

where ϕ is an angle variable, v is a transverse elastic displacement , $V_T \ll 1$ is the velocity of transverse (linear) elastic waves and a is the coupling parameter. Clearly, elasticity plays a secondary role in the phase-transition phenomena(temperature dependence could appear in the form of a temperature varying coefficient of the term (1+ cos ϕ)) since (1) simply is a forced linear elastic-wave equation. The system (1)-(2) admits exact, solitary-wave solutions that represent , for instance,moving electroacoustic domain walls in a ferroelectric crystal such as $NaNO_2$ or a moving magnetoelastic Néel domain wall in a thin film of ferromagnetic elastic yttrium-iron garnet . Equation (2), in the absence of coupling (a=0), is a sine-Gordon equation whose solitary-wave solutions are known to be true solitons. However, as clearly proved through analysis (inverse scattering method) and numerical simulation (both in Pouget and Maugin, 1985b) the solitary waves, solutions of (1)-(2) , are not exactly true solitons in that the nonlinear coupling of (2) with (1) - a linear wave equation - implies non-negligible linear radiations during the interaction of such waves of relatively important amplitude . This follows from the fact that the system (1)-(2) is only nearly integrable . It is of interest to see if such a property also holds true in the case of ferroelastic crystals . We give this case in greater detail by following method 1 in Figure 1 . The model was developed by D.Lebriez and J.Pouget.

THE CASE OF FERROELASTIC AND SHAPE-MEMORY ALLOYS

For the sake of definiteness we consider a one-dimensional lattice model with axis of the chain in the stacking direction of sheared layers - which are parallel to the (110) habit plane in a twinning material- in agreement with Figure 2. The change in longitudinal position of the atom sites is of higher order and, therefore, discarded. Only interactions of a nonlinear type between nearest neighbours are accounted for and we assume small relative transverse displacements. With rotational invariance of the potential energy and pair interactions we have

$$\phi = \sum_n \varphi(\vec{r}_n, \vec{r}_{n-1}) = \sum_n \varphi(||\vec{r}_n - \vec{r}_{n-1}||) \quad , \vec{r}_n = x_n \vec{e}_x + y_n \vec{e}_y \, , \tag{4}$$

with (a is the lattice spacing)

$$||\vec{r}_n - \vec{r}_{n-1}|| = a + \varepsilon_n \quad , x_n \simeq X_n = n\,a \, , \quad y_n = v_n \, ,$$

$$\varepsilon_n = a\left(\frac{1}{2} e_n^2 - \frac{1}{8} e_n^4 + \frac{1}{16} e_n^6 + \dots \right) , \tag{5}$$

$$e_n = (v_n - v_{n-1})/\,a \quad , \quad |e_n| \ll 1 \; , \; \forall n=1,2,\dots,N \, .$$

Figure 2. Lattice model for shape-memory alloy

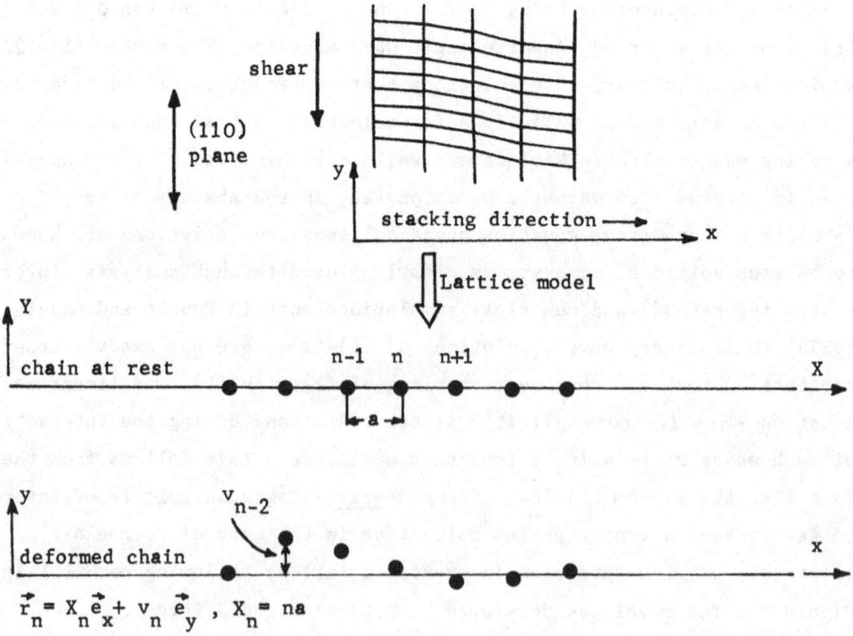

Expanding ϕ one gets an expansion in even powers as

$$\phi = \phi_o + \frac{aA}{2} \sum_n e_n^2 - \frac{aB}{4} \sum_n e_n^4 + \frac{aC}{6} \sum_n e_n^6 + \dots \tag{6}$$

with

$$\phi_o = \sum_n \varphi(a) = N\varphi(a),$$

$$A = \varphi'(a) > 0 \quad , \quad B = \frac{1}{2}[\varphi'(a) - a\varphi''(a)] \; , \tag{7}$$

$$C = \frac{3}{8}[\varphi'(a) - a\varphi''(a) + \frac{a^2}{3}\varphi'''(a)].$$

The discrete equations of motion are the Euler-Lagrange equations associated with the Lagrangian density $L = \sum_n \frac{1}{2} m(\partial v_n/\partial t)^2 - \phi$. We obtain thus

$$m \frac{d^2 v_n}{d t^2} = -[A(e_n - e_{n+1}) - B(e_n^3 - e_{n+1}^3) + C(e_n^5 - e_{n+1}^5)]. \tag{8}$$

On taking the continuum limit such that $v_{n+1} = v((n+1)a) = v(na) + a(\partial v/\partial x) + ..$, we evaluate $e_n = (\partial v/\partial x) - ..$, $e_n - e_{n+1}$, $e_n^3 - e_{n+1}^3$,etc . Being careful to retain all relevant terms, from eqns.(8) we obtain the following partial differential equation

$$\frac{\partial^2 v}{\partial t^2} - c_T^2[1 + \epsilon g(v_x) + \epsilon\delta^2 \frac{\partial^2}{\partial x^2}] \frac{\partial^2 v}{\partial x^2} = 0 \tag{9}$$

for the continuous transverse displacement v , with

$$c_T^2 = A/\rho a^2 > 0 \, , \qquad \epsilon \sim -3B/A \text{ or } 5C/A \; , \quad \delta^2 = a^2 A/\,36\,B,$$
$$g(v_x) = (\partial v/\partial x)^2 + \gamma(\partial v/\partial x)^4 \quad , \gamma = 5C/\,3B \quad , \rho = m/a^3. \tag{10}$$

Equation (9) is derivable from the following total Lagrangian

$$L = \int (\frac{1}{2}\rho\, a^2 (\frac{\partial v}{\partial t})^2 - [\frac{A}{2}\left\{ (\frac{\partial v}{\partial x})^2 + \frac{a^2}{4}(\frac{\partial^2 v}{\partial x^2})^2 + \frac{a^3}{3}\frac{\partial v}{\partial x}\frac{\partial^3 v}{\partial x^3}\right\}$$
$$- \frac{B}{4}(\frac{\partial v}{\partial x})^4 + \frac{C}{6}(\frac{\partial v}{\partial x})^6])dx \, , \tag{11}$$

which, following avenue 2 , can be deduced directly from the continuum limit of the discrete Lagrangian . Equations (9)-(10) deserve the following comments . First, eqn.(9) is a nonlinear hyperbolic dispersive equation that is very close to the Boussinesq equation for which $g(v_x) = \partial v/\partial x$; see Maugin,1985,p.91. Thus we call it a modified Boussinesq equation (MBE) . Second , taking the space derivative of eqn.(9), introducing the appropriate nondimensionalization and setting $e = v_x$, we can write the result as

$$\frac{\partial^2 e}{\partial t^2} = \frac{\partial^2}{\partial x^2}\left(\sigma - \frac{\partial \mu}{\partial x}\right) = \frac{\partial^2}{\partial x^2}\left(\frac{\delta \Psi}{\delta e}\right) \quad , \frac{\delta}{\delta e} = \frac{\partial}{\partial e} - \frac{\partial}{\partial x}\left(\frac{\partial}{\partial e_x}\right), \tag{12}$$

where

$$\sigma = \frac{\partial \Psi}{\partial e} = 2\,K\,e - 4\,e^3 + 6\,e^5 \;, \mu = -\,2\,\frac{\partial e}{\partial x} \quad , \Psi = \tilde{\Psi}(e, e_x, T). \tag{13}$$

Equation (12) is a one-dimensional version for pure shear motions of the second-gradient theory of elastic solids of Mindlin and Tiersten (1962), but with an energy density of the sixth order in e and quadratic in the strain gradient. Being sixth order, this may admit minimizers such as twins with two oppositely sheared configurations (martensites) and a zero-shear austenite phase of an obviously differing energy level and stability behaviour. The quantities σ and μ are intrinsic stresses of the first and second order as introduced in the full continuum approach using method 3 in which the density of virtual power of internal forces would a priori be written as a linear continuous form on the velocities of successive strain gradients [e.g., $p_{(i)} = -(\sigma\dot{e} + \mu\nabla\dot{e} + ..)$; cf .Germain, (1973)] . The <u>first-order</u> phase-transition picture of our elastic material is complete with A or K positive and other signs specified by (13) and K written as $K = K^{+}(T-T_c)$, $K^{+} > 0$, where T_c is the transition temperature . When this temperature dependence is discarded, then eqns. (12)-(13) are exactly those obtained by Falk(1982,1983a,b) in statics or in dynamics (Falk,1984,1986). However, we have here a means of evaluating all phenomenological coefficients from microscopic parameters through eqns. (7). Accompanying boundary conditions (here limit conditions at the ends of a finite x-interval) follow from method 3 and are consistent with (12).Falk (1984) has studied the static solutions of (12) on a finite interval as well as the solitary-wave solutions in the dynamical case on the complete real line.Non-propagating "solitary-wave"solutions of the front type may be interpreted as nuclei of one phase, one martensite (or austenite), in a matrix of the other martensite (or one martensite), while propagating kink (hump) solutions may be interpreted as domain walls. The kinks occur only at rest in the absence of external force . When driven by an external force, the domain-wall motion obeys a Rankine-Hugoniot equation (in close analogy with shock waves) while a generalized Maxwell area construction holds in the stress(σ)-strain(e) diagram (Falk,1986). Finally, the type of eqn. (9) allows one to answer the question whether the obtained solitary waves are true solitons or not . Indeed, depending on the expression of $g(v_x)$ and the type of fourth-order derivative, equations of the type of (9) are known in the literature as "improved" and "modified improved" Boussinesq equations (for short, IBE and MIBE). Such equations occur, for instance, in nonlinear electric transmission lines and in the study of solitary waves propagating along a nonlinear elastic rod of circular cross section (Soerensen et al, 1984). They are only nearly integrable in the sense that: (i) they exhibit solitary-wave solutions, (ii) they possess a finite number of conservation laws, (iii) they reduce to exactly integrable equations in the small-amplitude limit, and (iv) they numerically exhibit an almost soliton-like behaviour, only little linear radiation being created during the interaction of their solitary-wave solutions. This was shown numerically by Bogolubsky (1977) and Iskandar and Jain (1980). Equation (9) is of the same type and, therefore, its solitary-wave solutions are not true solitons, but they are not so far from such a behaviour. Obviously, any further coupling with another parameter such as an electric or a magnetic one, as indicated in the third column of Table 1, will certainly not improve the situation, but rather aggravate it. Nonetheless, it remains to perform the complete analytical and numerical study.

CONCLUSION

The qualitative result enunciated in the foregoing paragraph and that mentioned before are tantamount to saying that the solitary waves that can be shown to exist in the three ferroic states of Table 1 for deformable crystals and that are supposed to represent domain walls in motion, are never true solitons although the various cases are described by differing systems of equations. But these systems have the common feature of only nearly integrable. Consequently, the interaction (e.g., head-on collision, annihilation and reappearance, with a change of phase) of the domain-wall solutions is always accompanied by linear radiations and this results from the very form of the relevant equations and the inherent physics and is not an artifact of computations.

Acknowledgments. This work was supported by Theme "Nonlinear Waves in Electromagnetic Elastic Materials of the A.T.P. "Mathématiques Appliquées et Méthodes Numériques Performantes" (M.P.B., C.N.R.S., Paris, France).

References

Aizu K.(1970).Possible species of ferromagnetic, ferroelectric and ferroelastic crystals. Phys.Rev., B2 , 754-772.

Berdichevskii V. and Truskinovskii L.(1985). Energy structure of localization. In Local effects in the analysis of structures, ed. P. Ladevèze ,pp.127-158, Elsevier:Amsterdam.

Boccara N.(1976). Symétries brisées. Paris :Hermann.

Bogolubsky I.L.(1977). Some examples of inelastic soliton interaction. Comp.Phys.Comm.,13, 149-155.

Falk F.(1982).Landau theory and martensitic phase transitions. J.Phys.Coll.C4., 43 ,C4-3 - C4.15.

Falk F.(1983a).One-dimensional model of shape-memory alloys. Arch.Mech., 35 , 63-84.

Falk F.(1983b).Ginzburg-Landau theory and static domain walls in shape-memory alloys. Zeit.Phys.B.Condensed matter,51,177-185.

Falk F.(1984). Ginzburg-Landau theory and solitary waves in shape-memory alloys. Zeit.Phys.B.Condensed matter, 54, 159-167.

Falk F.(1986).Driven kinks in shape-memory alloys. In Trends in applications of Pure mathematics to mechanics, ed.E.Kröner and K. Kirchgassner,pp.164-167, Berlin:Springer-Verlag.

Germain P.(1973).La méthode des puissances virtuelles en mécanique des milieux continus.I.Théorie du second gradient. J.Mécanique, 12, 235-274.

Iskandar I. and Jain P.C.(1980).Numerical solutions of the improved Boussinesq equation. Proc.Indian Acad.Sci.(Math.Sci.), 89, 171-181.

Klassen-Naklyudova M.V.(1964).Mechanical twinning of crystals. New York: Consultants Bureau.

Magyari E.(1984).Mechanically generated solitons in nematic liquid crystals. Zeit.Phys.B.Condensed matter, 56, 1-3.

Maugin G.A.(1980).The method of virtual power in continuum mechanics:Application to coupled fields. Acta Mechanica, 35, 1-70.

Maugin G.A.(1985).Nonlinear electromechanical effects and applications. Singapore: World Scientific Publ..

Maugin G.A.(1986).Solitons and domain structure in elastic crystals with a microstructure:mathematical aspects. In Trends in applications of pure mathematics to mechanics, ed.E.Kröner and K.Kirchgassner , pp.195-211, Berlin:Springer-Verlag.

Maugin G.A.and Miled A.(1986a). Solitary waves in elastic ferromagnets. Phys.Rev., B33, 4830-4842.

Maugin G.A. and Miled A.(1986b).Solitary waves in micropolar elastic crystals . Int.J.Engng.Sci., 24 (in the press).

Maugin G.A. and Pouget J.(1986).Solitons in microstructured elastic materials-Physical and mechanical aspects. In Continuum models of discrete systems (6),ed.A.J.M.Spencer, Rotterdam:A.A.Balkema.

Maugin G.A.and Pouget J.(1980).Electroacoustic equations for one-domain ferroelectric bodies. J.Acoust.Soc.Amer., 68, 575-587.

Mindlin R.D. and Tiersten H.F.(1962). Effects of couple stresses in linear elasticity. Arch.Rat.Mech.Anal., 11, 415-448.

Pouget J.(1986).Ondes solitaires dans les milieux élastiques orientés. In Proc.Interdisciplinary workshop on Nonlinear excitations, ed. M.Remoissenet, pp.360-375, Dijon,France:Publ.L.O.R.C.,University of Dijon.

Pouget J., Askar A. and Maugin G.A.(1986).Lattice model for elastic ferroelectric crystals:Microscopic approach. Phys.Rev., B33, 6304-6319.

Pouget J. and Maugin G.A.(1984).Solitons and electroacoustic interactions in ferroelectric crystals.I.Single solitons and domain walls. Phys.Rev., B30, 5306-5325.

Pouget J. and Maugin G.A.(1985a).Influence of an external field on the motion of a ferroelectric domain wall. Phys.Lett., 109A, 389-392.

Pouget J. and Maugin G.A.(1985b).Solitons and electroacoustic interactions in ferroelectric crystals.II.Interactions between solitons and radiations. Phys.Rev., B31, 4633-4651.

Pouget J.P.(1981).Les transformations ordre-désordre. In Les transformations de phase dans les solides minéraux, ed.V.Gabis and M.Lagarde , pp.125-243,Paris: Société française de minéralogie et cristallographie.

Soerensen M.P., Christiansen P.L. and Lomdahl P.S.(1984). Solitary waves on nonlinear elastic rods.I. J.Acoust.Soc.Amer., 76, 871-879.

Tolédano J.C. and Tolédano P.(1986). The Landau theory of phase transitions. Singapore: World Scientific Publ..

Wadhawan V.K.(1982). Ferroelasticity and related properties of crystals. Phase transitions, 3 , 3-103.

Yomosa S.(1985). Solitary excitations in DNA double helices. In Dynamical problems in soliton systems, ed.S.Takeno, pp.242-247, Berlin: Springer-Verlag.

ON THE DYNAMICS OF STRUCTURAL PHASE TRANSITIONS IN SHAPE MEMORY ALLOYS

M. Niezgódka
Systems Research Institute, Polish Academy of Sciences, Newelska 6, 01-447 Warsaw, Poland; partially supported by the A. v. Humboldt Foundation while visiting Universität Augsburg

J. Sprekels
Institut für Mathematik, Universität Augsburg, Memminger Str. 6, 8900 Augsburg, West Germany;
partially supported by the Deutsche Forschungsgemeinschaft

1. Introduction.

In this paper we consider the nonlinear system of differential equations

$$u_{tt} - [\Psi_\varepsilon(\Theta,\varepsilon)]_x - \mu u_{xxt} = f(x,t), \tag{1.1a}$$

$$-\Theta[\Psi_\Theta(\Theta,\varepsilon)]_t - \mu u_{xt}^2 - k\Theta_{xx} - \alpha k\Theta_{xxt} = \lambda(x,t), \tag{1.1b}$$

$$\varepsilon = u_x, \tag{1.1c}$$

for $(x,t) \in \Omega \times (0,T]$ (where $\Omega = (0,1)$ and $T > 0$), together with the initial and boundary conditions

$$u(x,0) = u_0(x), u_t(x,0) = u_1(x), \Theta(x,0) = \Theta_0(x),\ x \in \overline{\Omega}, \tag{1.1d}$$

$$u = 0, \quad \text{on } \Gamma \times [0,T], \tag{1.1e}$$

$$k\Theta_\nu = k_1(\Theta_\Gamma - \Theta), \quad \text{on } \Gamma \times [0,T]. \tag{1.1f}$$

Here, $\Gamma = \{0,1\}$ and $\Theta_\nu = -\Theta_x$ at $x = 0$ and $\Theta_\nu = \Theta_x$ at $x = 1$, respectively. The positive constants k, α, k_1, μ and the functions f, λ, Θ_Γ, u_0, u_1, Θ_0 are given.

Problem (1.1) constitutes a onedimensional model for the structural phase transitions in shape memory alloys (Niezgódka & Sprekels 1985). The involved quantities have their usual physical meaning, i.e.: u – displacement, Θ – temperature, ε – (linearized) strain, f – loads, λ – heat sources; the Helmholtz free energy $\Psi(\Theta,\varepsilon)$ has to be specified. (1.1a) reflects the momentum balance and (1.1b) the energy balance. The mass density ρ is assumed constant and normalized to unity; accordingly, the equation of continuity is ignored. The terms $-\mu u_{xxt}$ in (1.1a) (and $-\mu u_{xt}^2$ in (1.1b)) and $-\alpha k\Theta_{xxt}$ in (1.1b) indicate the presence of viscous stresses and a short thermal memory, respectively.

2. Phenomenology and Free Energy.

The load-deformation ($\sigma - \varepsilon$) curves of shape memory alloys exhibit a strong dependence on temperature: There exists a critical temperature Θ_c such that the following shapes are observed:

a) $\Theta < \Theta_c$ b) $\Theta > \Theta_c$ c) $\Theta >> \Theta_c$

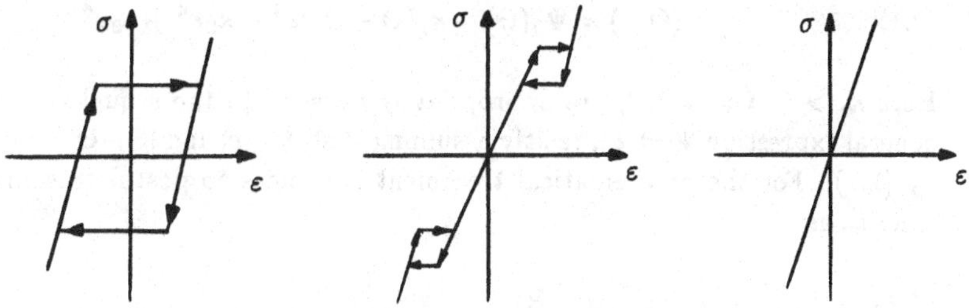

Fig. 1: temperature-dependent σ-ε-curves

Due to the form of the σ-ε–curves, a shape memory effect results: Loading above the yield load at $\Theta < \Theta_c$ results in a residual deformation after unloading; if the deformed material is then heated to a temperature above Θ_c, it creeps back to its virginal state (= original shape) since this is for $\Theta > \Theta_c$ the only possible unloaded configuration.

In a series of papers (Achenbach & Müller 1984, Müller 1979, Müller & Wilmański 1980; see also I. Müller's contribution to this volume), I. Müller and his co-workers constructed a macroscopic model, based on statistical mechanics. In particular, the hysteresis effects were ascribed to structural phase transitions between different stable configurations of the crystal lattice, highly symmetric austenite (A) and its two sheared versions, the martensitic twins (denoted M_+ and M_-).

The basic variable in this approach is the Helmholtz free energy $\Psi = \Psi(\Theta, \varepsilon)$, assumed to be a function of temperature and macroscopic strain alone. The σ-ε-curves resulting from the theory have the form

a) $\Theta < \Theta_c$ b) $\Theta > \Theta_c$ c) $\Theta >> \Theta_c$

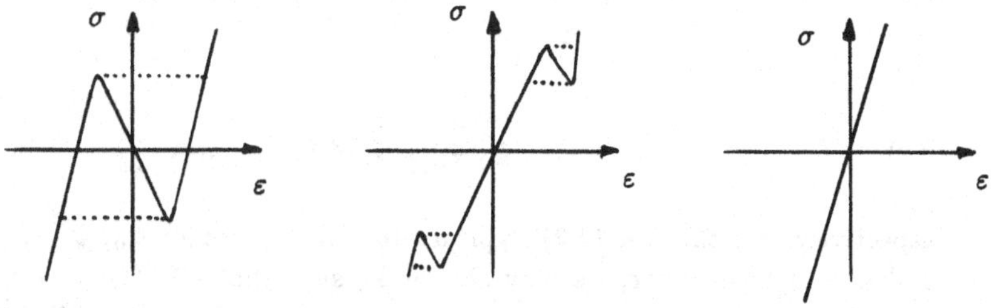

Fig. 2

The branches corresponding to $\Psi_{\varepsilon\varepsilon}(\Theta,\varepsilon) < 0$ are unstable. Thus the system follows the dotted lines and a hysteresis is observed.

In our initial-boundary value problem (1.1), the free energy function $\Psi(\Theta,\varepsilon)$ has still to be specified. The simplest polynomial expression such that Ψ_ε matches the behaviour depicted in Fig. 2, is the **LANDAU-DEVONSHIRE**-form, valid for small values of $|\varepsilon|$ and $|\Theta - \Theta_c|$:

$$\Psi(\Theta,\varepsilon) = \Psi_0(\Theta) + \kappa_1(\Theta-\Theta_c)\varepsilon^2 - \kappa_2\varepsilon^4 + \kappa_3\varepsilon^6. \tag{2.1}$$

Here $\kappa_i > 0$, $i = 1,2,3$, are appropriately chosen. In the sequel we work with the general expression $\Psi(\Theta,\varepsilon)$, tacitly assuming that, for $|\varepsilon|$ and $|\Theta-\Theta_c|$ small, Ψ is given by (2.1). For the mathematical treatment it suffices to postulate some asymptotic conditions:

$(H1)$ $\quad \Psi \in C^3(\mathbb{R}\times\mathbb{R});$

$(H2)$ $\quad \Psi_\Theta(\Theta,\varepsilon) = 0\,, \quad \text{on} \quad (-\infty,0]\times\mathbb{R};$

$(H3)$ $\quad -\Theta\Psi_{\Theta\Theta}(\Theta,\varepsilon) \geq 0, \quad \text{on} \quad \mathbb{R}\times\mathbb{R};$

$(H4)$ $\quad (1+\Theta^2)|\Psi_{\Theta\varepsilon}(\Theta,\varepsilon)| \leq c_1 + c_2|\varepsilon|\,, \quad \text{on} \quad \mathbb{R}\times\mathbb{R};$

$(H5)$ $\quad |\Psi_{\varepsilon\varepsilon}(\Theta,\varepsilon)| \leq c_3, \quad \text{on} \quad \mathbb{R}\times\mathbb{R};$

$(H6)$ $\quad |\Psi_\varepsilon(\Theta,\varepsilon)| \leq c_4 + c_5|\varepsilon|, \quad \text{on} \quad \mathbb{R}\times\mathbb{R};$

$(H7)$ $\quad |\int_0^\infty \xi^2\Psi_{\Theta\Theta\varepsilon}(\xi,\varepsilon)d\xi| \leq c_6 + c_7|\varepsilon|, \quad \text{for all} \quad \varepsilon\in\mathbb{R}.$

Here c_i, $i = 1,\cdots,7$, are fixed positive constants.

In Niezgódka & Sprekels (1985) and Niezgódka et. al. (1986) the hypotheses $(H1)$ - $(H7)$ have been formulated individually for the components Ψ_i of Ψ in the two cases

$$\Psi(\Theta,\varepsilon) = \Psi_0(\Theta) + \Psi_1(\Theta)\varepsilon^2 + \Psi_2(\varepsilon), \tag{2.2}$$

and

$$\Psi(\Theta,\varepsilon) = \Psi_0(\Theta) + \Psi_1(\Theta)\varepsilon^2 + \Psi_2(\Theta,\varepsilon), \tag{2.3}$$

respectively. For the case (2.2), typical admissible forms for Ψ_0, Ψ_1, Ψ_2 are depicted in Fig. 3. In particular, we may choose Ψ_0 such that $-\Psi_0(\Theta) \sim \Theta^4$ for $\Theta \to 0+$ and $-\Psi_0(\Theta) \sim \Theta$, for $\Theta \to +\infty$, respectively; this reflects the **DEBYE**-law for the temperature dependence of the specific heat.

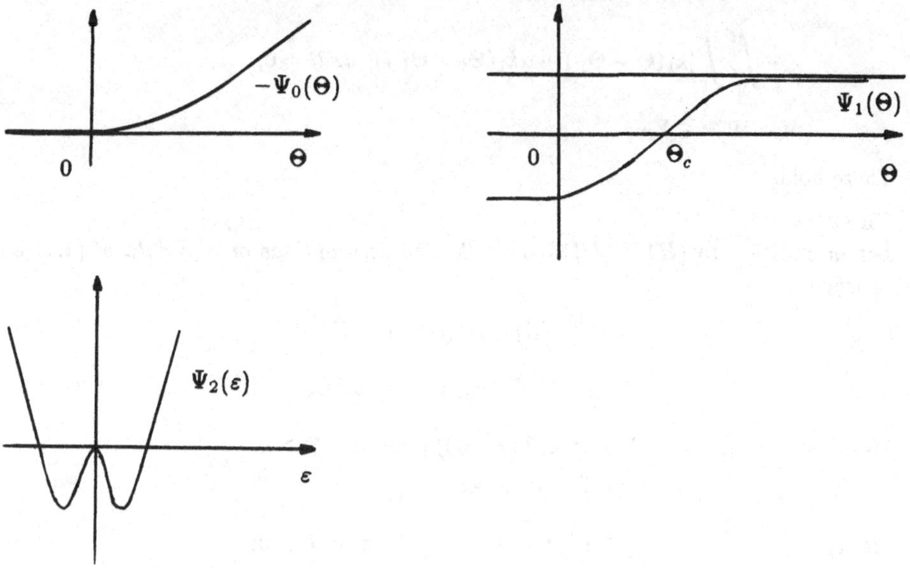

Fig. 3.: Admissible forms of Ψ_0, Ψ_1, Ψ_2.

3. Existence and Uniqueness.

For $t > 0$ we use the notations: $\Omega_t = \Omega \times (0,t)$, $\|\cdot\|$ = norm in $L^2(\Omega)$, $\|\Theta(t)\|_\Gamma^2 = \Theta^2(0,t) + \Theta^2(1,t)$, $Y_t = L^2(0,t;H^1(\Omega))$, $Z_t = L^2(0,t;L^2(\Omega))$. We define:

Definition:

A pair (u,Θ) is called a "weak solution" of (1.1) on $\overline{\Omega}\times[0,T]$ for $T > 0$, if the following is satisfied:

(3.1) $\quad u \in H^2(0,T;L^2(\Omega)) \cap H^1(0,T;\overset{\circ}{H}{}^1(\Omega) \cap H^2(\Omega))$;

(3.2) $\quad \Theta \in H^1(0,T;L^2(\Omega)) \cap H^1(0,T;L^2(\Gamma)) \cap L^2(0,T;H^1(\Omega))$;

(3.3) $\quad \Theta(x,0) = \Theta_0(x), u(x,0) = u_0(x), u_t(x,0) = u_1(x)$, a. e. on Ω;

(3.4) $\quad$ with $\varepsilon = u_x$ there hold

(i) $$\int_{\Omega_T}\int [u_{tt} - (\Psi_\varepsilon(\Theta,\varepsilon))_x - \mu u_{xxt} - f]\eta\,dx\,dt = 0, \quad \text{for all } \eta \in Z_T,$$

(ii) $$\int_{\Omega_T}\int [-\Theta(\Psi_\Theta(\Theta,\varepsilon))_t\zeta + \mu u_t(u_{xxt}\zeta + u_{xt}\zeta_x) +$$

$$+ k\Theta_x\varsigma_x + \alpha k\Theta_{xt}\varsigma_x + \lambda\varsigma]dxdt$$

$$+ \int_0^t \int_\Gamma [k_1(\Theta - \Theta_\Gamma) + \alpha k_1(\Theta_t - \Theta_{\Gamma,t})]\varsigma dxdt = 0,$$

for all $\varsigma \in Y_T$.

There holds:

Theorem

Let in addition to (H1) – (H7) the following assumptions on the data of (1.1) be satisfied:

(H8) $u_0 \in \mathring{H}^1(\Omega) \cap H^2(\Omega)$, $u_1 \in \mathring{H}^1(\Omega)$;

(H9) $\Theta_0 \in H^1(\Omega)$, $\Theta_0(x) > 0$ *on* $\overline{\Omega}$;

(H10) $\lambda \in L^2(0,T;L^2(\Omega))$, *for any* $T > 0$,
$\lambda(x,t) \geq 0$, a. e.;

(H11) $f \in L^2(0,T;L^2(\Omega))$, *for any* $T > 0$;

(H12) $\Theta_\Gamma \in H^1(0,T)$, *for any* $T > 0$
$\Theta_\Gamma(t) > 0$ *and* $\Theta_{\Gamma,t}(t) \geq 0$, a. e.

Then problem (1.1) has a global unique weak solution (u,Θ) *such that*

(3.5) $$\Theta, \varepsilon \in C(\overline{\Omega} \times [0,\infty)); \ \Theta(x,t) \geq 0, \quad on\ \overline{\Omega} \times [0,\infty).$$

Proof:

See Niezgódka & Sprekels (1985) and Niezgódka et. al. (1986) for the existence; the uniqueness proof is analogous to that in Hoffmann & Songmu (1986).

□

Remarks:

1. According to (3.5), ε does not jump across the phase transition front (which should be the case for the order parameter of a first order phase transition). This stems from the smoothing effects of viscosity and thermal memory and from the one-dimensionality.
2. Clearly $\Theta(x,t) > 0$ on $\overline{\Omega} \times [0,T]$, for some $T > 0$.
3. If the data are smooth, (u,Θ) is a classical solution.

4. Numerical Simulations.

We report about simulations carried out in Alt et. al. (1985) for the case that $\mu = 0$, $\alpha = 0$, $f \equiv 0$, $\lambda \equiv 0$, $\Omega = [-1,+1]$ and (1.1e, f) replaced by $\sigma = \Psi_\varepsilon = \sigma_\Gamma(t)$, $\Theta = \Theta_\Gamma(t)$, on the boundary. As potential we chose

(4.1) $$\Psi(\Theta,\varepsilon) = \varepsilon^2 + \frac{1}{2}(5 - \Theta^2)(1+\varepsilon^2)^{-1/2},$$

which exhibits the behaviour depicted in Fig. 1 with $\Theta_c = \sqrt{5}$. Moreover, $u_0(x) = \varepsilon_0 x$ ($\varepsilon_0 > 0$), $u_1 \equiv 0$, $\Theta_0 \equiv 1$; i.e., at $t = 0$ the material is completely in the M_+-phase.

We only display a case where $M_+ \to M_-$ and $M_\pm \to A$ transitions occured. The experiment extended over 800 time steps during which σ_Γ, Θ_Γ had the form:

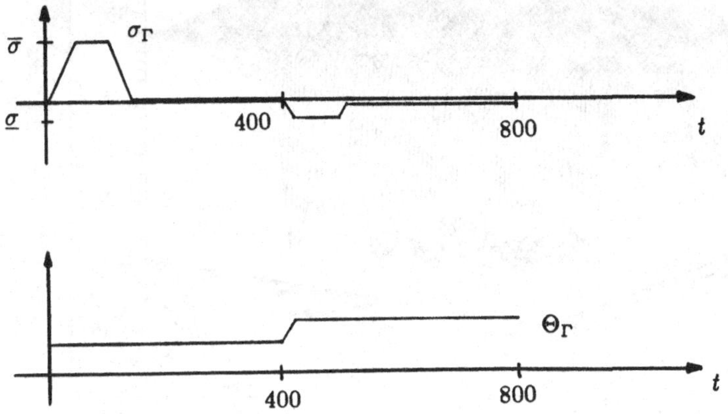

Fig. 4

Figs. 5 - 28 show the evolution of ε, Θ, u, where each picture shows 500 time steps. Since $\overline{\sigma}$ exceeds the yield limit, in Figs. 5 – 8 we see stress-induced $M_+ \to M_-$ transitions, accompanied by temperature outbursts due to released latent heat (Figs. 9 – 12) and the formation of sharp interfaces between M_+ and M_- (Figs. 13 – 16) along which ε is discontinuous. Upon unloading at $t = 150$, a superposition effect induces another transition nucleating from the center at $x = 0$. After the stabilization at $t = 400$, a load below the yield limit is applied, together with a sharp increase of Θ_Γ. This results (see Figs. 17 – 28) in $M_\pm \to A$ transitions, until at $t = 800$ a configuration is reached which corresponds to a state close to the origin in the σ-ε-plane; i.e., the material is a stressed austenite, and we observe shape memory.

7

DEFORMATION

Fig. 5

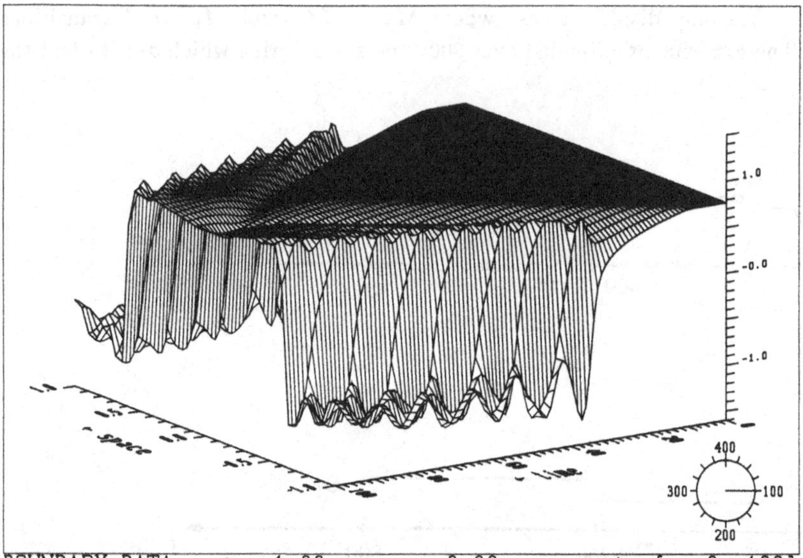

BOUNDARY DATA : θ = 1.00 , σ = 0.00 ; t ∈ [0, 100]

DEFORMATION

Fig. 6

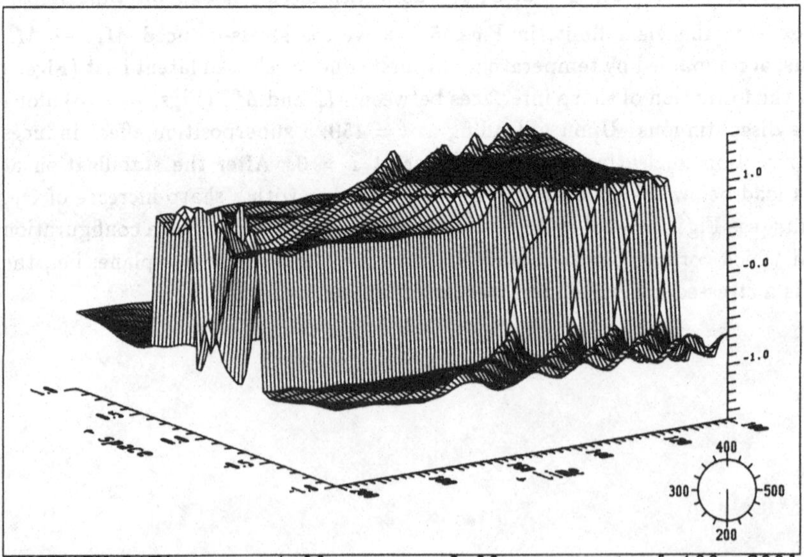

BOUNDARY DATA : θ = 1.00 , σ = 0.00 ; t ∈ [100, 200]

Fig. 7

DEFORMATION

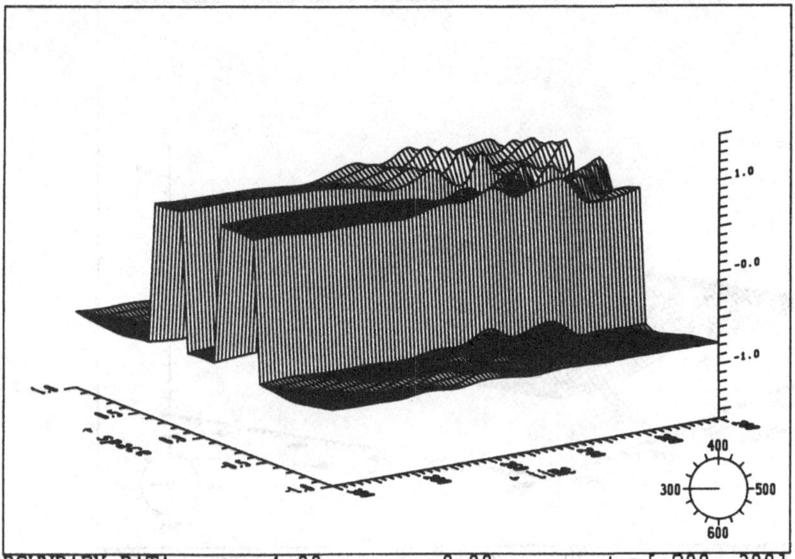

BOUNDARY DATA : θ = 1.00 , σ = 0.00 ; t ∈ [200, 300]

Fig. 8

DEFORMATION

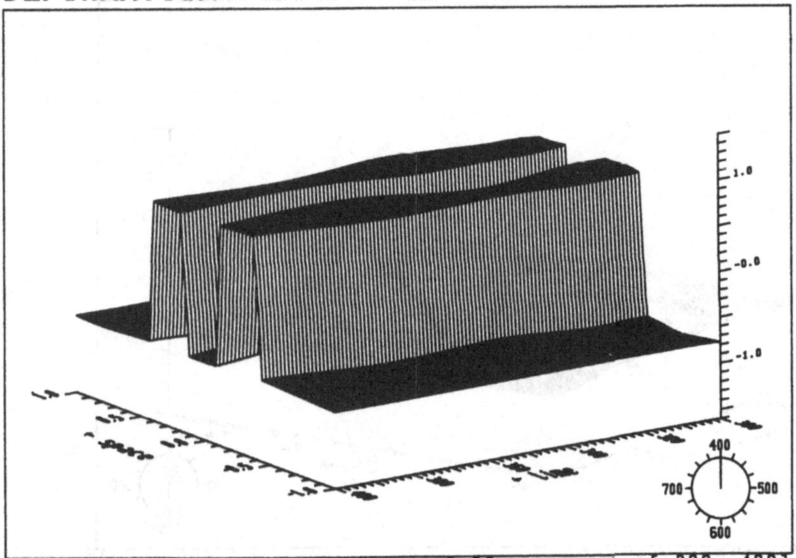

BOUNDARY DATA : θ = 1.00 , σ = 0.00 ; t ∈ [300, 400]

TEMPERATURE

Fig. 9

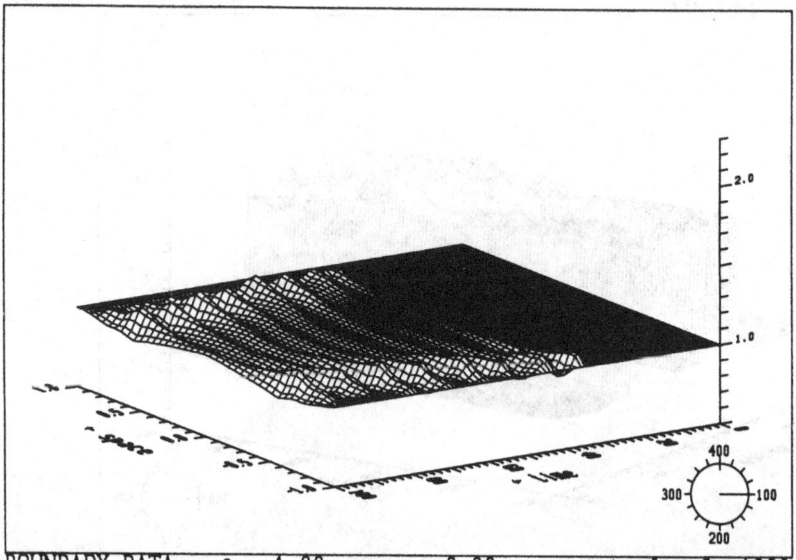

BOUNDARY DATA : θ = 1.00 , σ = 0.00 ; t ∈ [0, 100]

TEMPERATURE

Fig. 10

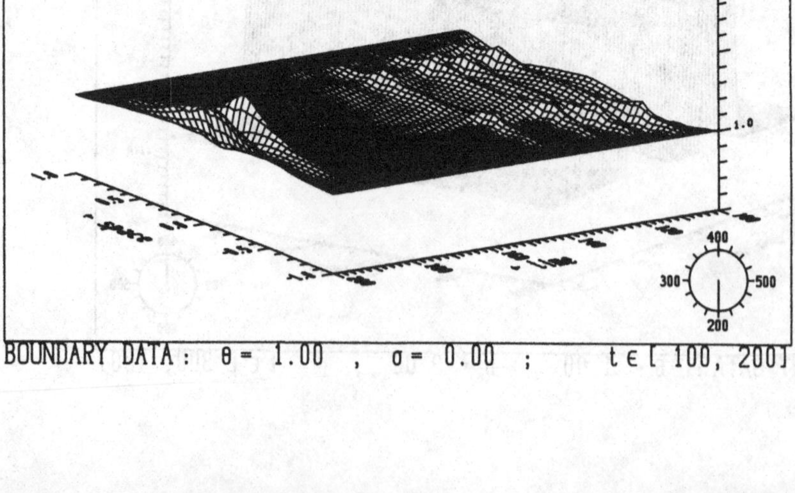

BOUNDARY DATA : θ = 1.00 , σ = 0.00 ; t ∈ [100, 200]

Fig. 11

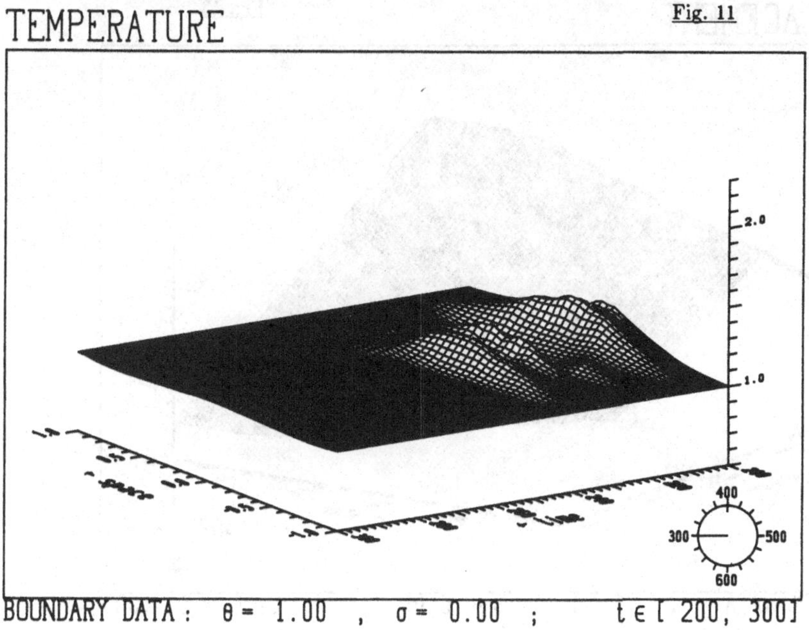
TEMPERATURE
2.0
1.0
400
300
500
600
BOUNDARY DATA : θ = 1.00 , σ = 0.00 ; t ∈ [200, 300]

Fig. 12

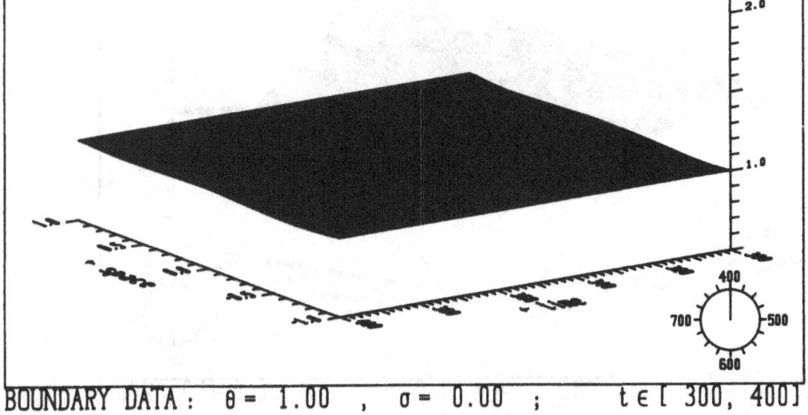
TEMPERATURE
2.0
1.0
400
700
500
600
BOUNDARY DATA : θ = 1.00 , σ = 0.00 ; t ∈ [300, 400]

DISPLACEMENT

Fig. 13

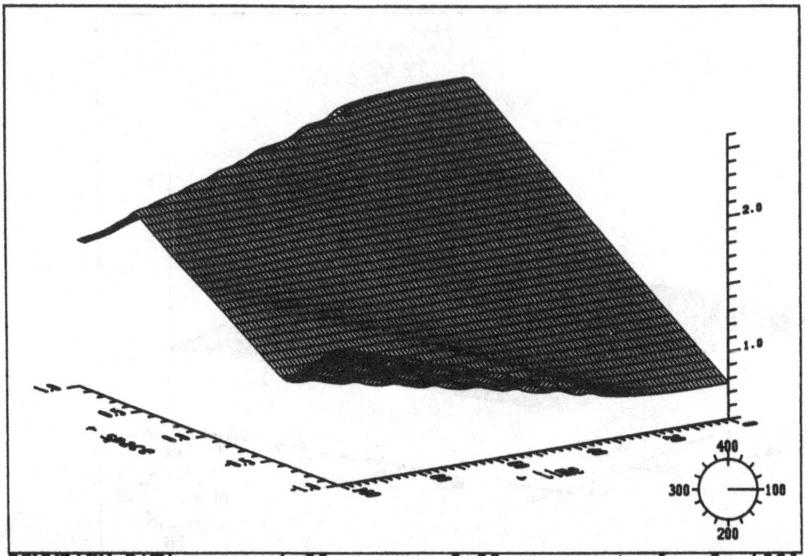

BOUNDARY DATA : θ = 1.00 , σ = 0.00 ; t ∈ [0, 100]

DISPLACEMENT

Fig. 14

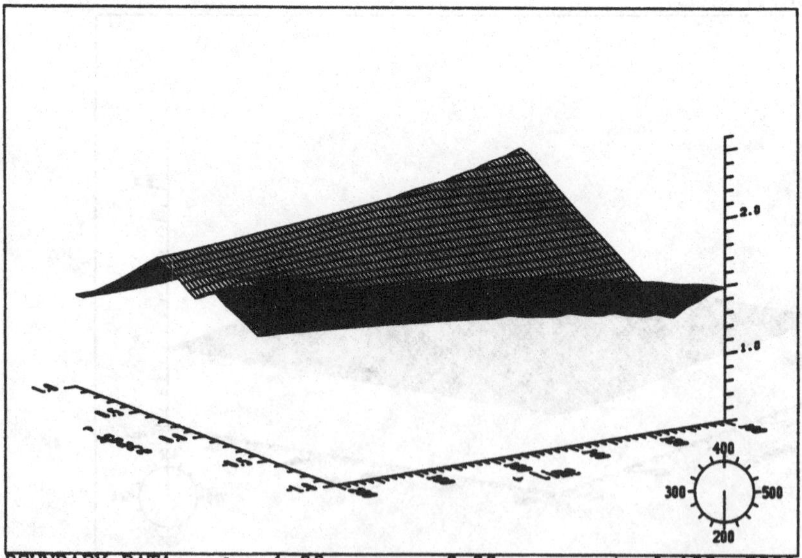

BOUNDARY DATA : θ = 1.00 , σ = 0.00 ; t ∈ [100, 200]

Fig. 15

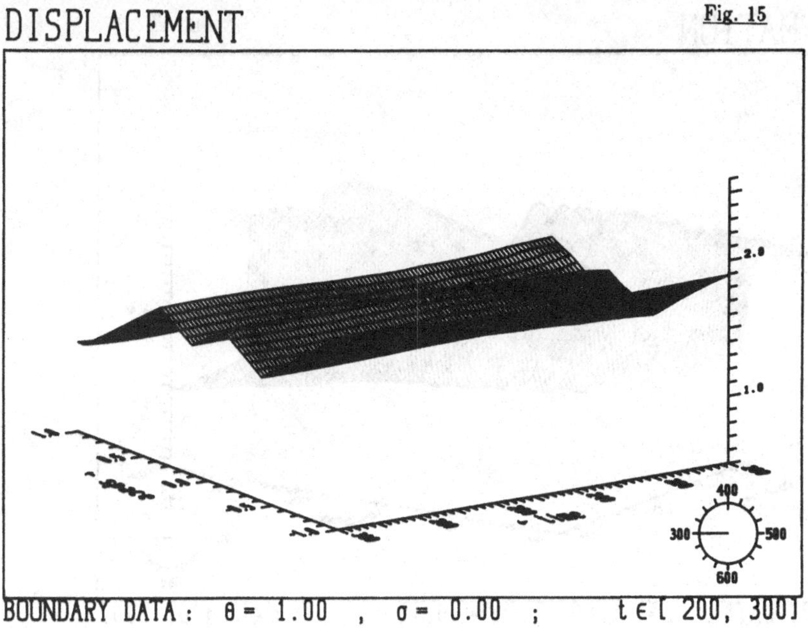
DISPLACEMENT
2.0
1.0
400
300
500
600
BOUNDARY DATA : θ= 1.00 , σ= 0.00 ; t ∈ [200, 300]

Fig. 16

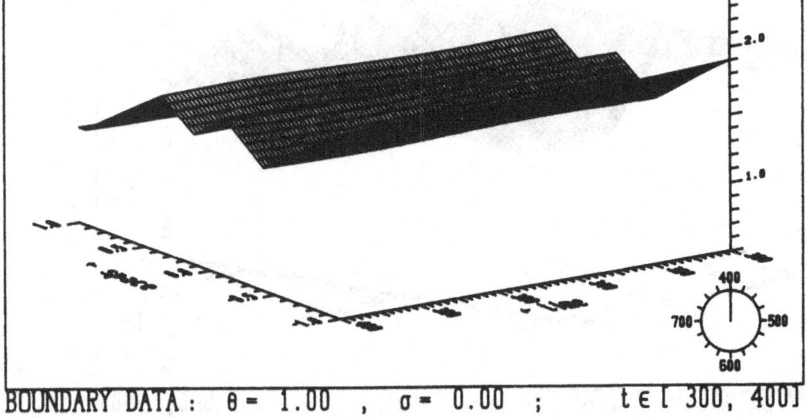
DISPLACEMENT
2.0
1.0
400
700
500
600
BOUNDARY DATA : θ= 1.00 , σ= 0.00 ; t ∈ [300, 400]

Fig. 17

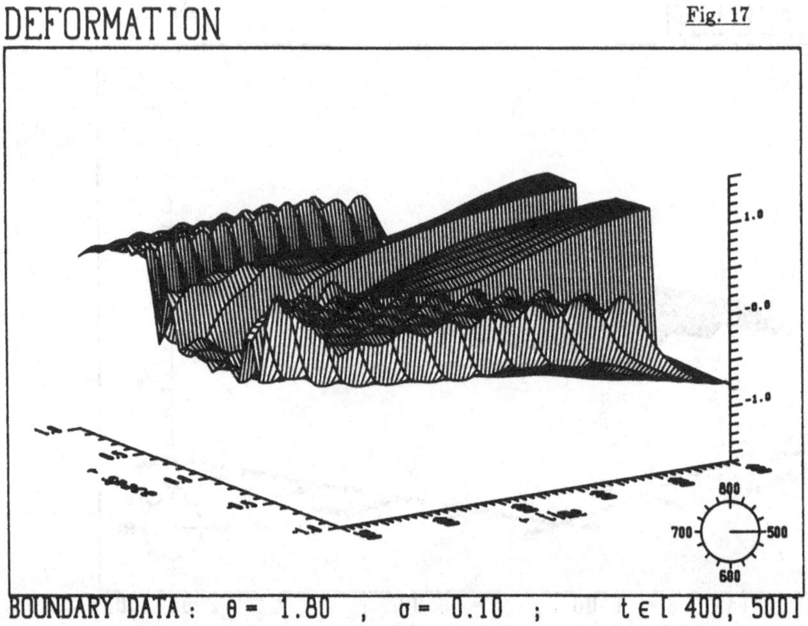
DEFORMATION
1.0
-0.0
-1.0
800
700
500
600
BOUNDARY DATA : θ = 1.80 , σ = 0.10 ; t ∈ [400, 500]

Fig. 18

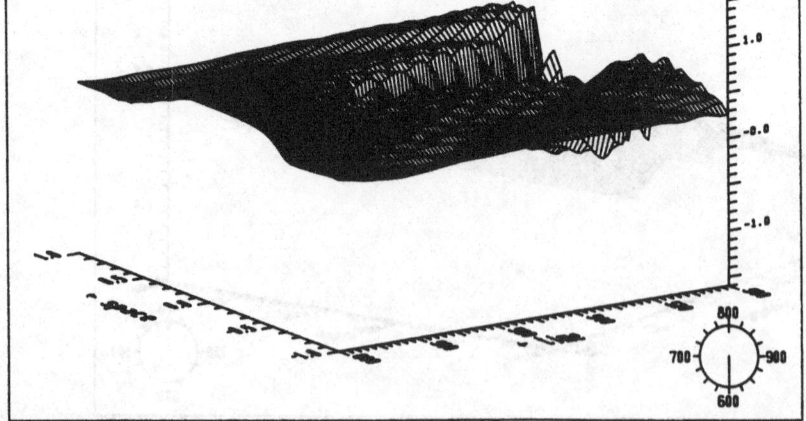
DEFORMATION
1.0
-0.0
-1.0
800
700
900
600
BOUNDARY DATA : θ = 1.80 , σ = 0.10 ; t ∈ [500, 600]

DEFORMATION

Fig. 19

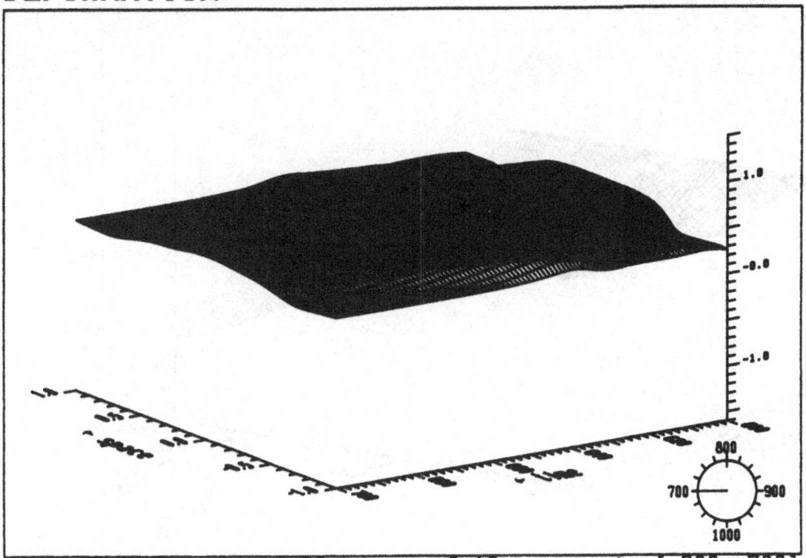

BOUNDARY DATA : θ = 1.80 , σ = 0.10 ; t ∈ [600, 700]

DEFORMATION

Fig. 20

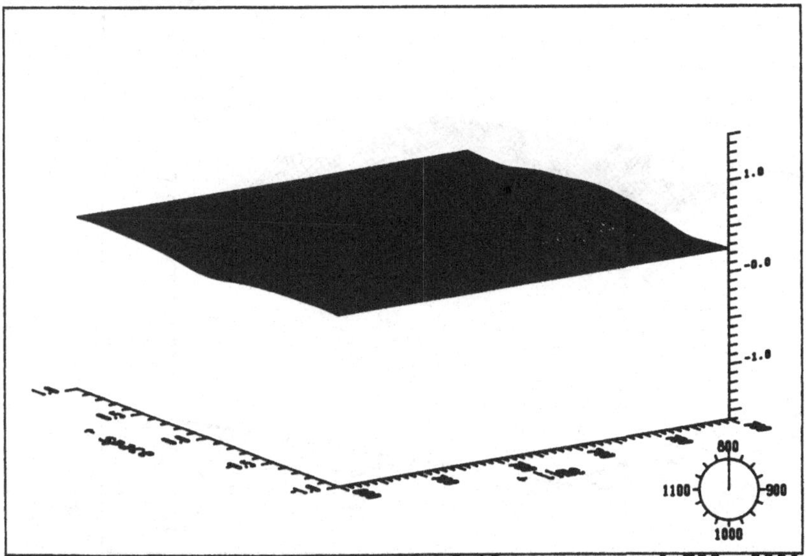

BOUNDARY DATA : θ = 1.80 , σ = 0.10 ; t ∈ [700, 800]

Fig. 21

TEMPERATURE

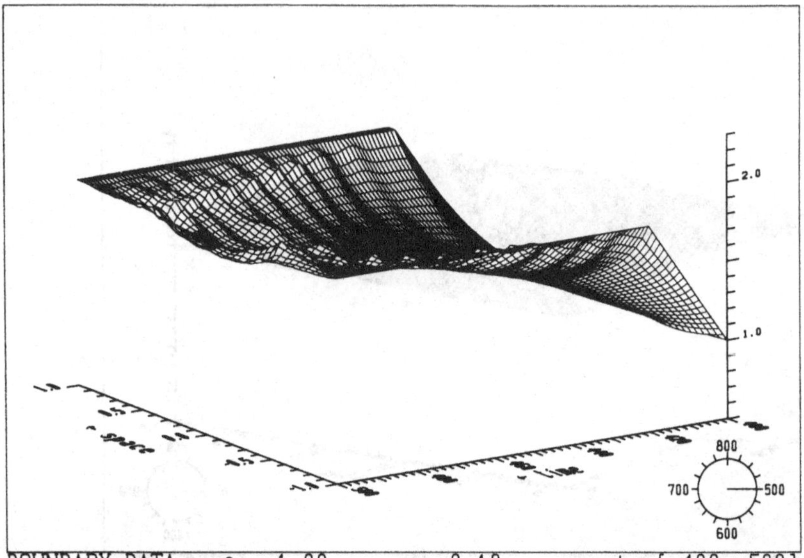

BOUNDARY DATA : θ = 1.80 , σ = 0.10 ; t ∈ [400, 500]

Fig. 22

TEMPERATURE

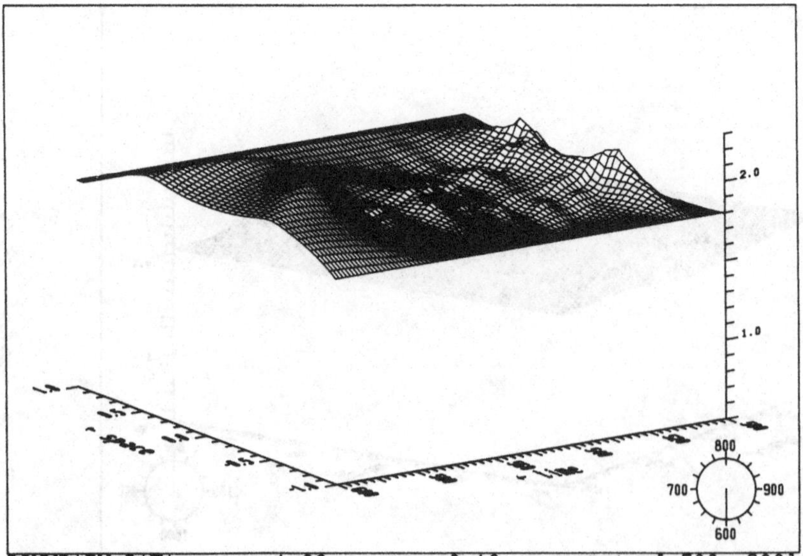

BOUNDARY DATA : θ = 1.80 , σ = 0.10 ; t ∈ [500, 600]

Fig. 23

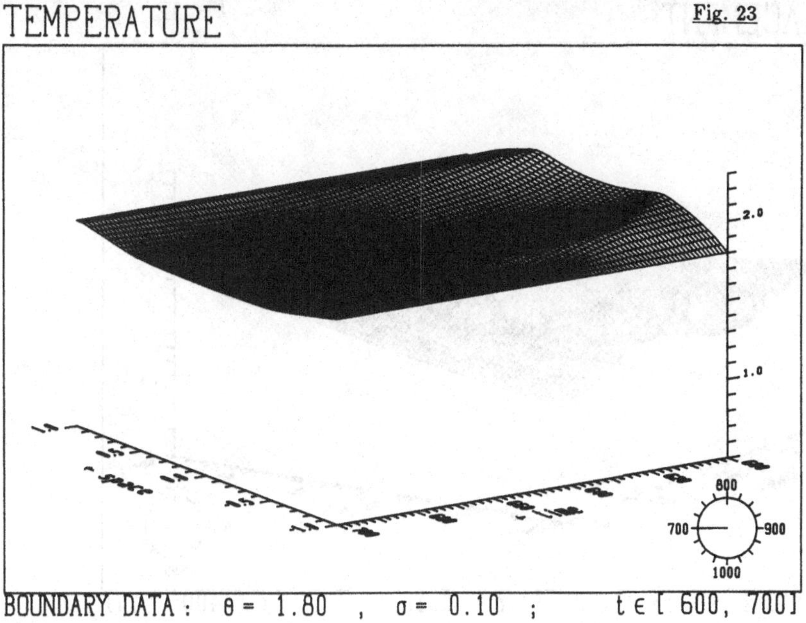
TEMPERATURE
2.0
1.0
800
700
900
1000
BOUNDARY DATA : θ = 1.80 , σ = 0.10 ; t ∈ [600, 700]

Fig. 24

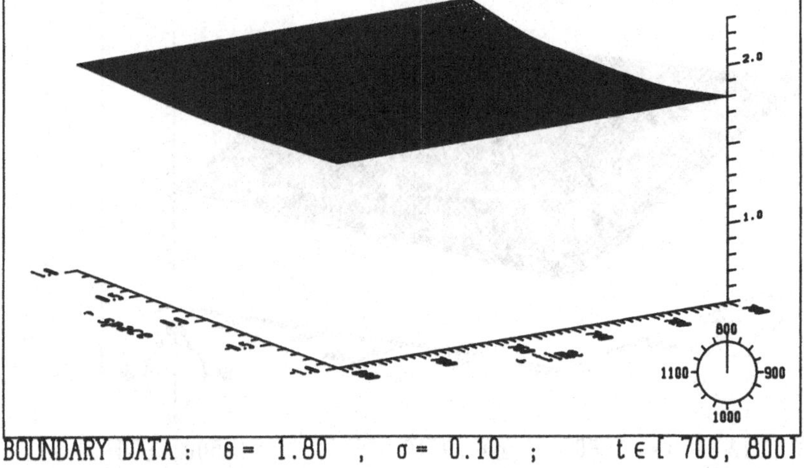
TEMPERATURE
2.0
1.0
800
1100
900
1000
BOUNDARY DATA : θ = 1.80 , σ = 0.10 ; t ∈ [700, 800]

DISPLACEMENT

Fig. 25

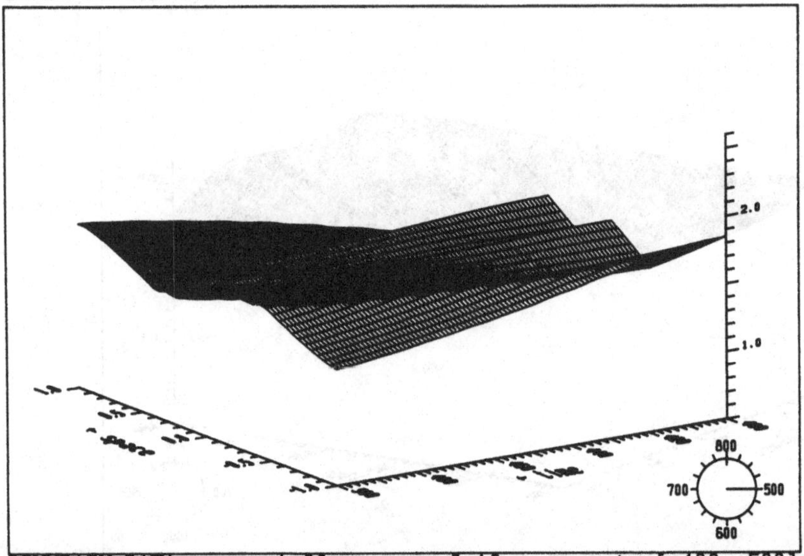

BOUNDARY DATA : θ = 1.80 , σ = 0.10 ; t ∈ [400, 500]

DISPLACEMENT

Fig. 26

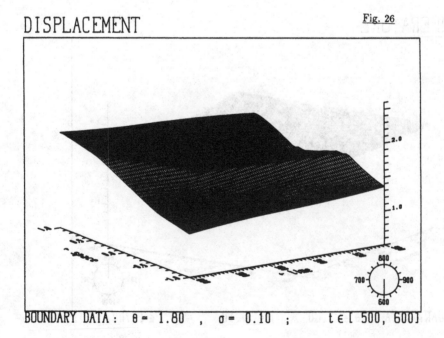

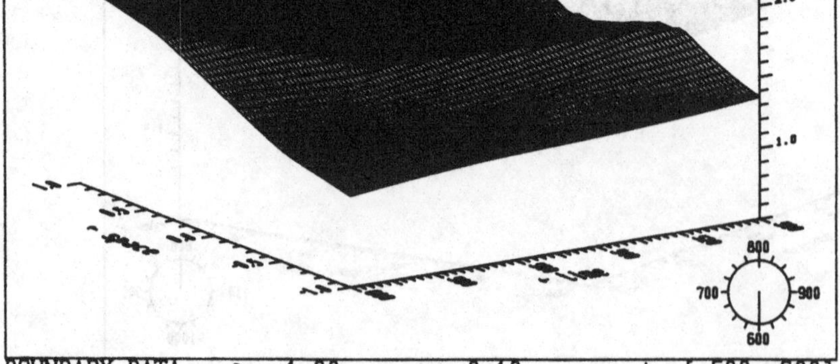

BOUNDARY DATA : θ = 1.80 , σ = 0.10 ; t ∈ [500, 600]

Fig. 27

DISPLACEMENT

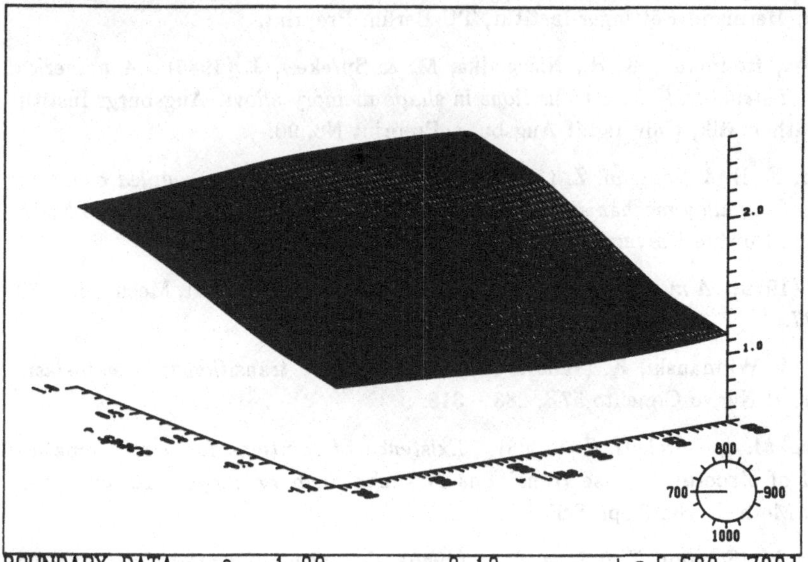

BOUNDARY DATA : θ = 1.80 , σ = 0.10 ; t ∈ [600, 700]

Fig. 28

DISPLACEMENT

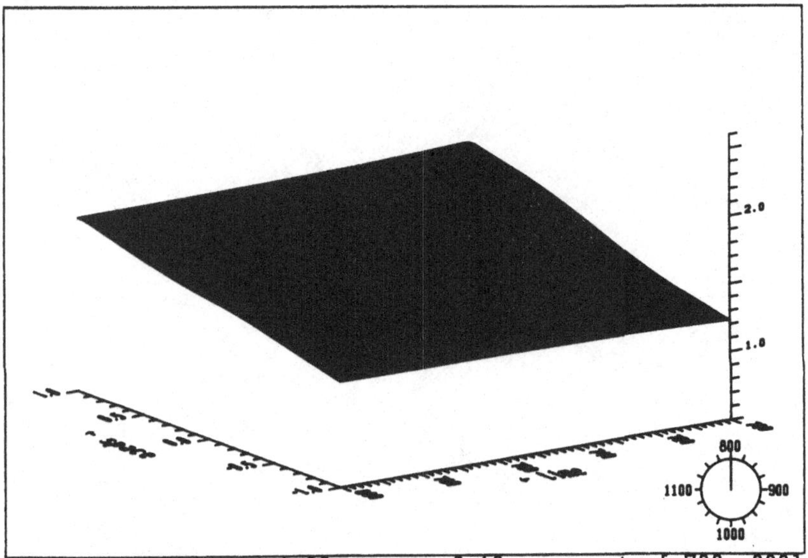

BOUNDARY DATA : θ = 1.80 , σ = 0.10 ; t ∈ [700, 800]

References:

Achenbach, M. & Müller, I. (1984). *Shape memory as a thermally activated process.* Berlin: Hermann-Föttinger-Institut, TU Berlin, Preprint.

Alt, H. W., Hoffmann, K.-H., Niezgódka, M. & Sprekels, J. (1985). *A numerical study of structural phase transitions in shape memory alloys.* Augsburg: Institut für Mathematik, Universität Augsburg, Preprint No. 90.

Hoffmann, K.-H. & Songmu, Z. (1986). *Uniqueness for nonlinear coupled equations arising from alloy mechanism.* West Lafayette, Indiana: Center for Applied Mathematics, Purdue University, Techn. Report No. 14.

Müller, I. (1979). *A model for a body with shape-memory.* Arch. Rat. Mech. Anal. 70, 61 - 77.

Müller, I. & Wilmanski, K. (1980). *A model for phase transition in pseudoelastic bodies.* Il Nuovo Cimento 57B, 283 - 318.

Niezgódka, M. & Sprekels, J. (1985). *Existence of solutions for a mathematical model of structural phase transitions in shape memory alloys.* To appear in: Math. Meth. in the Appl. Sci.

Niezgódka, M., Songmu, Z. & Sprekels, J. (1986). *Global solutions to a model of structural phase transitions in shape memory alloys.* To appear in: J. Math. Anal. Appl.

ON THE HOMOGENIZATION OF SOME FREE BOUNDARY PROBLEMS

José-Francisco Rodrigues
University of Lisbon and C.M.A.F.
Av. Prof. Gama Pinto, 2. 1699 LISBOA CODEX (PORTUGAL)

1. Introduction

In this article we present some results and open problems related to the stability, with respect to the homogenization, of the coincidence sets and free boundaries associated with elliptic and parabolic variational inequalities. They can be considered as partial contributions to an interesting and deep problem posed by J.L.Lions several years ago (see [L1,2]).

For each parameter $\varepsilon \geq 0$, we consider the Dirichlet unilateral problem

(1.1) $u^\varepsilon \geq 0,\ L^\varepsilon u^\varepsilon \geq f^\varepsilon$ and $u^\varepsilon(L^\varepsilon u^\varepsilon - f^\varepsilon) = 0$ a.e. in Q

(1.2) $u^\varepsilon = g^\varepsilon$ on Σ,

where, in the elliptic case, we denote (if $v=v(x)$, $x \in \Omega$, open and bounded)

(1.3) $L^\varepsilon v = -(a^\varepsilon_{ij} v_{x_j})_{x_i}$ and $Q = \Omega \subset \mathbf{R}^n$, $\Sigma = \partial\Omega$ Lipschitz;

and, in the parabolic case, (if $v=v(x,t)$),

(1.4) $L^\varepsilon v = v_t - (a^\varepsilon_{ij} v_{x_j})_{x_i}$ and $Q = \Omega \times]0,T[\subset \mathbf{R}^{n+1}$, $\Sigma = \partial\Omega \times]0,T[\cup \Omega \times \{0\}$.

In both cases, we assume the coefficients $a^\varepsilon_{ij} = a^\varepsilon_{ij}(x)$, $x \in \Omega$, satisfy

(1.5) $a^\varepsilon_{ij}\xi_i\xi_j \geq \alpha|\xi|^2$, $\forall \xi \in \mathbf{R}^n$ and $|a^\varepsilon_{ij}| \leq M$, a.e. $x \in \Omega$,

where $M \geq \alpha > 0$ are constants independent of $\varepsilon \geq 0$.

Under appropriate conditions on the data f^ε and g^ε ($g^\varepsilon \geq 0$), see [KS] or [BL] for instance, it is well known that the variational inequality (1.1)(1.2) admits a unique continuous solution u^ε ($u^\varepsilon=u^\varepsilon(x)$ or $u^\varepsilon=u^\varepsilon(x,t)$, respectively in the elliptic or parabolic cases). In general, the domain $\bar{Q}$ is "a posteriori" divided into two parts: an open (in $\bar{Q}$) subset, the continuation set, where

(1.6) $\quad L^\varepsilon u^\varepsilon = f^\varepsilon \quad$ a.e. in $\Lambda^\varepsilon \equiv \{u^\varepsilon > 0\}$,

and its complement in $\bar{Q}$, the coincidence set, separated in Q by the so-called free boundary denoted, respectively, by

(1.7) $\quad I^\varepsilon \equiv \{u^\varepsilon=0\}$ (closed) and $\Phi^\varepsilon \equiv \partial I^\varepsilon \cap Q$, $(\varepsilon \geq 0)$.

The problem (1.1)(1.2) is well-behaved for the homogenization of the coefficients, that is, if the matrices $\{a^\varepsilon_{ij}\}$ H-converge to $\{a^o_{ij}\}$ as $\varepsilon\to 0$ (see [L1], [BLP] or [T]) and $f^\varepsilon\to f^o$, $g^\varepsilon\to g^o$ under appropriate conditions, one has $u^\varepsilon\to u^o$. It is then natural to discuss in what sense and under which conditions one can expect $I^\varepsilon\to I^o$ and $\Phi^\varepsilon\to\Phi^o$.

In the next section we give general conditions in order to obtain the convergence of the coincidence sets in Hausdorff distance (see [CR1] and [R1]) and we discuss its applicability to the homogenization. In the Section 3 a different independent convergence is given in terms of their characteristic functions. It improves previous results of [M2][CR2] and [R1]. In the following section we apply the preceding results to a typical free boundary model, the one-phase Stefan problem. Finally one concludes with some related open questions, namely in the dam problem and in the two-phases Stefan problem.

2. Convergence in Hausdorff Distance

Recall the Hausdorff distance defined by

(2.1) $\quad h(I,J) = \inf\{\delta > 0 : I \subset B_\delta(J) \text{ and } J \subset B_\delta(I)\}$,

where $B_\delta(I) = \bigcup_{x\in I}\{y : |x-y| < \delta\}$ is the δ-parallel body of I, and notice that the family of all nonempty compact subsets of $\bar{Q}$ is a compact metric space for h.

THEOREM 1 Suppose the solutions to (1.1)(1.2) are such that

(2.2) $\quad u^\varepsilon \to u^o \quad$ in $C^o(\bar{Q})$ (i.e., uniformly in $\bar{Q}$)

(2.3) $\quad L^\varepsilon u^\varepsilon - f^\varepsilon \rightharpoonup L^o u^o - f^o$ in $\mathcal{D}'(Q)$ (i.e. in the sense of distributions) and in the limit problem one has the nondegeneracy conditions

(2.4) $\quad I^o = \overline{\mathrm{int}\,(I^o)}$ and

(2.5) $\quad f^o \neq 0$ a.e. in Q (assuming $f^o \in L^1_{loc}(Q)$).

Then the corresponding coincidence sets are such that

(2.6) $\quad I^\varepsilon \to I^o$ in Hausdorff distance.

Proof: We only describe the main ideas since the details are similar to [CR1] and [R1]. By compactness, there is a compact subset $I^* \subset \bar{Q}$ and a subsequence such that $h(I^{\varepsilon'}, I^*) \to 0$; by the continuity of u^o and (2.2), $I^* \subset I^o$ follows easily; then the convergence (2.3) implies that $\mathrm{int}(I^o \setminus I^*) = \emptyset$ by the assumption (2.5); hence $\mathrm{int}(I^*) = \mathrm{int}(I^o)$ and the condition (2.4) yields $I^o \subset I^*$, i.e., $I^* = I^o$ and (2.6) follows. □

REMARK 1 The restrictive condition (2.4) is a regularity assumption on the limit coincidence set and it precludes the existence of "pathological components in I^o (like isolated points, for instance) which are easily seen to be unstable. On the other hand, the nondegeneracy assumption (2.5) is, in some sense, also necessary since we can easily produce counter-examples from the extreme case where $u^o = f^o = 0$. Notice that the conclusion (2.6) can be localized in any nonempty compact subset $K \subset Q$. □

In order to apply this result to the homogenization problem, both in the elliptic and in the parabolic cases, we need to guarantee the convergence properties (2.2) and (2.3). Hence, we assume the H-convergence for the coefficients

(2.7) $\quad \{a^\varepsilon_{ij}\} \xrightarrow[H]{} \{a^o_{ij}\}$ (see [T]),

and the following convergences for the data

(2.8) $\quad f^\varepsilon \to f^o$ in $L^2(Q)$-strong,

(2.9) $\quad g^\varepsilon \to g^o$ in $H^{1/2}(\Sigma)$-weak and $L^\infty(\Sigma)$-strong,

under the additional conditions, uniformly in $\varepsilon \geq 0$,

(2.10) $\quad \|f^\varepsilon\|_{L^\infty(Q)} \leq C$ and $\|g^\varepsilon\|_{C^{o,\gamma}(\Sigma)} \leq C$,

and, in the parabolic case, also

(2.11) $\quad \|f^\varepsilon_t\|_{L^\infty(Q)} \leq C$ and $\|g^\varepsilon_t\|_{L^\infty(\Sigma)} \leq C$,

for some fixed constants $C > 0$, $0 < \gamma < 1$.

These conditions are slightly stronger than what is necessary to apply the homogenization techniques to (2.1)(2.2) (see [BLP], [M1]) but they are convenient for our purposes. In particular, (2.10) provides the De Giorgi-Nash estimate

(2.12) $\|u^\varepsilon\|_{C^{0,\gamma'}(\bar{Q})} \leq C$ $(0 < \gamma' \leq \gamma$ and $C > 0$ independent of $\varepsilon)$

and (2.11), by a maximum principle argument for the parabolic problem, also

(2.13) $\|u_t^\varepsilon\|_{L^\infty(Q)} \leq C$ $(C > 0$ independent of $\varepsilon)$.

Then, using the theory for homogenization of variational inequalities (see [L1], [BLP], [M1]), and taking Theorem 1 into account, one can state in both cases (elliptic and parabolic):

<u>THEOREM 2</u> Under the preceding assumptions one has

(2.14) $u^\varepsilon \to u^0$ in $H^1(Q)$-weak and $C^{0,\gamma''}(\bar{Q})$-strong $(0<\gamma''<\gamma')$

(2.15) $L^\varepsilon u^\varepsilon \rightharpoonup L^0 u^0$ in $L^\infty(Q)$-weak-*.

If, in addition, (2.4) and (2.5) are satisfied the coincidence sets converge in Hausdorff distance, i.e., (2.6) holds. □

<u>REMARK 2</u> This last conclusion is due to [CR1] in the elliptic case and to [R1] in the parabolic one. It is only a partial contribution since, in general, the condition (2.4) is difficult to establish. A special case due to H. Lewy and G.Stampacchia, where (2.4) holds, can be found in [KS]: in the elliptic case, let $\Omega \subset \mathbb{R}^2$ be a convex smooth domain, a_{ij}^0 be constant coefficients, $g^0 = -\psi > 0$ on $\partial\Omega$ and $f^0 = L^0\psi$, where $\psi \in C^2(\bar{\Omega})$ is a strictly concave function verifying $\max_\Omega \psi > 0$. □

<u>REMARK 3</u> The result of Theorem 2 was applied in [CR2] to the homogenization of a steady flow through a rectangular porous dam with a permeability independent of the vertical direction (for instance, a dam with vertical layers). In the variational inequality approach to this problem the coincidence set represents the dry part of the dam and, since for each $\varepsilon \geq 0$, the free boundary is given by a monotone decreasing graph, the conclusion (2.6) can be used to show the (local) uniform convergence of the free boundaries with respect to the homogenization. □

REMARK 4. The uniform convergence (2.2) alone is sufficient in order to obtain an upper semi-continuity general property for the coincidence sets I^ε. In an equivalent way, this can be stated in terms of lower semi-continuity of the characteristic functions $\chi_{\Lambda^\varepsilon}$ of the continuation sets $\Lambda^\varepsilon=\bar{Q}\setminus I^\varepsilon=\{u^\varepsilon>0\}$:

(2.16) $$\liminf_{\varepsilon\to 0} \chi_{\Lambda^\varepsilon} \geq \chi_{\Lambda^o} \quad \text{everywhere in } \bar{Q}.$$

In fact, if a point $P\in\Lambda^o(\chi_{\Lambda^o}(P)=1)$ one has $u^o(P)>0$ and by the uniform convergence $u^\delta(P)>0$, for all $\delta>0$, $0<\delta<\varepsilon_p$ for ε_p small enough. Hence $P\in\Lambda^\delta$, $\forall\delta<\varepsilon_p$, and (2.16) follows since $P\in\liminf_{\varepsilon\to 0}\Lambda^\varepsilon = \bigcup_{\varepsilon>0}\bigcap_{\varepsilon>\delta}\Lambda^\delta$. □

3. Convergence in Lebesgue Measure

A different, and in fact independent, result on the stability of the coincidence sets (defined now up to a null set) can be given using the Lebesgue measure μ (in the elliptic case, $\mu=\mu_n$ is n-dimensional, and, in the parabolic one, $\mu=\mu_{n+1}$ is (n+1)-dimensional).

From the well-known Lewy-Stampacchia's inequalities (see [BL], for instance), if each $f^\varepsilon\in L^2(Q)$, the variational solution $u^\varepsilon(\varepsilon\geq 0)$ to (1.1)(1.2) satisfies

(3.1) $$f^\varepsilon \leq L^\varepsilon u^\varepsilon \leq \sup(f^\varepsilon,0) \quad \text{a.e. in } Q.$$

Then there exists a function $\lambda^\varepsilon=\lambda(u^\varepsilon)\in L^\infty(Q)$, satisfying

(3.2) $$L^\varepsilon u^\varepsilon = f^\varepsilon\lambda^\varepsilon, \quad \text{a.e. in } Q, \text{ and}$$

(3.3) $$0\leq 1-\chi_{I^\varepsilon}=\chi_{\Lambda^\varepsilon}\leq\lambda^\varepsilon\leq 1, \quad \text{a.e. in } Q.$$

This follows easily from (1.6) and (3.1), and if $f^\varepsilon\neq 0$ a.e. in Q we immediately see that λ^ε is uniquely determined up to a set of measure zero. Actually, under some additional regularity assumption one has $\lambda^\varepsilon=\chi_{\Lambda^\varepsilon}$ a.e. in Q (see Remark 5, below).

Now we want to let $\varepsilon\to 0$, assuming (2.3) and (2.8). Since $0\leq\lambda^\varepsilon\leq 1$, there exists a subsequence $\varepsilon'\to 0$ and some $\lambda^*\in L^\infty(Q)$ such that

(3.4) $$\lambda^{\varepsilon'} \rightharpoonup \lambda^* \text{ in } L^\infty(Q)\text{-weak* and } 0\leq\lambda^*\leq 1 \text{ a.e. in } Q.$$

Since $f^{\varepsilon'}\to f^o$ in $L^2(Q)$-strong, one has $f^{\varepsilon'}\lambda^{\varepsilon'} \rightharpoonup f^o\lambda^*$ in $L^2(Q)$-weak. Using (3.2) for the limit case, one finds $f^o\lambda^*=$

$=f^o\lambda^o$ a.e. in Q and clearly the nondegeneracy condition (2.5) implies $\lambda^*=\lambda^o$ a.e. Hence, by uniqueness of λ^o, the following convergence holds

(3.5) $$\lambda^\varepsilon \rightharpoonup \lambda^o \quad \text{in} \quad L^\infty(Q)\text{-weak*}.$$

This very weak result is particularly interesting under the additional mild regularity assumption on the limit problem: $\lambda^o=\chi_{\Lambda^o}$, that is, if

(3.6) $$L^o u^o = f^o(1-\chi_{I^o}) \quad \text{a.e. in } Q.$$

THEOREM 3 Assume $u^\varepsilon \to u^o$ in $L^1(Q)$ under the convergence assumptions (2.3) and (2.8) and the nondegeneracy conditions (2.5) and (3.6). Then the characteristic functions of their associated coincidence sets are such that

(3.7) $$\chi_{I^\varepsilon} \to \chi_{I^o} \quad \text{in} \quad L^p(Q)\text{-strong}, \quad \forall p,\ 1 \le p < \infty.$$

Proof: From (3.3) one can select some $q^* \in L^\infty(Q)$ and a subsequence $\varepsilon' \to 0$, such that

(3.8) $$\chi_{I^{\varepsilon'}} \rightharpoonup q^* \quad \text{in} \quad L^\infty(Q)\text{-weak*}.$$

From (3.3) and (3.5), it follows that

(3.9) $$0 \le 1-q^* \le \lambda^o = 1-\chi_{I^o} \le 1 \quad \text{a.e. in } Q,$$

by the assumption (3.6). Hence $q^* \ge \chi_{I^o}$ a.e.

On the other hand, since $u^\varepsilon \to u^o$ in $L^1(Q)$ one has

$$0 = \int_A u^\varepsilon \chi_{I^\varepsilon} = \int_Q (u^\varepsilon \chi_A)\chi_{I^\varepsilon} \to \int_Q (u^o\chi_A)q^* = \int_A u^o q^* = 0,$$

for any measurable subset $A \subset Q$. Then $u^o q^* = 0$ a.e. in Q, which means that $q^* = 0$ if $u^o > 0$, i.e., $\chi_{I_o} \ge q^*$ a.e.. Therefore $q^*=\chi_{I^o}$ and the convergence (3.7) follows, first weakly and, afterwards, also strongly since they are characteristic functions. □

The application of this result to the homogenization problem can be done without the assumption (2.10) since now the uniform convergence of the solutions is not necessary. Again from [L1], [BLP] [M1] and the Theorem 3 one can state.

THEOREM 4 Under the assumptions (2.7), (2.8), (2.9) (and (2.11) in the parabolic case), as $\varepsilon \to 0$ the solutions to to (2.1)(2.2) satisfy

(3.10) $u^\varepsilon \to u^o$ in $H^1(Q)$-weak and $L^2(Q)$-strong

(3.11) $L^\varepsilon u^\varepsilon \rightharpoonup L^o u^o$ in $L^2(Q)$-weak.

If, in addition, (2.5) and (3.6) are satisfied then (3.7) holds, or, equivalently, one has

(3.12) $I^\varepsilon \to I^o$ in the sense $\mu(I^\varepsilon \div I^o) \to 0$,

where $\div$ denotes the symmetric difference of the coincidence sets. □

REMARK 5 This result extends analogous conclusions of [M2] and [CR1,2] for the elliptic case and [R1] for the parabolic one. As before the conclusion can be localized in any measurable subset $A \subset Q$ where (2.5) and (3.6) are a.e. satisfied, yielding now $\mu((I^\varepsilon \div I^o) \cap A) \to 0$.

The regularity assumption (3.6) is satisfied, for instance,

(3.13) if $\mu(\partial I^o \cap Q) = 0$ or if $a^o_{ij} \in C^{o,1}(Q)$,

being trivially satisfied this last condition in the case of periodic homogenization, since then a^o_{ij} are constants (see [BLP]). □

4. Application to the One-Phase Stefan Problem

The results of the preceding sections can be applied to a simple model describing the melting of ice (at zero temperature) in a nonhomogeneous medium, whose thermal conductivity is given by $\{a^\varepsilon_{ij}\}$, in contact with the liquified region at temperature $\theta^\varepsilon = \theta^\varepsilon(x,t) > 0$, where

(4.1) $\eta\theta^\varepsilon_t - (a^\varepsilon_{ij}\theta^\varepsilon_{x_j})_{x_i} = 0$ in $\{\theta^\varepsilon > 0\} \subset Q$.

Here η denotes the specific heat: if $\eta > 0$ we have a parabolic problem and if $\eta = 0$ an elliptic one, corresponding to a quasi-steady model where the time is just a parameter.

On the moving boundary $\Phi^\varepsilon(t)$, the liquid-solid free interface, the double condition holds

(4.2) $\theta^\varepsilon = 0$ and $a^\varepsilon_{ij}\theta^\varepsilon_{x_j} n^\varepsilon_i = -\lambda(\nu^\varepsilon . n^\varepsilon)$

being the normal $n^\varepsilon = \{n^\varepsilon_i\}$ to $\Phi^\varepsilon(t)$, ν^ε its velocity of propagation and $\lambda > 0$ a constant representing the latent heat.

We can extend θ^ε by 0 on $Q \setminus \{\theta^\varepsilon > 0\}$ and on the fixed boundary we assume

$$(4.3) \qquad \theta^{\varepsilon} = \theta_D(x,t), \quad \text{for} \quad (x,t) \in \partial\Omega \times]0,T[\qquad (\theta_D \geq 0)$$

and at the initial instant

$$(4.4) \qquad \theta^{\varepsilon} = h(x), \quad \text{for} \quad x \in \Omega, \; t = 0, \qquad (h \geq 0)$$

where θ_D is assumed to be Hölder continuous and $h \in L^{\infty}(\Omega)$ with $J = \bar{\Omega} \setminus \mathrm{int}(\mathrm{supp}\, h)$, $J \neq \emptyset$, representing the initial frozen region. In the case $\eta = 0$, (4.4) is not necessary and the initial condition is given only by J.

Then, introducing the transformation

$$(4.5) \qquad u^{\varepsilon}(x,t) = \int_0^t \theta^{\varepsilon}(x,\tau)d\tau \;, \quad (x,t) \in \bar{Q}_T = \bar{\Omega} \times [0,T],$$

the one-phase Stefan problem reduces to the variational inequality (1.1)(1.2), where for all $\varepsilon \geq 0$,

$$(4.6) \qquad f^{\varepsilon} = f(x) = \eta h(x) - \lambda \chi_J(x), \quad x \in \Omega,$$

$$(4.7) \qquad g^{\varepsilon} = g(x,t) = \int_0^t \theta_D(x,\tau)d\tau, \; (x,t) \in \partial\Omega \times]0,T[\text{ and } g^{\varepsilon}(x,0)=0, \; x \in \bar{\Omega}.$$

In the quasi-steady case (η=0) L^{ε} is given by (1.3) and in the parabolic case ($\eta > 0$) L^{ε} is given by (1.4) with v_t replaced by ηv_t. In both cases we can apply the Theorems 2 and 4, considering in the elliptic case ($\eta = 0$) the solution $u^{\varepsilon}(t)$ at each fixed time $t > 0$ and the corresponding coincidence sets $I^{\varepsilon}(t)$.

The variational solutions u^{ε} are continuous functions and it is not difficult to verify that $\{u^{\varepsilon}(t) > 0\} = \{\theta^{\varepsilon}(t) > 0\}$, for $t > 0$, where $\theta^{\varepsilon} = u_t^{\varepsilon}$ denotes the generalized temperature. In general, we have

$$(4.8) \qquad 0 \leq u_t^{\varepsilon}(x,t) \leq C \quad \text{for} \quad (x,t) \in \Omega \times]0,T[\;, \quad \eta \geq 0,$$

so that the "ice region" I^{ε} is decreasing in time:

$$(4.9) \qquad I^{\varepsilon}(t) \equiv \{x \in \bar{\Omega} : u^{\varepsilon}(x,t) = 0\} \subset I^{\varepsilon}(t') \subset J, \text{ for } 0 < t' < t,$$

and it can be represented in the form

$$(4.10) \qquad I^{\varepsilon} = \bigcup_{o<t<r} I^{\varepsilon}(t) = \{(x,t) : t \leq S^{\varepsilon}(x), \; x \in \bar{\Omega}\}$$

where the bounded functions S^{ε} are defined by

$$(4.11) \qquad S^{\varepsilon}(x) = \inf\{t \in [0,T] : u^{\varepsilon}(x,t) > 0\} \quad x \in \bar{\Omega}.$$

Notice that $S^{\varepsilon}(x)=0$ if $x \in \overline{\Omega \setminus J}$, by (4.9). In the degenerate case

$\eta=0$ the application of the Theorem 2 and 4 must be localized in the compact subset J where (2.5), by the definition (4.6), is everywhere satisfied.

THEOREM 5 Under the preceding assumptions, in the homogenization of the one phase Stefan problem one has the convergence of the free boundaries in mean, namely

(4.12) $$S^{\varepsilon} \to S^{0} \quad \text{in} \quad L^{1}(\Omega),$$

provided (3.6) is satisfied (for instance, if (3.13) holds).

Proof: As we have remarked, all the assumptions of the Theorem 4 are naturally fulfilled (with the exception of (3.6) which was added), so that (3.12) holds in both cases $\eta > 0$ and $\eta = 0$. Now (4.12) follows immediately by remarking that from (4.10) one has

$$\mu_{n+1}(I^{\varepsilon} \div I^{0}) = \int_{\Omega}\int_{0}^{T} |\chi_{I^{\varepsilon}} - \chi_{I^{0}}| \, dt dx = \int_{\Omega} |S^{\varepsilon} - S^{0}| \, dx . \quad \Box$$

REMARK 6. This result was essentially established in [R1] for the special case of periodic homogenization and it gives an answer to a problem posed in [L1]. More references to the one-phase Stefan problem can be found in the survey article [R3]. □

REMARK 7. In the case of one-dimensional space variable, if $\Omega =]\alpha,\beta[$, under appropriate assumptions (see [KS] and [R1]) one can show that the free boundary Φ^{ε} is a Lipschitz curve in the coordinates obtained from (x,t) by rotating through $\pi/4$. Then the sets I^{ε} satisfy an exterior cone property uniformly in ε, which together with $I^{\varepsilon} \to I^{0}$ (Hausdorff or in measure) implies the uniform convergence of the free boundary. □

REMARK 8. In higher dimensions, under certain conditions and assuming a_{ij}^{ε} independent of one direction, for instance x_1, it can be shown than the free boundary $\Phi^{\varepsilon}(t) = \partial I^{\varepsilon}(t) \cap \Omega$ admits a local representation

$$\Phi^{\varepsilon}(t) : x_1 = \phi^{\varepsilon}(x',t), \quad t > 0, \quad x' = (x_2,\ldots,x_n), \quad (\varepsilon \geq 0).$$

In this case, as in Theorem 5 one also obtains the convergence of $\phi^{\varepsilon}(t)$ to $\phi^{0}(t)$ in L^1. This holds easily for the elliptic case

$\eta=0$ and it can be extended to the parabolic one $\eta>0$ by refining the conclusion (3.12) in the form $\mu_n(I^\varepsilon(t) \div I^o(t))\to 0$, for fixed $t>0$ (see [R1]). A similar application was done in [CR2] for the variational inequality approach to the dam problem with horizontal layers. □

5. Related Open Problems

Without being complete let us mention the following

PROBLEM 1. In [L2], J.L.Lions reported some numerical experiments of Bourgat (Laboria/INRIA, France) on the approximation of the solution and the free boundary of an elliptic obstacle problem in composite media by using homogenization techniques, which suggested a good convergence, and he has raised the problem of not only the convergence of the free boundaries $\Phi^\varepsilon\to\Phi^o$, but also if, in some appropriate sense, it could be estimated in terms of the parameter ε. This problem seems to be still open. □

PROBLEM 2. Analogously, the same question has been formulated for the boundary obstacle problem, or Signorini problem, which consists of

$$L^\varepsilon u^\varepsilon = f^\varepsilon \quad \text{in} \quad \Omega \tag{5.1}$$

$$u^\varepsilon \geq 0, \quad \partial u^\varepsilon/\partial\nu^\varepsilon \geq g^\varepsilon, \quad u^\varepsilon(\partial u^\varepsilon/\partial\nu^\varepsilon - g^\varepsilon) = 0 \quad \text{on } \partial\Omega \tag{5.2}$$

where $\partial/\partial\nu^\varepsilon = a^\varepsilon_{ij} n_i \partial/\partial x_j$ denotes the conormal derivative. In the elliptic case, as well as in the corresponding parabolic one, there remains open the question of the convergence of the boundary coincidence sets

$$J_\varepsilon \to J_o \; , \quad \text{where} \quad J_\varepsilon = \{x \in \partial\Omega : u^\varepsilon(x) = 0\}. \quad □$$

PROBLEM 3. The similar problem to (1.1)(1.2) for higher order elliptic (or parabolic) operators can also be posed. For instance, if

$$L^\varepsilon v = (a^\varepsilon_{ijk\ell} v_{x_k x_\ell})_{x_i x_j} \; , \quad (i,j,k,\ell = 1,2) \tag{5.3}$$

the unilateral problem (1.1) with appropriate boundary conditions, corresponds to a nonhomogeneous plate constrained against a plane. The convergence of the contact region $\{u^\varepsilon=0\}$ with respect to homogenization is an open problem. □

PROBLEM 4. In the already mentioned steady dam problem

with a rectangular geometry and a two dimensional permeability in the special isotropic case $a^\varepsilon_{ij}=\alpha^\varepsilon\delta_{ij}$ with $\alpha^\varepsilon(x_1,x_2) = \alpha^\varepsilon_1(x_1)\alpha^\varepsilon_2(x_2)$, the problem for the pressure $p^\varepsilon=p^\varepsilon(x_1,x_2)$ is transformed into a variational inequality of the form (1.1)(1.2) for the Baiocchi's transformation $u^\varepsilon(x_1,x_2)=\int_y^b \alpha^\varepsilon_2(\eta)p^\varepsilon(x_1,\eta)d\eta$. The corresponding homogenization can be found in [CR2] and it corresponds to the physical situation whenever $\{p^\varepsilon>0\}=\{u^\varepsilon>0\}$. However in the case of horizontal layers it may happen that $\{p^\varepsilon>0\}\subsetneq\{u^\varepsilon>0\}$ and we must use a direct and more general formulation due to Alt (see [A] or [R2]). In general, if $\partial\Omega=\Gamma_1\cup\Gamma_2\cup\Gamma_3$, the hydrostatic pressure is given by $h>0$ on Γ_3 and $h=0$ on Γ_2, $e=(0,1)$ denotes the vertical direction and a^ε the matrix $\{a^\varepsilon_{ij}\}$, the problem consists of:

"Find a pair $(p^\varepsilon,\gamma^\varepsilon)\in H^1(\Omega)\times L^\infty(\Omega)$, such that,

(5.4) $\quad p^\varepsilon\geq 0$ in Ω, $p^\varepsilon=h$ on $\Gamma_2\cup\Gamma_3$,

(5.5) $\quad 0\leq\chi_{\{p^\varepsilon>0\}}\leq\gamma^\varepsilon\leq 1$ a.e. in Ω,

(5.6) $\quad \int_\Omega a^\varepsilon(\nabla p^\varepsilon+\gamma^\varepsilon e).\nabla\zeta\geq 0,\quad \forall\zeta\in H^1(\Omega):\ \zeta|_{\Gamma_2}\leq 0,\ \zeta|_{\Gamma_3}=0$".

Since in general we may have $\gamma^\varepsilon\neq\chi_{\{p^\varepsilon>0\}}$, $(\varepsilon\geq 0)$ the general form for the homogenized problem (5.4)(5.5)(5.6) must be relaxed and there exists a counterexample for the non-convergence of the functions γ^ε to γ^o (see [R2]).

The homogenization of the dam problem leads to other open problems: (i) the requirement of a new and more general modelling; (ii) the characterization of the situations where the above formulation is closed for the H-convergence.

Partial answers to the last question have been given in [R2], where in the special case $a^\varepsilon=\{a^\varepsilon_{ij}(x_2)\}$ (including horizontal layers) one has $p^\varepsilon\to p^o$ in $H^1(\Omega)$-weak and $C^{o,\beta}(\bar\Omega)$-strong and $\gamma^\varepsilon\rightharpoonup\gamma^o$ in $L^\infty(\Omega)$-weak-*. Under the additional regularity assumption $\gamma^o=\chi_{\{p^o>0\}}$ (for instance, if a^o_{ij} are constants) then $\chi_{\{p^o>0\}}=\lim_{\varepsilon\to o}\chi_{\{p^\varepsilon>0\}}=\lim_{\varepsilon\to o}\gamma^\varepsilon$ strongly in $L^2(\Omega)$. □

PROBLEM 5. The homogenization of the two-phase Stefan problem have been considered by Damlamian [Da] in the case of a ε-periodic structure composed of two different substances. The temperature $\theta^\varepsilon = \theta^\varepsilon(x,t)$ $(x,t) \in \bar{Q}$, $\varepsilon \geq 0$ can be seen as the unique weak solution of the following singular parabolic equation

(5.7) $$[\beta(\theta^\varepsilon)]_t - (a^\varepsilon_{ij}\theta^\varepsilon_{x_j})_{x_i} \ni f \quad \text{in } Q,$$

with a nonhomogeneous Dirichlet data $\theta^\varepsilon = g$ on the parabolic boundary $\Sigma = \partial Q \setminus (\Omega \times \{T\})$. The enthalpy $\beta(\theta^\varepsilon)$ is given in terms of a maximal monotone graph of the form ($\lambda > 0$ is the latent heat)

$$\beta(\tau) = \beta_-\tau, \text{ if } \tau<0, \ \beta(0) = [0,\lambda] \quad \text{and} \quad \beta(\tau) = \beta_+\tau + \lambda \text{ if } \tau > 0$$

where $\lambda > 0$ is the latent heat and $\beta_\pm$ are positive constants.

If $\{a^\varepsilon_{ij}\} \underset{H}{\to} \{a^0_{ij}\}$ one has $\theta^\varepsilon \to \theta^0$ weakly in appropriate Sobolev spaces [Da], and, by an equicontinuity property (depending on f,g and $\partial\Omega$, see, for instance [Db]) also uniformly on $\bar{Q}$. Hence, by the argument of the Remark 4, one has a lower semicontinuity property for the convergence of the liquid region $\Lambda^\varepsilon_+ = \{\theta^\varepsilon > 0\}$ and of the solid one $\Lambda^\varepsilon_- = \{\theta^\varepsilon < 0\}$, $\varepsilon \geq 0$:

(5.8) $$\liminf_{\varepsilon \to 0} \Lambda^\varepsilon_\pm \supset \Lambda^0_\pm \quad .$$

However, no other results seem to be known for this more delicate two-phase problem, in particular, what can be said about the structure and the stability of the free boundary $\Phi^\varepsilon = \partial\Lambda^\varepsilon_+ \cap \partial\Lambda^\varepsilon_- = \{\theta^\varepsilon = 0\}$. The general situation is still more complicated since the enthalpy is in fact of the form $\beta^\varepsilon = \beta^\varepsilon(x,\tau)$ and the limit medium undergoes changes of phase at all the change of phase temperatures of the initial media (see [Da]). □

REFERENCES

[A] H.W.ALT - "The dam problem", in Research Notes in Math (Pitman, Boston) 78 (1982), 52-68.

[BL] A.BENSOUSSAN & J.L.LIONS - "Contrôle impulsionnel et inéquations quasi-variationnelles", Dunod, Paris, 1982.

[BLP] A.BENSOUSSAN, J.L.LIONS & G.PAPANICOLAOU - "Asymptotic Analysis for Periodic Structures", North-Holland, Amsterdam,1978.

[CR1] M.CODEGONE & J.F.RODRIGUES - "Convergence of the coincidence set in homogenization of the obstacle problem",Ann. Fac.Sci.Toulouse 3(1981), 275-285.

[CR2] M.CODEGONE & J.F.RODRIGUES - "On the homogenization of the rectangular dam problem" Rend.Sem.Mat.Univ.Pol.Torino, 39 nº2 (1981), 125-136.

[Da] A.DAMLAMIAN - "Homogénéisation du problèm de Stefan", C.R.Acad.Sc.Paris, 289-A (1979), 9-11.

[Db] E.DIBENEDETTO - "A boundary modulus of continuity for a class of singular parabolic equations", J.Differential Equations, 63 (1986), 418-447.

[KS] D.KINDERLEHRER & G.STAMPACCHIA - "An introduction to variational inequalities and their applications", Academic Press, New York, 1980.

[L1] J.L.LIONS - "Asymptotic behaviour of solutions of variational inequalities with highly oscillating coefficients", Lect. Notes Maths (Springer-Verlag, Berlin), 503 (1976), 30-55.

[L2] J.L.LIONS - "Sur l'approximation de problèmes à frontière libre dans les matériaux inhomogènes", Lect.Notes. Maths. (Springer-Verlag, Berlin), 606 (1977), 194-203.

[M1] F.MURAT - "Sur l'homogéneisation d'inéquations elliptiques du 2ème ordre relative an convexe $K(\psi_1,\psi_2)$". Thèse d'État, Univ.Paris VI, 1976.

[M2] F.MURAT - Private communication, Paris (1980).

[R1] J.F.RODRIGUES - "Free boundary convergence in the homogenization of the one phase Stefan problem", Trans.Amer. Math.Soc.274(1982), 297-305.

[R2] J.F.RODRIGUES - "Some remarks on the homogenization of the dam problem", manuscripta math. 46(1984), 65-82.

[R3] J.F.RODRIGUES - "The variational inequality approach to the one phase Stefan problem" Acta Appl.Math. (1986).

[T] L.TARTAR - "Cours Peccot, Collège de France, Paris, 1977 (see also these Proceedings).

THE POINT INTERACTION APPROXIMATION, VISCOUS FLOW THROUGH POROUS MEDIA, AND RELATED TOPICS

J. Rubinstein
Department of Mathematics
Stanford University
Stanford, CA 94305

I. INTRODUCTION TO THE P.I.A.

The Point Interaction Approximation is a technique to analyse boundary value problems in domains with many tiny obstacles. It was introduced by Foldy [4] for the problem of multiple scattering of waves. The general framework is as follows: We take a smooth domain D in R^3 and cut out of it N holes B_1, B_2, B_N. In the domain of interest

$$D^N = D - \bigcup_1^N B_i \tag{1.1}$$

we consider the problem

$$\begin{aligned} L[u^N] &= f \quad \text{in} \quad D^N \\ M[u^N] &= 0 \quad \text{on} \quad \partial B_i \\ M^0[u^N] &= 0 \quad \text{on} \quad \partial D \end{aligned} \tag{1.2}$$

If G^N is the Green's function of (1.2), we can write

$$u^N = G^N[f].$$

The essence of the P.I.A. is to approximate

$$G^N(x,y) \approx G(x,y) + \sum_1^N q_i^y \, G(x,w_i^N) \equiv H^N(x,y) \tag{1.3}$$

where G is Green's function of D (with respect to L), q_i^y are appropriately chosen "charges", and w_i^N are the centers of the holes. In fact Foldy assumed (1.3) as a good representation of the medium, and proceeded to look for smooth (macroscopic, effective) approximation for it. There are 3 questions one has to answer while applying (1.3):

(i) Evaluate q_i^y from the microstructure. (1.4)

(ii) Prove $\|G^N - H^N\| < < 1$

(iii) Show that there is a smooth operator J s.t.

$$\|J-H^N\| << 1.$$

The convergence estimates in (ii) and in particular in (iii) are relevant when $N >> 1$. Note that we have as yet left the norms unspecified. Rigorous results in the sprit of (1.4) were recently obtained by Figari, Orlandi and Teta [2] for the Laplace equation, using estimates derived by Ozawa [6]. Although we formulated the P.I.A. for linear operators, it can be used successfully for nonlinear problems as well. Caflisch et. al [1] applied it (formally) to study wave propagation in bubbly liquids, and the problem of diffusion in regions with many melting holes was recently studied by Figari, Papanicolaou and Rubinstein [3]. In the next section we apply the P.I.A. to Stokes equations and derive rigorously the equations associated with Brinkman and Darcy. Detailed proofs are given in [7] and [8]. Some related problems are discussed at section III.

II. APPLICATION TO FLOW IN POROUS MEDIA

Using the notation of the preceding section, we let L be the stokes operator in D^N, $\underset{\sim}{u}^N$ the local velocity field and $M[f] = f$, i.e.

$$\mu\Delta \underset{\sim}{u}^N - \lambda \underset{\sim}{u}^N = \nabla P + \underset{\sim}{f} \qquad (2.1)$$

$$\underset{\sim}{\nabla} \cdot \underset{\sim}{u}^N = 0$$

$$\underset{\sim}{u}^N = 0 \text{ on } \partial B_j \text{ and on } \partial D.$$

Here μ is the viscosity and λ is a nonnegative parameter. Each hole has radius $R = \frac{\alpha_0}{N}$. (In the context of homogenization theory, this corresponds to $R \sim \varepsilon^3$ where ε is the size of the cell). $\underset{\approx}{G}(\underset{\sim}{x},\underset{\sim}{y})$ is as before the Green's function (in fact it is called here Oseen's tensor) in D, and we let $\underset{\approx}{\psi}$ be the $3N\times 3N$ matrix which is composed of $N\times N$ blocks of $\underset{\approx}{G}$, s.t. the (i,j) block is $\underset{\approx}{G}(\underset{\sim}{w}_i, \underset{\sim}{w}_j)$ for $i\neq j$ and zero otherwise. We also denote by $\underset{\approx}{\phi}$ the matrix

$$\phi(x,\cdot) = \begin{matrix} \underset{\approx}{G}(\underset{\sim}{x},\underset{\sim}{w}_1) \\ \cdot \\ \cdot \\ \cdot \\ \cdot \\ \underset{\approx}{G}(\underset{\sim}{x},\underset{\sim}{w}_N) \end{matrix} \quad .$$

We write (1.3) in matrix form:

$$\underset{\approx}{H}^N(\underset{\sim}{x},\underset{\sim}{y}) = \underset{\approx}{G}(\underset{\sim}{x},\underset{\sim}{y}) - \frac{\alpha}{N} \underset{\approx}{\phi}^T(\cdot,\underset{\sim}{y})(\frac{\alpha}{N}\underset{\approx}{\psi} + \underset{\approx}{I})^{-1}\underset{\approx}{\phi}(\underset{\sim}{x},\cdot), \qquad (2.2)$$

where $\alpha = 6\pi\mu\alpha_0$.

Geometrical assumptions:

1) $\{\underset{\sim}{w}_i^N\}$ are bounded away from ∂D (uniformly in N).

2) $\min_{i\neq j} |\underset{\sim}{w}_i^N - \underset{\sim}{w}_j^N| > CN^{-1+\nu}$, $\quad \nu < 1/3$

3) $N^{-2} \sum_{\substack{i,j \\ i\neq j}} |\underset{\sim}{w}_i^N - \underset{\sim}{w}_j^N|^{-3+\xi} < c < \infty$, $\quad \xi > 0$.

(Note that assumption (2) guarantees that there is no overlapping).

Theorem 1 [7]

Let $\underset{\sim}{f} \in C_0(D)$, $\{\underset{\sim}{W}_i^N\}$ satisfy (1) - (3), and $\lambda > 0$. Then for every $e > 0$, there is N_0 s.t. for every $N > N_0$

$$\|(\underset{\approx}{H}^N - \underset{\approx}{G}^N)(\underset{\sim}{f})\|_2 < c\, N^{-\frac{1}{6}}.$$

Now, under the conditions of Theorem 1 we can show that $\|\frac{\alpha}{N} \underset{\approx}{\psi}\| < 1$ (matrix norm), and hence we can expand $(\frac{\alpha}{N} \underset{\approx}{\psi} + I)^{-1}$ in (2.2). We let $\{\underset{\sim}{w}_i^N\}$ be i.i.d. random variables with probability density function $\rho(\underset{\sim}{x})$. Then we can analyse the statistics of (2.2) and prove

<u>Theorem 2 [7]</u>

Let $\rho(\underset{\sim}{x}) \in C(D)$, f and λ as in Theorem 1. Then

$$\lim_{N\to\infty} P_\Omega^N \{N^\sigma \|\underset{\sim}{v} - \underset{\sim}{u}^N\|_2 < \epsilon\} = 1$$

$\forall\ \epsilon > 0$, $\sigma < \frac{1}{6}$. P_Ω^N is the probability measure induced by $\rho(\underset{\sim}{x})$ on the space of configurations and $\underset{\sim}{v}$ solves

$$\mu\Delta\underset{\sim}{v} - \lambda\underset{\sim}{v} - 6\pi\mu\alpha_0\, \rho(\underset{\sim}{x})\underset{\sim}{v} = \nabla p + \underset{\sim}{f} \qquad (2.3)$$

$$\nabla \cdot \underset{\sim}{v} = 0 \quad \text{in} \quad D$$

$$\underset{\sim}{v} = 0 \quad \text{on} \quad \partial D,$$

where (2.3) is Brinkman's equation. It holds in the capacity scaling, i.e., for very dilute systems - the porosity is $\sim 1 - O(N^{-2})$. It is well known that flow in dense materials is macroscopically described by Darcy's equation. While the P.I.A. surely breaks down at high concentrations, it still holds slightly "above" the capacity scaling:

Let $R = \alpha_0\, N^{-1+\gamma}$, $\gamma \in (0,1/60)$, and set $\lambda = N^\gamma\Gamma$. We denote by $\underset{\sim}{U}(\underset{\sim}{x}, \Gamma)$ the solution of Darcy's equation

$$\Gamma\, \underset{\sim}{U} + 6\pi\mu\alpha_0\rho\, \underset{\sim}{U} = -\nabla P - \underset{\sim}{f}, \qquad (2.4)$$

$$\nabla \cdot \underset{\sim}{U} = 0 \qquad \text{in} \quad D$$

$$\underset{\sim}{U} \cdot \hat{n} = 0 \quad \text{on} \quad \partial D,$$

and expand

$$\underset{\sim}{u}^N = N^{-\gamma}\, \underset{\sim}{u}^{N,0} + N^{-2\gamma}\, \underset{\sim}{u}^{N,1} + \ldots .$$

Then [8]

$$\lim_{N\to\infty} P^N_\Omega \{\|\underset{\sim}{u}^{N,0} - \underset{\sim}{U}\|_Q < e\} = 1 \tag{2.5}$$

for every $e > 0$, where

$$Q = \{\underset{\sim}{v} \in L_2(D);\ \underset{\sim}{\nabla} \cdot \underset{\sim}{v} = 0,\ \underset{\sim}{v} \cdot \hat{n}|_{\partial D} = 0\} . \tag{2.6}$$

(See also the related work of Sanchez-Palencia [10])

III. LOW ORDER CONNECTIONS

There is a large body of work on this difficult problem, and all the results here so far are formal. Hoping to get better understanding of the origin for such corrections and rigorous estimates, we suggest to consider a simpler problem: we start at the level of the P.I.A. and continue to the averaged equation. To first order we needed to know only the distribution density of $\{\underset{\sim}{w}_j\}$. It is natural to conjecture that the 2 point correlation function will be needed to evaluate further terms. We proceed formally:

Let $\{\underset{\sim}{w}_j\}$ be distributed uniformly in D

$$P_r(\underset{\sim}{w}_j \in d\underset{\sim}{x}) = \frac{1}{V} d\underset{\sim}{x} \qquad V = \text{vol } (D) \tag{3.1}$$

$$P_r\,(\underset{\sim}{w}_j \in \underset{\sim}{r}_j + d\underset{\sim}{x} \,|\, \underset{\sim}{w}_i = \underset{\sim}{r}_i) = \frac{1}{V}[1 - F(\underset{\sim}{r}_j - \underset{\sim}{r}_i)]d\underset{\sim}{x}$$

We assume $F(\underset{\sim}{x}) = F(|\underset{\sim}{x}|)$, and that $F(\underset{\sim}{x})$ decays over a length scale ℓ (the correlation length). $\langle\cdot\rangle$ will denote averaging with respect to the configurations C and $\langle\cdot\rangle_x$ means averaging given a sphere centre at $\underset{\sim}{x}$. (i.e. configurations C^x). Let G be the Green function of D (w.r. to Δ), and consider in D^N the microscopic problem:

$$\Delta u = -f \quad \text{in} \quad D^N$$

$$u^N = 0 \text{ , for } |\underset{\sim}{x} - \underset{\sim}{w}_j| = \delta \text{ , } \quad \underset{\sim}{x} \in \partial D \text{ . } \quad \delta << 1.$$

Then

$$u^N = \int_{D^N} G(\underset{\sim}{x}-\underset{\sim}{x}')f(\underset{\sim}{x}')d\underset{\sim}{x}' - \sum_j \int_{|\underset{\sim}{x}'-\underset{\sim}{w}_j|=\delta} G(\underset{\sim}{x}-\underset{\sim}{x}') \frac{\partial u^N}{\partial n}(\underset{\sim}{x}')dS' \text{ .}$$

Set

$$\gamma_j^N(\underset{\sim}{w}_1,, \underset{\sim}{w}_N) = \int_{|\underset{\sim}{x}-\underset{\sim}{w}_j| = \delta} \frac{\partial u^N}{\partial n}(\underset{\sim}{x}) \, ds.$$

The P.I.A. is written as

$$\Delta w^N = -f + \sum_j \gamma_j^N \, \delta(\underset{\sim}{x}-\underset{\sim}{w}_j) \quad \text{in} \quad D \tag{3.2}$$

where the γ_j^N are determined by

$$\int_{|\underset{\sim}{x}-\underset{\sim}{w}^j|=\delta} w(\underset{\sim}{x})dS = 0. \tag{3.3}$$

Averaging (3.2) gives

$$\Delta\langle w\rangle = -f + \langle \sum_j \gamma_j^N(C)\ \delta(\underset{\sim}{x}-\underset{\sim}{w}_j)\rangle = \tag{3.4}$$

$$= -f + \sum_j \underbrace{\int \cdots \int}_{N-1} \int_j \gamma_j^N(C)\ \delta(\underset{\sim}{x}-\underset{\sim}{w}_j)dP(\underset{\sim}{w}_j)dP(C^{w_j})$$

$$= \frac{N-1}{V} \underbrace{\int \cdots \int}_{N-1} \gamma_x(C^x)dP(C^x) \simeq n\langle\gamma_x\rangle_x$$

where we use $N \gg 1$, and $n = \frac{N}{V}$ is the number density of the obstacles.

Let $w^N = \langle w\rangle + \tilde{w}$, then

$$\Delta\tilde{w} = \sum_j \gamma_j^N(C)\ \delta(\underset{\sim}{x}-\underset{\sim}{w}_j) - n\langle\gamma_x\rangle_x,$$

so

$$\langle\tilde{w}(x)\rangle_y = \int G(\underset{\sim}{x}-\underset{\sim}{x}')[\langle\sum_j \gamma_j^N(C)\delta(\underset{\sim}{x}'-\underset{\sim}{w}^j)\rangle_y - n\langle\gamma_{x'}\rangle_{x'}]d\underset{\sim}{x}' =$$

$$= G(\underset{\sim}{x}-\underset{\sim}{y})\langle\gamma_y\rangle_y + \int G(\underset{\sim}{x}-\underset{\sim}{x}')\ \{n[1-F(\underset{\sim}{x}'-\underset{\sim}{y})]\langle\gamma_{x'}\rangle_{x',y} - n\langle\gamma_{x'}\rangle_{x'}\}\ d\underset{\sim}{x}'$$

We apply now the Quasi Crystallian Approximation

$$\langle\gamma_{x'}\rangle_{x',y} \simeq \langle\gamma_{x'}\rangle_{x'}, \tag{3.5}$$

and condition (3.3) to get

$$\overline{w}(\underset{\sim}{y}) = \langle\gamma_y\rangle_y\ G(\delta) + n\int G(\underset{\sim}{y}-\underset{\sim}{x}')F(\underset{\sim}{y}-\underset{\sim}{x}')\ \langle\gamma_{x'}\rangle_{x'}d\underset{\sim}{x}'\ .$$

Since we assumed that F decays fast we get

$$\langle\gamma_y\rangle_y \sim [G(\delta)]^{-1}\ \overline{w}(\underset{\sim}{y})/[1 - \frac{n}{G(\delta)}\int G(\underset{\sim}{y}-\underset{\sim}{x}')F(\underset{\sim}{y}-\underset{\sim}{x}')d\underset{\sim}{x}']$$

and thus we close (3.4)

$$\Delta\overline{w} = \theta\overline{w}\ \frac{1}{1-b\theta\ell^2} \tag{3.6}$$

where we scale as usual $\theta = \frac{n}{G(\delta)}$, and

$$b = \ell^{-2}\int G(\underset{\sim}{x})F(\underset{\sim}{x})d\underset{\sim}{x} \qquad (b = O(1),\ \theta = O(1)).$$

This is the "random analog" of the Hasimoto result for periodic structures. (Hasimoto [5], Saffman [9]). If we set

$$\ell = O(\delta^q),$$

then the correction is $O(\beta^q)$ where β is the volume fraction of the obstacles.

We stated precisely the approximations we used. The most crucial one (recall that we <u>start</u> at the P.I.A. level) is (3.5). Since this is exact for periodic structures it seems natural to look first for a rigorous justification of (3.6) in that case.

Acknowledgement

This research was supported in part by the Institute for Mathematics and its Applications with funds provided by the NSF and the ARO.

IV. REFERENCES

[1] R.E. Caflisch, M.J. Mikisis, G.C. Papanicolaou and L. Ting, J. Fluid Mech 153 (1985) 259-273.

[2] R. Figari, E. Orlandi and S. Teta, J. Stat. Phys. 41 (1985), 465.

[3] R. Figari, G. Papanicolaou and J. Rubinstein (In preparation)

[4] L.L. Foldy, Phys. Rev. 67 (1945), 107.

[5] H. Hasimoto, J. Fluid Mech., 5 (1959), 317.

[6] S. Ozawa, Comm. Math. Phys., 91 (1983), 473.

[7] J. Rubinstein, J. Stat. Phys., 1986, (to appear).

[8] J. Rubinstein, Proceedings of the IMA Workshop on Infinite particle systems and hydrodynamics limits, Springer-Verlag (to appear).

[9] P.G. Saffman, Stud. Appl. Math., 52 (1973), 115.

[10] E. Sanchez-Palencia, Int. J. Eng. Sci. 20 (1982), 1291.

THE VANISHING VISCOSITY-CAPILLARITY APPROACH TO THE RIEMANN PROBLEM FOR A VAN DER WAALS FLUID

M. Slemrod*
Department of Mathematical Sciences
Rensselaer Polytechnic Institute
Troy, New York 12180-3590

0. INTRODUCTION

In this note we sketch a proof for the solvability of the Riemann problem for the isothermal motion of a van der Waals fluid. The problem is of physical interest in that it simulates the behavior in a shock tube of a compressible fluid which may exhibit two different phases, say liquid and vapor. The main tools of the analysis are (i) Dafermos's [1] formulation of the vanishing viscosity method for the Riemann problem and (ii) resolution of the viscous system via bifurcation ideas suggested in the paper of Hale [2]. Rigorous proofs will appear elsewhere.

1. PROBLEM FORMATION

We wish to solve the Riemann initial value problem for the isothermal motion of a van der Waals fluid. We denote by u the velocity, w the specific volume, p the pressure, x the Lagrangian mass variable, and t the time. The equations of one dimensional inviscid motion are then

$$\begin{aligned} &u_t + p(w)_x = 0 \ , \\ &w_t - u_x = 0 \ . \end{aligned} \tag{1.1}$$

In addition we prescribe Riemann initial data

$$u(x,0) = \begin{matrix} u_R^- \\ u_R^+ \end{matrix} \ , \quad w(x,0) = \begin{matrix} w_R^- \\ w_R^+ \end{matrix} \ , \quad \begin{matrix} x < 0 \\ x > 0 \end{matrix} \ .$$

Here u_R^-, u_R^+, w_R^-, w_R^+ are constants.

---

*This research was sponsored in part by the Air Force Office of Scientific Research, USAF, Contract/Grant No. AFOSR-85-6239. The U. S. Government's right to retain a nonexclusive royalty free license in and to copyright this paper for government purposes is acknowledged.

The graph of p is given in Figure 1*. It resembles a typical van der Waals isotherm when the temperature is below critical. In this case $p' > 0$ on (α,β) when (1.1) is <u>elliptic</u> and $p' < 0$ on (b,α), (β,∞) when (1.1) is <u>hyperbolic</u>. We denote the two values A,B as the two values of specific volume that yield the Maxwell equal area rule

$$\int_A^B (p(s) - p(A))ds = 0 . \tag{1.3}$$

We then choose our Riemann data as

$$\begin{aligned} u_R^- &= \mu u^- , \qquad w_R^- = A + \mu w^- , \\ u_R^+ &= \mu u^+ , \qquad w_R^+ = B + \mu w^+ , \end{aligned} \tag{1.4}$$

where u^-, u^+, w^-, w^+ are constants and $\mu > 0$ is a small parameter.

We now sketch a proof which shows that this two phase Riemann problem has a solution for small $\mu > 0$ which is obtained via viscosity-capillarity limits.

In order to resolve (1.1), (1.4) we imbed (1.1) in the "viscous" system ($\varepsilon > 0$)

$$\begin{aligned} u_t + p(w)_x &= \varepsilon\mu\, t u_{xx} , \\ w_t - u_x &= \varepsilon\mu\, t w_{xx} . \end{aligned} \tag{1.5}$$

The terms on the right hand side simultaneously attempt to capture the effects of viscosity and capillarity (see [3]) and preserve the structure of solutions of the Riemann problem in terms of the independent variable $\xi = \frac{x}{t}$.

The program is to show that for every fixed $\varepsilon > 0$ (1.5) admits a solution $u^\epsilon(\xi)$, $w^\epsilon(\xi)$ satisfying the ordinary differential equations

$$\begin{aligned} -\xi u' + p(w)' &= \varepsilon\mu\, u'', \\ -\xi w' - u' &= \varepsilon\mu\, w'', \end{aligned} \tag{1.6}$$

and boundary conditions

$$u(-\infty) = u_R^-,\ w(-\infty) = w_R^-, \quad u(+\infty) = u_R^+,\ w(+\infty) = w_R^+ . \tag{1.7}$$

If, in addition, we can show $u^\epsilon(\xi)$, $w^\epsilon(\xi)$ have total variation bounded independently of $\epsilon > 0$ it will follow from Helly's theorem that $u^\epsilon(\xi)$,

---

*See page 335

w (ξ) possess a subsequence which converges boundedly a.e. as $\epsilon \to 0^+$ to functions $u(\xi)$, $w(\xi)$ which are solutions to the Riemann problem, (1.1), (1.4).

2. PRELIMINARY ANALYSIS

It is convenient to introduce the new independent variable

$$\theta = \frac{\xi}{\mu} - \alpha$$

where α is constant yet to be determined. In this case (1.6) becomes

$$\begin{aligned} -\mu(\theta+\alpha)u' + p(w)' &= \varepsilon u'', \\ -\mu(\theta+\alpha)w' - u' &= \varepsilon w'', \end{aligned} \tag{2.1}$$

and ' denotes differentiation with respect to θ. As $\mu > 0$ the boundary conditions in ξ go into identical boundary conditions in θ

$$u(-\infty) = u_R^-, \quad u(-\infty) = w_R^-, \quad u(+\infty) = u_R^+, \quad w(+\infty) = w_R^+ . \tag{1.2}$$

When $\mu = 0$ (2.1) becomes

$$\begin{aligned} p(w)' &= \varepsilon u'', \\ -u' &= \varepsilon w'', \end{aligned} \tag{2.3}$$

which may exhibit both homoclinic and heteroclinic orbits. In particular (2.3) exhibits the classical van der Waals solution

$$u_m^{(0)}(\theta), \quad w_m^{(0)}(\theta)$$

which is the heteroclinic connection between the equilibrium points

$$u = 0, \quad w = A \quad \text{and} \quad u = 0, \quad w = B$$

of (2.3) i.e., $u_m^{(0)}(\theta)$, $w_m^{(0)}(\theta)$ satisfies (2.3) and boundary conditions

$$u_m^{(0)}(-\infty) = 0, \; w_m^{(0)}(-\infty) = A, \; u_m^{(0)}(+\infty) = 0, \; w_m^{(0)}(+\infty) = B. \tag{2.4}$$

On the other hand (2.3) exhibits other non-oscillating solutions, namely homoclinic orbits. We denote the homoclinic orbit which connects the point

$$u = 0, \quad w = A + \mu v$$

to itself by $u_m(\xi;\mu)$, $w_m(\xi;\mu)$. Again, this means that $u_m(\xi;\mu)$, $w_m(\xi;\mu)$ satisfies (2.3) and boundary conditions

$$u_m(-\infty;\mu) = 0 \ , \qquad w_m(-\infty,\mu) = A + \mu v \ . \tag{2.5}$$

Here v is as yet undetermined scalar. (Since v = 0 is a possibility, we don't specify conditions at $\theta = +\infty$. Of course if v = 0, $u_m(+\infty;\mu) = 0$, $w_m(+\infty;\mu) = B$ while if $v \neq 0$, $u_m(+\infty;\mu) = 0$, $w_m(+\infty;\mu) = A + \mu v$.)

3. <u>THE FINITE DOMAIN PROBLEM</u>

We consider (2.1) on the finite domain (-L,L) with boundary conditions

$$\begin{aligned} u(L) &= u_m^{(0)}(L) + \mu u_+ \ , \\ w(L) &= w_m^{(0)}(L) + \mu w_+ \ , \\ u(-L) &= u_m^{(0)}(-L) + \mu u_- \ , \\ w(-L) &= u_m^{(0)}(-L) + \mu w_- \ . \end{aligned} \tag{3.1}$$

We shall show that (2.1), (3.1) can be solved for $u(\theta;L)$, $w(\theta,L)$ which are bounded independently of L. In this case the functions $u(\theta;L)$, $w(\theta;L)$ can be extended to the infinite domain $(-\infty,\infty)$ by setting

$$\begin{aligned} &u(\theta;L) = u_m^{(0)}(L) + \mu u_+ \\ &w(\theta;L) = w_m^{(0)}(L) + \mu w_+ \end{aligned} \quad \text{for } \xi > L \ ;$$

$$\begin{aligned} &u(\theta;L) = u_m^{(0)}(-L) + \mu u_- \\ &w(\theta;L) = w_m^{(0)}(-L) + \mu w_- \end{aligned} \quad \text{for } \xi < L \ . \tag{3.2}$$

The argument of Dafermos given after Theorem 3.1 of [1] then yields a sequence $\{L_n\}$, $L_n \to \infty$ so that $u(\theta;L_n) \to \tilde{u}(\theta)$, $w(\theta;L_n) \to \tilde{w}(\theta)$ uniformly on $(-\infty,\infty)$. The limit functions $\tilde{u}(\theta)$, $\tilde{w}(\theta)$ are then seen to satisfy (2.1) and the boundary conditions (2.2).

To resolve the finite domain problem we write

$$\begin{aligned} u(\theta) &= u_m(\theta;\mu) + U(\theta;\mu) \ , \\ w(\theta) &= w_m(\theta;\mu) + W(\theta;\mu) \ , \end{aligned} \tag{3.3}$$

where we recall u_m, w_m is a solution of the $\mu = 0$ problem (2.3).

Substitution of (3.3) into (2.1) yields

$$\begin{aligned} -\mu(\theta + \alpha)(u_m' + U') + (p(w_m + W) - p(w_m))' &= \varepsilon U'' , \\ -\mu(\theta + \alpha)(w_m' + W') - U' &= \varepsilon W'' , \end{aligned} \tag{3.4}$$

and the boundary conditions (3.1) become

$$\begin{aligned} u_m(L;\mu) + U(L;\mu) &= u_m^{(0)}(L) + \mu u_+ , \\ w_m(L;\mu) + W(L;\mu) &= w_m^{(0)}(L) + \mu w_+ , \\ u_m(-L;\mu) + U(-L;\mu) &= u_m^{(0)}(-L) + \mu u_- , \\ w_m(-L;\mu) + W(-L;\mu) &= w_m^{(0)}(-L) + \mu w_- . \end{aligned} \tag{3.5}$$

The solution of (3.4), (3.5) can formally be envisaged in terms of matched asymptotic expansions. (A rigorous proof follows from the Liapunov - Schmidt alternative method as given in in Hale [2].)

Write

$$\begin{aligned} u_m(\theta;\mu) &= u_m^{(0)}(\theta) + \mu\, u_m^{(1)}(\theta) + \ldots ; \\ w_m(\theta;\mu) &= w_m^{(0)}(\theta) + \mu\, u_m^{(1)}(\theta) + \ldots ; \\ U(\theta;\mu) &= \mu\, U_1(\theta) + \mu^2\, U_2(\theta) + \ldots ; \qquad -L \le \theta \le L \\ W(\theta;\mu) &= \mu\, W_1(\theta) + \mu^2\, W_2(\theta) + \ldots ; \\ \alpha &= \alpha_0 + \mu\, \alpha_1 + \ldots ; \end{aligned}$$

and substitute these expressions into (3.4), (3.5). If we introduce the new independent variable

$$\zeta = \frac{\theta}{\epsilon}$$

and dependent variables

$$u^{(j)}_{M}(\zeta) = u^{(j)}_{m}(\epsilon\zeta) , \quad j = 0,1 \ldots$$

$$w^{(j)}_{M}(\zeta) = w^{(j)}_{m}(\epsilon\zeta) ,$$

$$\underline{U}_j(\zeta) = U_j(\epsilon\zeta) ,$$

$$\underline{W}_j(\zeta) = W_j(\epsilon\zeta) ,$$

and equate orders of μ we find equations and boundary conditions of $\underline{U}_j(\zeta)$, $\underline{W}_j(\zeta)$. For $\underline{U}_1(\zeta)$, $\underline{W}_1(\zeta)$ we find

$$\begin{aligned} \ddot{\underline{U}}_1 &= -(\epsilon\zeta + \alpha_0)(u^{(0)}_M)^{\cdot} + (p'(w^{(0)}_M)W_1)^{\cdot}, \\ \ddot{\underline{W}}_1 &= -(\epsilon\zeta + \alpha_0)(w^{(0)}_M)^{\cdot} - \dot{\underline{U}}_1, \end{aligned} \tag{3.6}$$

and boundary conditions

$$\begin{aligned} u^{(1)}_M(-\tfrac{L}{\varepsilon}) + \underline{U}_1(-\tfrac{L}{\varepsilon}) &= u_- , \\ w^{(1)}_M(-\tfrac{L}{\varepsilon}) + \underline{W}_1(-\tfrac{L}{\varepsilon}) &= w_- , \\ u^{(1)}_M(-\tfrac{L}{\varepsilon}) + \underline{U}_1(-\tfrac{L}{\varepsilon}) &= u_+ , \\ w^{(1)}_M(-\tfrac{L}{\varepsilon}) + \underline{W}_1(-\tfrac{L}{\varepsilon}) &= w_+ , \end{aligned} \tag{3.7}$$

where $\cdot = \frac{d}{d\zeta}$. If we integrate (3.6) from $-\infty$ to ζ (3.6) is equivalent to solving

$$L\begin{pmatrix}\underline{U}_1\\ \underline{W}\end{pmatrix} = \begin{matrix} -\int_{-\infty}^{\zeta} (\epsilon\zeta + \alpha_0)\, \dot{u}^{(0)}_M(\zeta)d\zeta + \gamma_1 , \\ -\int_{-\infty}^{\zeta} (\epsilon\zeta + \alpha_0)\, \dot{w}^{(0)}_M(\zeta)d\zeta + \gamma_2 , \end{matrix} \tag{3.8}$$

where γ_1, γ_2 are constants and

$$L\begin{pmatrix}\underline{U}_1\\ \underline{W}_1\end{pmatrix} \overset{\text{def}}{=} \begin{matrix} \dot{\underline{U}}_1 - (p'(w^{(0)}_M(\zeta))\, \underline{W}_1 , \\ \dot{\underline{W}}_1 + \underline{U}_1 . \end{matrix}$$

L is a Fredholm operator $C_b^1(-\infty,\infty) \times C_b^1(-\infty,\infty)$ where $C_b^1(-\infty,\infty)$ denotes the space of bounded continuous functions on $(-\infty,\infty)$ possessing bounded continuous derivatives. (A proof of this fact maybe found in Hale [2].) So the existence of $C_b^1(-\infty,\infty) \times C_b^1(-\infty,\infty)$ solutions of (3.6) requires the right hand side of (3.8) to be orthogonal to the null space of L^* (the adjoint of L). L^* has a one dimensional null space proportional to

$$\begin{pmatrix} -\dot{w}_M^{(0)}(\zeta) \\ \dot{u}_M^{(0)}(\zeta) \end{pmatrix}.$$

So taking the $L^2(-\infty,\infty) \times L^2(-\infty,\infty)$ inner product of the above vector with the right hand side of (3.8) yields

$$\alpha_0 = -\frac{\epsilon\int_{-\infty}^{\infty} \zeta\, \dot{w}_M^{(0)}(\zeta)^2 d\zeta + \frac{\epsilon}{4}(B-A)^2 - \frac{1}{2}\gamma_1(B-A)}{\int_{-\infty}^{\infty} \dot{w}_M^{(0)}(\zeta)^2 d\zeta}. \qquad (3.9)$$

Formula (3.9) determines the <u>propagation speed</u> of the interfacial wave to lowest order in terms of known quantities and the still undetermined constant γ_1. For convenience set

$$R = -\frac{\int_{-\infty}^{\infty} \zeta\, \dot{w}_M^{(0)}(\zeta)^2 d\zeta + (B-A)^2}{\int_{-\infty}^{\infty} \dot{w}_M^{(0)}(\zeta)^2 d\zeta},$$

$$\lambda = +\frac{(B-A)}{2\int_{-\infty}^{\infty} \dot{w}_M^{(0)}(\zeta)^2 d\zeta},$$

so that

$$\alpha_0 = -\lambda\gamma_1 + \epsilon R,$$

where

λ, R are known quantities.

Having enforced the orthogonality condition (3.9) we can

write the expression for the general $C_b^1(-\infty,\infty) \times C_b^1(-\infty,\infty)$ solution of (3.8). This solution will depend on γ_1, γ_2 in an affine manner. (Remember that the term α_0 in (3.8) should be replaced with its value from (3.9).) So we can write

$$\begin{aligned} \underline{U}_1(\zeta) &= p_1(\zeta)\gamma_1 + p_2(\zeta)\gamma_2 + \epsilon\underline{P}(\zeta) + a\,\dot{u}^{(0)}(\zeta)\ , \\ \underline{W}_1(\zeta) &= q_1(\zeta)\gamma_1 + q_2(\zeta)\gamma_2 + \epsilon\underline{Q}(\zeta) + a\,\dot{u}^{(0)}(\zeta)\ . \end{aligned} \qquad (3.10)$$

The terms $p_1(\zeta)$, $p_2(\zeta)$, $q_1(\zeta)$, $q_2(\zeta)$, $\underline{P}(\zeta)$, $Q(\zeta)$ are determined in terms of known quantities. The terms $a\,\dot{u}^{(0)}(\zeta)$, $a\,\dot{w}^{(0)}(\zeta)$ enter (3.10) as solutions of the homogeneous equation

$$L\begin{pmatrix}\underline{U_1}\\ \underline{W_1}\end{pmatrix} = 0$$

where a is as yet undetermined constant.

Next we note that the $\mu = 0$ equation (2.3) implies $u_M(\zeta;\mu) = u_M(\frac{\theta}{\epsilon};\mu)$, $w_M(\rho;\mu) = w_M(\frac{\theta}{\epsilon};\mu)$, satisfy

$$\begin{aligned} p(w)^{\cdot} &= u^{\cdot\cdot}\ , \\ -u^{\cdot} &= w^{\cdot\cdot}\ , \end{aligned} \qquad (3.11)$$

where the boundary conditions (2.5) are preserved as

$$u_M(-\infty;\mu) = 0,\quad w_M(-\infty;\mu) = A + \mu v\ . \qquad (3.12)$$

If we integrate (3.11) explicitly we ascertain

$$\begin{aligned} u_M^{(1)}(\zeta) &= \frac{2p'(A)}{u_0(\rho)}\,(w_M^{(0)}(\zeta) - A)v\ , \\ w_M^{(1)}(\zeta) &= \frac{-2p'(A)\,(w_M^{(0)} - A)v}{p(w_M^{(0)}(\zeta)) - p(A)}\ . \end{aligned} \qquad (3.13)$$

If we substitute (3.10), (3.13) into boundary conditions (3.7) we have four linear equations for the four unknowns v, a, γ_1 γ_2. From (3.8) we see

$$
\begin{aligned}
-p'(B)\underline{W}_1(+\infty) &= \gamma_1 + O(\varepsilon) , \\
\underline{U}_1(+\infty) &= \lambda\gamma_1(B-A) + \gamma_2 , \\
-p'(A)\underline{W}_1(-\infty) &= \gamma_1 , \\
\underline{U}_1(-\infty) &= \gamma_2 .
\end{aligned}
\tag{3.14}
$$

From (3.14) we can pick off the approximate values for p_1, p_2, q_1, q_2 for large values of L, e.g., $p_1(\frac{L}{\varepsilon}) \approx \gamma(B-A)$, $p_2(\frac{L}{\varepsilon}) \approx 1$, etc. Similarly, for large values of L (3.13) gives approximate relations for $u_M^{(1)}(\frac{L}{\varepsilon})$, $w_M^{(1)}(\frac{L}{\varepsilon})$, $u_M^{(1)}(-\frac{L}{\varepsilon})$, $w_M^{(1)}(-\frac{L}{\varepsilon})$,

$$
\begin{aligned}
u_M^{(1)}(\tfrac{L}{\varepsilon}) &\approx \frac{-2\, p'(A)(B-A)v}{w_M^{(0)}(\frac{L}{\varepsilon})} , \\
w_M^{(1)}(\tfrac{L}{\varepsilon}) &\approx \frac{2\, p'(A)(B-A)v}{w_M^{(0)}(\frac{L}{\varepsilon})} , \\
w_M^{(1)}(-\tfrac{L}{\varepsilon}) &\approx \frac{-2\, p'(A)\left(w_M^{(0)}(-\frac{L}{\varepsilon})-A\right)v}{w_M^{(0)}(-\frac{L}{\varepsilon})} , \\
w_M^{(1)}(-\tfrac{L}{\varepsilon}) &\approx -2v .
\end{aligned}
\tag{3.15}
$$

Substitution of (3.14), (3.15) into (3.7) shows that the approximations if treated as exact would yield a simple set of four linear equations in the four unknowns v, a, γ_1 γ_2. We evaluate the determinant of the relevant coefficient matrix. If $-p'(B) > -p'(A)$ we can pick out the dominant term of this determinant which is not zero when $\lambda(B-A) \neq (-p'(A))^{-1/2} - (-p'(B))^{-1/2}$. However, if we switch the integrations in deriving (3.8) from $+\infty$ to ζ instead of $-\infty$ to ζ, we are led to the requirement $-\lambda(B-A) \neq (-p'(A))^{-1/2} - (p'(B))^{-1/2}$ so if $-p'(B) > -p'(A)$ we need only require $\lambda(B-A) \neq 0$ which we know to be true. On the other hand if $-p'(A) > -p'(B)$ we set the boundary conditions $w_M(\infty,\mu) = B + \mu v$ and argue in a similar fashion. By continuity we assert that the invertibility of the approximate system implies invertibility of the exact system for large L and small ε. This shows solveability of (3.6), (3.7). Similar arguments work for successive

$$\underline{U}_j, \underline{W}_j, \quad j = 2,3,\ldots .$$

4. THE INFINITE DOMAIN PROBLEM

In the previous section we outlined the resolution of the finite problem (3.4), (3.5). Remember however, that for each $\underline{U}_j(\zeta)$, $\underline{W}_j(\zeta)$ is in $C_b^1(-\infty,\infty) \times C_b^1(-\infty,\infty)$ so for small μ $u_m(\theta;\mu) + U(\theta;\mu)$, $w_m(\theta,\mu) + W(\theta;\mu)$ will be bounded uniformly in ε and L. Hence the extension argument of Dagermos suggested earlier may be applied to yield a solution $\tilde{u}(\theta)$, $\tilde{w}(\theta)$ of (2.1), (2.2). Finally since $u_M^{(0)}(\zeta)$, $w_M^{(0)}(\zeta)$ connects (0,A) to (0,B) monotonically in $w_M^{(0)}$ with $u_M^{(0)}(\zeta)$ possessing only one critical point $w_M^{(0)}(\zeta)$ lies in the region $p'(w_M^{(0)}(\zeta)) > 0$ on some finite domain $-M < \zeta < M$. For small μ (3.3) then says $\tilde{u}(\theta)$, $\tilde{w}(\theta)$ will have $p'(\tilde{w}(\theta)) > 0$ on a fininte domain $-\varepsilon M < \theta < \varepsilon M$. So the total variation of $\tilde{u}(\theta)$ on $[-\varepsilon M, \varepsilon M]$ is formally estimated by

$$\operatorname*{Var}_{[-\epsilon M,\epsilon M]} \tilde{u}(\theta) < \operatorname*{Var}_{[-\epsilon M,\epsilon M]} u_m(\theta;\mu) + \operatorname*{Var}_{[-\epsilon M,\epsilon M]} \mathbb{U}(\theta);\mu)$$

$$< \text{Constant} + \int_{-\epsilon M}^{\epsilon M} (\frac{d}{d\theta} U(\theta;\mu))d\theta$$

$$< \text{Constant} + 2\epsilon M \cdot \frac{(\text{Constant})}{\epsilon}$$

since

$$\frac{d}{d\theta} U(\theta;\mu) = \frac{d}{d\zeta} U(\zeta;\mu) \frac{d\zeta}{d\theta} = \frac{1}{\epsilon} \frac{dU}{d\zeta}(\zeta,\mu)$$

and each term in the expansion for $\underline{U}(\zeta,\mu)$ is bounded in $C_b^1(-\infty,\infty) \times C_b^1(-\infty,\infty)$. A similar argument for $\tilde{w}(\theta)$ shows $\operatorname*{Var}_{[-\epsilon M,\epsilon M]} \tilde{u}(\theta)$, $\operatorname*{Var}_{[-\epsilon M,\epsilon M]} \tilde{w}(\theta)$ are bounded independent of ε. Outside of $[-\epsilon M,\epsilon M]$ the orbit $\tilde{u}(\theta)$, $\tilde{w}(\theta)$ is in the hyperbolic $p' < 0$ regime where Dafermos's estimates guarantee uniform in ε bounded total variation. Thus, our goal is accomplished and passage to the limit as $\varepsilon \to 0+$ produces a solution of the Riemann problem (1.1), (1.2) for special data (1.4) with μ small.

REFERENCES

1. C. M. Dafermos, Solution of the Riemann problem for a class of hyperbolic systems of conservation laws by the viscosity method, Archive for Rational Mechanics and Analyses. 52, 1-9 (1973).

2. J. K. Hale, Introduction to dynamic bifurcation, in Bifurcation Theory and Application, Springer Lecture Notes in Mathematics, No. 1057, ed. L. Salvadori, Springer Verlag, Berlin-Heidelberg - New York - Toyko (1984).

3. M. Slemrod, Dynamics of first order phase transitions, Proc. Mathematics Research Center, University of Wisconsin conference on "Material Instabilities and Phase Transformation". ed., M. Gurtin, Academic Press (1984).

Figure 1.

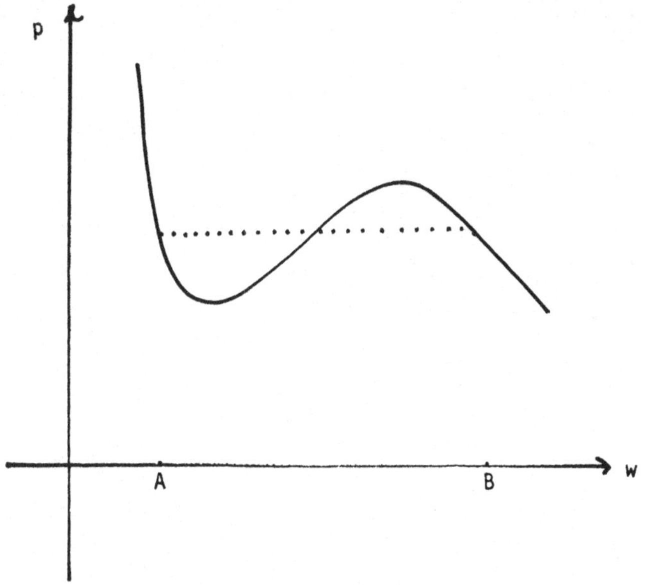

Printed in the United States
By Bookmasters